AIRPORT SLOTS

Airport Slots
International Experiences and Options for Reform

Edited by

ACHIM I. CZERNY
Berlin University of Technology, Germany
PETER FORSYTH
Monash University, Australia
DAVID GILLEN
University of British Columbia, Canada
HANS-MARTIN NIEMEIER
University of Applied Science, Germany

ASHGATE

Published by
Ashgate Publishing Limited
Gower House
Croft Road
Aldershot
Hampshire GU11 3HR
England

Ashgate Publishing Company
Suite 420
101 Cherry Street
Burlington, VT 05401-4405
USA

Ashgate website: http://www.ashgate.com

British Library Cataloguing in Publication Data
Airport slots : international experiences and options for
   reform
   1. Airport slot allocation 2. Airlines - Timetables
   3. Airport capacity
   I. Czerny, Achim I.
   387.7'4042

Library of Congress Cataloging-in-Publication Data
Airport slots : international experiences and options for reform / edited by Achim I. Czerny ... [et al.].
   p. cm.
   International bibliographical references and index.
   ISBN: 978-0-7546-7042-1
   1. Airport slot allocation. 2. Airport capacity. 3. Airports--Management. 4. Runway capacity--Costs. 5. Competition. I. Czerny, Achim I.

   HE9797.4.S56A39 2007
   387.7'364--dc22

2007027813

ISBN: 978-0-7546-7042-1

Printed and bound in Great Britain by TJ International Ltd, Padstow, Cornwall.

Contents

List of Figures

List of Tables

Acknowledgements

This book is a compilation of selected papers presented at five workshops organized by the German Aviation Research Society. We would like to thank the Berlin University of Technology, the HWWA – Institute of Economic Research, Hamburg, and the University of Applied Sciences Bremen for acting as hosts. We are also grateful to Australian Research Council, Hamburg Airport, Lufthansa and the Wolfgang-Ritter-Stiftung for providing financial support.

John Hindley and Guy Loft of Ashgate Publishers were encouraging with their active promotion of GARS.

Our thanks also go to Natalie Chantal-McCaughey for carefully revising the English style and to Harald Wiese for assistance in editing.

We would also like to thank Helen Huang, Christiane Müller-Rostin and Stephan Zass for their excellent research assistance in gathering the data for the appendix.

Contributors' Acknowledgements

Chapter 3   Economics Perspectives on the Problem of Slot Allocation
Batool Menaz and Bryan Matthews

This chapter has benefited from comments received following two presentations of the paper on which it was originally based, one to the German Aviation Research Society (GARS) workshop in Bremen in November 2003 and the other to colleagues at the Institute for Transport Studies in March 2004. In developing that original paper into this chapter, Bryan Matthews has also benefited through having cause to think and read further on the subject in the course of supervising Joseph Kendal's student research project into 'Stakeholder Views on Slot Allocation'. We hereby acknowledge our thanks to the different parties concerned, though the views we express, along with any remaining errors, are our own.

Chapter 5   Setting the Slot Limits at Congested Airports
Hans-Martin Niemeier and Peter Forsyth

We are grateful to Axel Claasen, Milan Janic, Christiane Müller-Röstin, David Starkie, and Dieter Wilken for very helpful suggestions and comments on an earlier draft. All errors are our responsibility.

Chapter 7   Managing Congested Airports Under Uncertainty
Achim I. Czerny

Many thanks to Peter Forsyth and Cornelia Templin for very helpful comments.

Chapter 8   Prices and Regulation in Slot Constrained Airports
Hans-Martin Niemeier and Peter Forsyth

We are grateful to David Starkie and Christiane Müller-Röstin for valuable comments on an earlier draft.

Chapter 11   The Dilemma of Slot Concentration at Network Hubs
David Starkie

I benefited greatly when preparing this chapter from comments on an early draft by participants at a CTS workshop, Sauder School of Business, UBC, Vancouver. In particular, special thanks are due to David Gillen who made arrangements for my stay and provided resources to enable me to complete the chapter. My gratitude also goes to Helen Huang who helped to compile the Appendix, and to Jan Brueckner and Leon Baros for comments on a later draft. The usual caveats, of course, apply.

Chapter 17   Formal Ownership and Leasing Rules for Slots
Erwin von den Steinen

The original paper was presented at the Workshop on Slot Markets conducted by the German Aviation Research Society at Bremen, 7–8 November 2003. I have drawn on the thoughtful commentary by Professor Jaap de Wit as well as a variety of other constructive suggestions and comments received in developing this amended version.

Chapter 20   Airport Slots: Perspectives and Policies
Peter Forsyth

I am grateful to Neelu Seetaram and Nathalie McCaughey for helpful research assistance. I am also grateful for perceptive comments from reviewers. Any errors are my own.

Editors and Contributors

Jörg Bauer is head of the slot management team at Deutsche Lufthansa AG. He holds a degree in Business and Engineering (Wirtschaftsingenieurwesen) from the University of Darmstadt. During his studies, he focused on transportation (airline and airport management) as well as aircraft engine design.

Kenneth Button is a University Professor at George Mason University where he is Director of the Center for Aerospace Policy Research and the Center for Transportation Policy, Operations, and Logistics. He holds degrees in economics from the University of East Anglia, the University of Leeds, and Loughborough University. He is editor of the *Journal of Air Transport Management* and of *Transportation Research Series D: Transport and Environment*. He is currently a visiting professor at the University of Bologna and Porto University. He has written or edited over a 100 books and authored over 400 articles and book chapters.

Nathalie-Chantal Mc Caughey is currently a PhD student at the department of economics at Monash University (Melbourne, Australia). Her PhD research is in the areas of applied microeconomics and industrial organization, with a focus on the impacts of frequent flier programs on competition in the global aviation industry. She is an assistant lecturer in subjects such as 'Competition and Regulation' and 'International Economics'. She has an honours degree in economics from Monash University and is an academically certified interpreter in Spanish and French, which she studied in Heidelberg, Germany.

Achim I. Czerny is a PhD student at the Workgroup for Infrastructure Policy (WIP) an economic teaching and research unit in the School of Economics and Management at Berlin University of Technology (Germany). His main activities include research in the area of Industrial Organization with focus on transport economics as well as utility and competition policy. He provides teachings in 'Transport Economics', 'Infrastructure Economics and Economic Policy', 'Advanced Microeconomics', and 'Game Theory'. Participation in research and consulting projects commissioned, e.g., by government agencies such as the European Commission (Directorate-General Competition), is also part of his work.

Peter Forsyth has been Professor of Economics at Monash University, Australia since 1997. Prior to this he held posts at Australian National University and the University of New England. He holds degrees form the University of Sydney and the University of Oxford. He has specialized in the economics of transport, especially aviation, privatization and regulation, and the economics of tourism.

Most recently, he has been paying particular attention to the privatization and regulation of airports, and to the use of computable general equilibrium models in evaluating the economic impacts of tourism. He has recently published an edited volume of classic articles in the economics of air transport (Edward Elgar), and, with Larry Dwyer, an edited *International Handbook on the Economics of Tourism* (Edward Elgar).

David Gillen graduated in 1975 from the University of Toronto with a PhD in Economics. He joined University of British Columbia in 2005 and currently holds the positions of YVR Professor of Transportation Policy in the Sauder School of Business and is Director, Centre for Transportation Studies, University of British Columbia. He is also Research Economist at the Institute of Transportation Studies at the University of California, Berkeley where he taught from 1990 to 1998. He has published over 100 articles and books dealing with a variety of topics in transportation economics and business. His current research includes evaluating investment in Intelligent Transportation Systems, pricing and auction mechanisms for roadways and runways, measuring performance of transportation infrastructure, vertical contracts in aviation and evolving network strategies and business models in airlines, airports, ports and gateways.

Milan Janic is an air transport and traffic engineer and planner. He is currently Senior Researcher in the air transport and research programme leader of the section Transport and Infrastructure at OTB Research Institute of Delft University of Technology (Delft, Netherlands). He was formerly Senior Research Fellow at Manchester Metropolitan University (Department of Physics, Geography, and Environment) (Manchester, UK), Research Fellow at Loughborough University (Civil and Building Engineering Department, Transport Studies Group) (Loughborough, UK) and Chief Researcher at the Institute of Traffic and Transport (Ljubljana, Slovenia). He has been involved in the aviation research and planning of the national and international projects for about 25 years. He has also published many papers in the scientific and professional journals, and conferences, as well as the scientific book *Air Transport System Analysis and Modelling: Capacity, Quality of Services and Economics*. Another book, *The Sustainability of Air Transportation: A Quantitative Approach*, is to be published in the year 2007. He is a member of ATRS (Air Transport Research Society) and of the US TRB (Transportation Research Board) Committee 'Airfield and Airspace Capacity and Delay'.

Daniel M. Kasper is a Managing Director of LECG, LLC (formerly The Law and Economics Consulting Group) and head of the firm's transportation practice. Prior to joining LECG, Dan was head of the Transportation Industry Program for Coopers & Lybrand, L.L.P., and vice-president of Harbridge House, Inc. In 1993, he was appointed as one of 15 voting members of the US National Airline Commission, a body created by Congress to evaluate the state of the US airline industry. In addition, he has served as a consultant to the US Departments of Transportation, State and Defense. Previously, he has held senior level positions at the United States Civil Aeronautics Board, first as a Special Assistant (chief of staff) for the Board's vice-chairperson, Elizabeth Bailey and subsequently as Director of the Bureau of International Aviation. He has authored two books on aviation as well as numerous articles, case studies, and research papers on various aspects of transportation and government policy. He frequently testifies as an expert on transportation industry matters before courts and federal administrative agencies, as well as legislative bodies and antitrust authorities both in the United States and abroad. Prior to joining the CAB, he served on the faculties of the Harvard Business School and the University of Southern California School of Business Administration. He holds MBA and JD degrees from the University of Chicago.

Matthias Kilian is senior research fellow at the Institute of Employment and Business Law in Cologne. He received his PhD in law at the University of Cologne where he has held several research posts. His research interests include the regulation of professional services, socio-legal studies, European law and Business law (including air transport law), areas in which he has published more than 150 articles. He is also a member of the Bar and partner at the Cologne office of a multidisciplinary law firm.

Bryan Matthews is a Senior Research Fellow at the Institute for Transport Studies (ITS) at the University of Leeds. He has a BA in Economics, an MA in Transport Economics and over ten years experience in consultancy and academia. His main focus is on transport pricing and costing issues. He recently co-edited a volume in the Research in Transport Economics series on the measurement of the marginal social costs of transport, and is currently Project Manager for a major EU research project (GRACE) to further this area of work. His teaching activities include lecturing on airline economics and leading the Air Transport systems module at ITS.

Nathalie-Chantal McCaughey is currently undertaking her PhD in Economics at Monash University (Melbourne, Australia). Her PhD research lies in the fields of applied microeconomics and industrial organisation, with a focus on the impacts of frequent flier programs on competition in the global aviation industry. She is an assistant lecturer in third year subjects such as 'Competition and Regulation' and 'International Economics' at Monash University and lectured 'International Economics' at third year level at Deakin University (Melbourne-Australia). She has an honours degree in economics from Monash University and is an academically certified interpreter in Spanish and French, which she studied in Heidelberg, Germany.

Batool Menaz is a Research Officer at the Institute for Transport Studies (ITS) at the University of Leeds. She has a BSc in Economics and an MA in Transport Economics. Her main research interests include transport pricing, transport demand and supply. She has worked on numerous European and UK projects, including IMPRINT-Europe, IMPRINT-NET, RRUK, ASSESS, EXTR@WEB,

DISTILLATE and DIFFERENT. She has recently been involved in a CfIT project looking at the demand elasticities for air travel.

William G. Morrison is an Associate Professor of Economics at the School of Business and Economics, Wilfrid Laurier University, Waterloo, Ontario. His academic publications include articles on predatory pricing in air transportation, airline competition and airport policy. Dr Morrison's research in air transportation also includes government reports on Canada's airport system and a study of demand-price sensitivity estimates for air travel.

Hans-Martin Niemeier is Professor of Transportation Economics and Logistics at the University of Applied Sciences, Bremen. He received his PhD in economics at the University of Hamburg and worked in the aviation section of the State-Ministry of Economic Affairs of Hamburg. His research focuses on airport regulation and management.

Frederik Sørensen holds a degree in transport economics from the University of Copenhagen in 1962. Since 2001 he has established his own counselling firm, FRESAIR in the areas of air transport regulatory matters in particular related to the EU. He joined the European Commission in 1973 after a career in Denmark, which went from local authority administration to a consultancy in transport planning and in particular airport planning. Mr. Sørensen retired from the European Commission on 1 October 2001. While at the European Commission he became responsible for the Economic Regulation and Air Transport Agreements Unit of the Directorate General for Energy and Transport of the European Commission. As such he was responsible for the development of an air transport policy for the EU. Mr. Sørensen therefore has been involved in this activity right from the start and in particular with the development of the first, second and third aviation packages which resulted in Community wide liberalization of air transport by 1 January 1993. In addition he has developed rules on CRS, slot allocation and passenger protection etc. After 2001 his time has been much occupied by the discussion on aviation relations with third countries (Central Europe and US), liberalization of air transport in Western Africa etc. Recently he wrote a guide on the implementation of EU legislation in air transport and associated matters.

David Starkie is a Senior Associate with Case Associates, London, and editor of The Journal of Transport Economics and Policy. He has held positions at the Universities of Leeds, Reading and Adelaide and at the Institute for Fiscal Studies. He has also worked in the Western Australian Public Service and been a director of a number of private sector companies. A former adviser to UK House of Commons Select Committees, the UK CAA and Office of Fair Trading, he currently advises the Irish Commission for Aviation Regulation. His degrees are from the London School of Economics.

William Spitz is Senior Economist at GRA, Inc., an aviation consulting firm based in Jenkintown, Pennsylvania, USA. He received his PhD in Economics from Yale University in 1988. His primary work focuses on the development and implementation of econometric and statistical models applied to economic issues in aviation. Dr. Spitz is the lead developer of GRA's proprietary network planning model that has been used extensively to analyze air carrier mergers and codesharing agreements, and as a modeling and planning tool for airlines. In addition, he has co-authored many transportation studies for the US Federal Aviation Administration and other government and private clients.

Erwin von den Steinen. Following government service as a US diplomat and senior air transport negotiator, Erwin established a consultancy practice in 1988 specializing in issues of market access, public infrastructure and traffic rights in international aviation. He has completed some 40 studies or projects in the field of aviation policy for public or private sector clients working independently or in teams. Erwin von den Steinen is also the author of a number of articles and most recently of the book *National Interest and International Aviation,* published by Kluwer Law and Aspen Publishers (2006), which provides a systematic survey of policy issues affecting international aviation, especially with respect to the North Atlantic market. He lives and works in Bonn, Germany.

Despina Tudor graduated in 2006 from Wilfrid Laurier University with a Bachelor Degree in Honours Economics. She held a Research Position at the University of British Columbia in 2005, under the supervision of Dr David Gillen, where she also worked on a demand management project for the YVR Airport in conjunction with InterVISTAS consulting company. As of September 2007, she will begin her graduate studies in Economics at Erasmus University in Rotterdam, Netherlands. She currently holds a position as a Forecasting Analyst for Unilever Canada.

Claus Ulrich is Airport Coordinator of the Federal Republic of Germany. Claus Ulrich was born in 1945 in Bremen, Germany. He joined Lufthansa German Airlines in 1968 for a three years training course. After holding various posts within Lufthansa and after joining the German Scheduling Coordination, Mr. Ulrich was appointed Airport Coordinator of the Federal Republic of Germany in 1986. The Airport Coordinator and his team of approximately 20 persons are responsible for the allocation and coordination of planned arrival and departure slots at presently 17 major German airports, with a total of over 2 million commercial flight movements per year. Additionally, the German Airport Coordinator handles the slot allocation of all other flights operated under IFR conditions serving these airports. At the same time the Airport Coordinator is responsible for the slot monitoring at the coordinated airports. Mr. Ulrich became Chairman of the European Union Airport Coordinators Association (EU-ACA) in 1994/95 and was elected Chairman again in 1997, 2000, and 2003. He also was elected Chairman of the Worldwide Airport Coordinators Group (WWACG) in 2003, and was re-elected to this position in 2005.

Chapter 1

Introduction and Overview

Achim I. Czerny, Peter Forsyth, David Gillen
and Hans-Martin Niemeier

The growth of air transport during the last decades indicates that this industry plays a key role for regional economies and the integration of the world economy as a whole. However, this development poses a great challenge to airlines, airports, regulators, and politicians. A particular problem is that demand growth for air transport services does not go along with a respective growth of airport capacity. For that reason many airports all around the world are short of capacity in relation to demand.

There are different ways to deal with limited airport capacity or, respectively, excess demand. One possibility is to allocate capacity on a first-come-first-serve basis with airlines queuing for runway access. Most airports' capacity in the US is allocated by first-come-first-serve which has its advantages, since it requires a minimum degree of regulatory intervention saving public resources. Furthermore, airlines are not discriminated by other criteria than by time of arrival. Consequently, competition between incumbent and newcomer airlines is possible. However, the airline's decision on airport usage does not take into account the additional queuing times imposed to other airlines. An excessive use of airports leading to serious congestion problems is often the consequence.

Congestion and congestion costs can be managed by different measures to reduce demand. In principle, these measures can be of two types: quantitative restrictions, such as the airport slot system, or pricing. Pricing systems involve setting prices at levels which adjusts likely demand according to the airport's capacity. An important difference between these two approaches is that with set prices, the allocation of airport capacity is always based on the willingness to pay of airlines. In contrast, when quantitative restrictions are used, some methods of allocation must be employed. These methods can be based on willingness to pay but in practice they are usually not.

The International Air Transport Association (IATA) is the trade association of international airlines. Around the world, the resolution of the excess demand problem is generally based on the use of the IATA slot system, with administrative allocation of the available slots. The most important principle is the 'grandfather'-principle which allocates airport capacity by historical use. Airlines are granted to use runways at specified times in the future which they already have used in the (recent) past. In other words airlines continuously retain the possibility to use a specific airport at a specified time, which is called a slot. Other administrative procedures applied in the air transport industry include lotteries which allocate runway capacity by chance or allocation by market share. The latter is often employed in order to stimulate competition between airlines by providing new entrants with a preferred status compared to incumbents.

It follows that the current system for the allocation of airport capacity strongly relies on administrative rules, which are frequently criticized by economists. Alternatively the move to more market oriented instruments is recommended in order to force an allocation relying on the airlines' willingness to pay. Why though do economists consider the willingness to pay to be a critical measure on which slot allocation should be primarily based on?

The answer is: the airline's willingness to pay for a slot is taken to be a valid indicator of the contribution from using the slot to social welfare. In other words, allocation of slots to an airline with a high willingness to pay for airport capacity is normally considered to generate the largest benefits to society. If that is true then allocation of airport resources on the basis of the airlines' willingness to pay would automatically allocate airport capacity to those airlines that make the best use of it from a welfare perspective.

The task of his book is to address the most relevant topics which need to be considered and analyzed in order to achieve an 'ideal' system of allocating airport capacity. For the reasons mentioned in the previous paragraph, the focus is often on market oriented instruments that allocate airport capacity according to the airlines' willingness to pay as basic elements for a new allocation system. Some of the key questions addressed in this book are summarized in the following.

An important issue addressed in this book is the comparative analysis of different systems used across the world for the allocation of scarce airport capacity. These include systems of airport slots as applied, for instance, in the EU and the first-come-first-serve approach which is common practice in the US. Another issue is the design of airport slot systems: could allocation of airport capacity be enhanced by trading and auctions? Is it possible to design an efficient auction mechanism which addresses the specific requirements of the airline industry? Furthermore, a fundamental element of airport slot systems is the determination of slot limits. This, however, is a complex problem which requires balancing the individual airline's needs and the effects additional use of airport capacity has on all other airlines and passengers due to congestion.

An alternative option to allocate airport capacity is the use of posted prices that adequately mirror scarcity. In spite of their potential advantages, such prices are very rarely used to ration scarce airport capacity. What are the economic effects and difficulties from moving to a greater reliance of posted prices instead of slot systems or a first-come-first-serve approach? Furthermore, airports are usually considered to have market power and, thus they face economic regulation. In this case, what are the implications of the economic airport regulation on capacity allocation? Instead of dealing with scarce airport resources one could also consider investments into new airport capacity as possible means to address congestion problems. The challenges here are to identify the optimal extent of airport investments and to build an environment that is in support to necessary and useful airport investments. Finally, lessons are drawn from current experiences and theoretical as well as empirical analysis for future reforms.

The book consists of six parts.

Part A, 'The Current Slot Allocation System' focuses on describing how the slot system works. The contribution by *Ulrich* includes a detailed description of the current systems to allocate runway slots among airlines. It is written from the perspective of an airport coordinator and depicts the different milestones that are passed until the final allocations of airport slots are reached. The contribution by *Menaz and Matthews* provides an overview of the different economic perspectives on airport slot allocation. It also includes an assessment of the current European system in allocating slots and examines various other types of allocation methods. In particular, the relative merits of pricing, auctioning, secondary trading and administrative arrangements are considered. Furthermore, the authors draw on their knowledge of the literature and experience of the allocation of scarce capacity within the rail sector. Contribution three is a primer on airport slots by *Gillen*. The purpose of this paper is to provide an introduction to and basic functioning of airport slots; what they are – definitions, how they are allocated and the various types of slot transactions. A brief description of the major factors that have contributed to the status quo of slot allocation is also presented. Furthermore, a comparison between the US and EU markets as well as differences that exist within the EU, such as between the UK and the rest of EU, is provided.

Part B, 'Congestion, Slots, and Prices' pays particular attention to the problem that the slot system seeks to address, namely that of reducing congestion. The first contribution is on slot constraints which are imposed at many busy airports to curb demand for airport capacity thereby reducing congestion costs. However, while much attention is given to the problem of slot allocation, little attention is paid to ensuring that slot capacities are set efficiently. In view of this *Forsyth and Niemeier* present a theory of how slot capacities can be set optimally. Furthermore, a test of whether slot capacities/congestion levels are set optimally at several airports is developed and applied to US and European airports. This shows that authorities on the two continents appear to make very different choices between slot availability and delays. The third contribution, by *Janic*, investigates the potential to model the effects of charging airport congestion fees to airlines. The author develops a model capable of estimating congestion as a matter of charging for it. Moreover, the model can be used to estimate the feasibility of additional aircraft/flight operation pricing conditions under congestion. The former two papers considered slot constraints or congestion pricing exclusively. The contribution by *Czerny* compares slot constraints and congestion pricing under uncertainty. In principal, both instruments can lead to the same efficient allocation of scarce airport capacity. However, the author demonstrates that uncertainty about demand and congestion costs has a significant impact on the welfare performance of the two different regulation instruments. Furthermore, it is argued that the network character of the air transport industry tends to favour slot constraints while a possible negative correlation between demand and congestion costs favours the use of congestion pricing. The choice of regulation instruments is, therefore, not straightforward. Alternative measures especially designed to deal with demand uncertainty are also presented. The subsequent contribution by *Forsyth and Niemeier* focuses on the role of slots in reconciling price regulation, as often implemented, with the efficient allocation of the airport's capacity. It outlines the characteristics of efficient price structures and explores the question of how regulation can set up incentives airports to choose either efficient or inefficient price structures. It also provides empirical evidence on price structures and regulation, showing that the actual price structures of most busy airports induce an inefficient allocation of slots.

Part C, 'Airline Strategies and Competition' investigates how airlines act in a slot constrained environment. The contribution by *Bauer* explores the slot usage behaviour of airlines. It discusses the four commonly alleged malpractices: overbidding, late-handback, seasonality and no-shows. Moreover, it shows that alleged malpractices often arise from operational necessities forcing airlines to use the flexibilities that the current slot regime offers in order to operate profitably. The contribution by *Gillen and Morrison* provides a theoretical approach to framing the competition issues arising from airport slots from an economic perspective. An international survey of how competition issues have been addressed by regulatory authorities is also presented. The authors conclude that the provisions in competition law dealing with the abuse of dominant position, and merger guidelines to protect against the anti-competitive effects of slot concentration appear adequate to protect competition.

A number of the world's major airports have a high proportion of their capacity utilized by a single airline, or alliance of airlines. In the third contribution to this chapter, *Starkie* discusses the tension between the advantages and disadvantages of slot concentration. The analysis incorporates recent developments in the theory of congestion pricing that, it is suggested, have an important bearing on the balance of the argument. It is also argued that higher fares at slot constrained airports do not constitute a prima facie case for abuse of market power. A part from the exploitation of a dominant position, a number of other possible reasons, which may lead to higher air ticket prices at major hubs, may in fact exist. Contribution four by *Gillen and Tudor* presents an event study methodology to assess the effect of monetary slot transactions on investors' reactions. In general, the authors find that slot transactions involving monetary compensation do not have much impact on stock prices. However, they conclude that this does not necessarily indicate efficiency of current slot allocation systems and that new industry developments could be steered in the wrong direction should loyalty to the current system be left intact. *Spitz* analyzes the value airlines put on flights at congested airports. His contribution contains a discussion of factors affecting conventional slot valuations, provides an economic analysis of how slot controls may affect carrier profitability, and presents estimates of valuations from recent sales and valuation studies. It also presents a new method to value flights that is based on network opportunity costs. Additionally, the results from applying this method at LaGuardia Airport in New York are compared to conventional slot valuations.

Part D, 'International Experiences' concentrates on the regulatory environment in the EU and the US. The first contribution by *Kilian* assesses the legal aspects of slot trading by outlining the regulatory regime the EU introduced in 1993. The paper also turns to the challenges EU policymakers have faced since and addresses the ongoing reform discussion and the legal problems a market based approach to slot allocation could create. It demonstrates that from a legal point of view, many questions remain open. Contribution two by *Kasper* analyzes the current as well as the potential future role of secondary markets for the allocation of take-off and landing slots at US airports. As measures to reduce airport delays, increases in available airport capacity and the use of market based demand management systems are suggested. Three alternative options are analyzed in more detail: congestion based landing fees, auctions, and secondary slot markets. It is suggested that the most effective measure for the allocation of scarce runway capacity would be the extended use of secondary slot markets which allows airlines to trade slots permanently.

Part E, 'Auctions and Alternatives' analyzes allocation regimes not or only rarely used in practice. In the first contribution by *Button* the potential benefits of auctions and the experiences with auctions in other industries are evaluated. Moreover, it provides a comparative analysis of slot auctions and alternative allocative mechanisms, such as modifications of administrative rules, capacity expansion, congestion charging, and secondary markets. The potential forms of slot auctions are also explored. Different topics regarding the design of a slot auction are discussed in more detail including the 'ownership' of slots, interdependence of slot values, frequency of slot auctions, design of slots, duration of the slot auctioning process, and the coordination with terminal and stand capacity. The second contribution by *v.d. Steinen* explores the issues of slot ownership and disposition rights. It is proposed to change the legal status of slots from that of public entitlements to private leases. To accomplish change and transition the introduction of a fee for making or holding a reservation of runway capacity reflecting its cost and value of operating time is suggested.

Part F 'Reforming the slots system' provides lessons that can be drawn from current experiences or from theoretical as well as empirical analysis regarding changes of regulation regimes. The first contribution by *Button* focuses on how airport take-off and landing slot allocation procedures have implications affecting the levels and the distributions of economic rents. The actors in the game are the airlines, airports and politicians (including the executive as well as legislators). It shows that the ultimate distribution of economic rents is essentially a political decision, as is any form of allocation, but that the form of specific procedures influences the nature of the benefits created and who enjoys them. Contribution two by *Sørensen* describes the legislative process concerning slot allocation and on air transport in general since 1977 and until 2001 in the EU. It provides a detailed picture of the development of legislation in this area and defines the key issues that influenced the way the legislation was developed and adopted in Europe. The discussion also includes possible ways of reforming the existing system. The final paper by *Forsyth* rounds off the discussion of the book, and comments on issues not covered in earlier papers. It begins by identifying a range of efficiency problems which need to be resolved in the context of busy airports, and follows this by making an assessment of how well current slots systems handle these. Then some of the choices between options are reviewed: a slot system versus the US delay system, slots versus prices, and alternative ways of improving slot allocation. The slot system has significant implications for the ways in which airline and airport markets work, and these are sketched out in a section on the political economy of slots. In the final sections some unresolved issues, worthy of further study, are identified, and some key policy conclusions are outlined.

PART A
The Current Slot Allocation System

Chapter 2

How the Present (IATA) Slot Allocation Works

Claus Ulrich

Introduction

The demand for air travel has been growing rapidly in most parts of the world, but the capacity of airports to handle this demand has been expanding less rapidly. The result is growing pressure on this capacity. Sometimes this is met by increasing delays – this is the case with many airports in the US. Alternatively, administrative means can be designed which keep actual demand for an airport at a level which can be handled at moderate delay. This is an approach which has been in operation in most countries other than the US. The system which most countries choose to adopt is the IATA Slot Allocation system. In this chapter, we focus on how this system works.

The chapter begins with a brief background to the system. It then outlines the parameters under which the system works, and the criteria used for slot allocation. After, this the system is explained through an example of the allocation process as it unfolds in real time. The next section discusses the enforcement of the system as it operates. The chapter concludes with a brief evaluation of the IATA slot system.

Background

Sometimes, when travelling by air, one encounters long delays at the airport. Fortunately, this does not happen too often. While waiting at the airport, one might question why airlines sometimes offer departure or arrival times which seem to be far from ideal, imposing rather long connecting times at these hub airports on passengers or forcing people to arrive or depart extremely early in the morning or extremely late at night. After all, why do airlines create their schedules that way, respectively is there anyone forcing them to do so? Finally, looking at all the crowds and activities at such a buzzling airport one might well ask – or be happy – that all this does not result in an absolute chaos.

Large airports, hub airports in particular, are quite congested and probably will stay that way. In Europe and elsewhere the development of air traffic demand has out-paced the development of capacities over the last years. One cannot expect that this situation might undergo any substantial change at major airports in the foreseeable future, in spite of some impressive expansion programmes we can see at airports such as Madrid Barajas today and probably at Frankfurt and Munich in a few years from now.

Of course, there are other airports, which still have capacities available, some of them a lot, so that one even can talk about an underusage of capacities at these airports. Today, the discussion goes even further than that, namely that countries such as Germany might have too many airports in operation. A large portion of such airports is not profitable and will never have a chance to reach black figures.

Such discussions may be extremely useful. However, they cannot alleviate the tight capacity situation at those airports which have a large catchment area or serve as a hub or are functioning both ways. Actually, the problem of congested airports is not new and of course the airlines as well as other participants in the air transport system had to come up with some kind of remedy, which at least would make regular air traffic possible.

The problem of demand outstripping capacity is one which is encountered across the world, and it is not one which will go away. Thus a worldwide approach is appropriate. Hence the emergence, over time, of the 'IATA System' This has been developed by the International Air Transport Association (IATA), which represents the interests of airlines all over the world.

The Slot Allocation System: Parameters and Criteria

The 'IATA System' has proved to be extremely successful over the years. It is a worldwide approach which is adopted, with some variations, in most regions of the world. Since for Europe, the European Commission took up these procedures and created a European Slot Allocation Regulation, which used the IATA practices in many aspects as a blue print. Of course, the authors of this EU Regulation had some ideas of their own which were integrated into the European procedures. These ideas on the other hand were more or less adopted by the 'IATA System', so that presently the air transport industry is following procedures, which match each other in the vast majority of activities. One indication of the success of the IATA procedures is the fact that legally they are only 'Recommendations', while of course the European Slot Allocation Regulation is legally binding in the EU member states. In many parts of the world, however, the IATA Procedures, as laid down in the WSG (World Scheduling Guidelines) have a binding character.

Setting the Parameters

The first issue which needs to be settled is the choice of the coordination parameters.

On the first glimpse the answer is quite simple: for example, at Frankfurt Airport the number of hourly slots which may be used by the airport coordinator is 80, out of which not more than 43 arrivals or 45 departures may be given to airlines and other airport users.

The immediate question then is why the coordinator may not accept 78 or 82 movements per hour, and, secondly what about the fact that the airport might be able to accommodate 85 movements during good weather conditions or just 70 or less during strong wind or bad visibility?

The second point is quite easy to explain. Indeed the actual capacity tends to change constantly, depending on items such as weather or the composition of the aircraft flow (e.g. a light aircraft behind a heavy aircraft needs more distance than two light aircraft behind each other etc.). Other factors of all kinds may also contribute to a change of the actual capacities. However, we know that we discuss figures for the coming summer season, which is more than six months away. So much in advance nobody would have an idea about any variable factors contributing to the actual capacity. On the other hand, the allocation process for the coming summer season 2004 will begin, as we will see a bit later, during the fall of 2003. Since this process does imply that an airline will have to accept arrival – or departure times for its regular schedules substantially differing from the original requests, that airline must be certain that the times, as allocated, are for sure at least under normal circumstances.

Consequently, coordination parameters are of course based on capacities, but they are not necessarily identical to them.

The first question we asked, namely why the coordination parameters at Frankfurt Airport lie with 80 movements per hour, is a bit more complex to answer. While, in the Frankfurt example, it is clear that the capacity bottleneck is the runway system, the amount of 'normal', hence tolerable delay rates must be considered. Generally speaking, a certain amount of delay will be accepted by the entire industry, provided, that the acceptance would lead to some higher numbers of slots, which can be distributed to airlines. We could even say that a zero delay tolerance, which automatically would reduce the number of slots available for allocation, is a waste of capacity. The big question then is what the industry will tolerate, and one can image that this question is an almost permanent item for debating at the meetings of the coordination committees.

The runway capacity is only one aspect, which has to be considered. Other capacity elements could be the number of parking positions, waiting room capacities, the number of gates and their respective capacities, which again have an influence on the number of flights, which might be handled simultaneously. This again depends on the size of aircraft planned for which flights (!) etc. Specific items such as the fact that flights into certain directions (e.g. to the US or to Israel) might require specific security handling could add to the complexity of the set of coordination parameters. These categories might be the 'hard facts'.

In many cases, this is not the end of the parameter framework: for instance, most airports have some kind of night flying restrictions, or, if they have a full night ban, there are exceptions from the ban. One of the more complex night regulations is the noise point scheme for Frankfurt Airport where each night flight would use a certain number of noise points, depending on the planned aircraft category. Since, for an entire scheduling period, only a certain amount of noise points is available for the slot allocation at night, this scheme contributes to the set of restrictions at this airport. On the other hand, the noise point system is quite an incentive for all airlines to use quiet aircraft, which would give to them some perspectives to operating more movements at night. Other night flying restrictions deal with home base clauses, e.g. home-based carriers might be able to operate later at night than other airlines or even for 24 hours. Another possibility could be to introduce 'bonus list aircraft' which would mean that certain aircraft categories have a longer operating time at the airport than other aircraft types. Outside Germany, we find even a benchmarking scenario, which would reduce the number of possible night movements for the noisiest airline or airlines.

Apart from night flying restrictions there may be other restrictions to observe. For instance, due to legal reasons, associated with environmental impacts, the number of hourly movements for Düsseldorf Airport has been restricted to a maximum number of movements (currently 45 per hour). The technical capacity of the runway system lies somewhere between 50 and 60 movements per hour. Similar restrictions apply for Paris Orly Airport, just to name another example.

So much for the set of coordination parameters. I think it is very easy to understand that this potentially complex issue might trigger vivid and sometimes extremely controversial debates among the members and attendees of the coordination committees.he European Slot Allocation Regulation even requests the member state to have carried out a capacity analysis in regular intervals.

After all these debates, and hopefully after having reached a general agreement, the Federal Ministry of Transport would decide on the coordination parameters for that airport and inform the coordinator.

Slot Allocation Criteria

The basics principle of the IATA slot allocation system is that airlines are allocated slots on the basis of their previous use of the airport. This is necessary so that they can operate services in the long term. This approach is often called one of 'Grandfather Rights'.

Of course, there are other priorities besides the Grandfather Rights. Some of them are:

- 'Changed Historics' – e.g. an air carrier wants to change the departure time of a flight from 10:00 to 11:30, with all other schedule elements unchanged. Such a flight would have a higher priority than simple expansion programmes.
- 'New Entrants' – this priority segment was introduced by the European Slot Regulation and was adopted by IATA's WSG. Basically it says that an air carrier which is new at that airport or a new schedule itinerary planned by an incumbent air carrier also have a higher priority than the already mentioned expansion programme.
- 'Year Round Services' – a carrier has started a new service during a winter period and wants to continue this service throughout the coming summer. In the interest of schedule stability such flights again would have a higher priority than the mentioned expansion programmes.

Other, more 'soft' criteria might also play a role in the set of priorities, IATA's WSG in particular requests coordinators to consider such factors.

So, in short, the coordination parameters frame work or the 'resource pool' at an airport would basically include the following elements.

- airport and ATC (technical capacities);
- environmental restrictions (e.g. night flying restrictions, night bans);
- administrative restrictions (examples: e.g. Düsseldorf, Paris-Orly);
- the priority system;
- additional (local) rules – if any.

Above all the coordinator has to stick to the principles of transparency, neutrality and non-discrimination.

I think it came out clearly by now that a 'slot' may not be such a simple thing. Rather, it may become quite difficult to exchange slots, considering all the different capacity elements and other potentially limiting factors.

The Slot Allocation Process: An Example in Real Time

We should now have a more detailed look into these procedures in order understand them properly, and we should focus on the EU. Keeping in mind that I am responsible for the coordination and allocation of arrival and departure slots in Germany I would like to take the activities in this country as an example, however emphasizing, that there are not many differences – if any – to what might be going on in other EU member states at the same time.

To start with I think it is a good idea to follow the procedures for a scheduling period from its first preparational stage throughout its duration.

Let's take summer 2004 as an example.

July 2003

The Airport Coordinator receives the official invitation to attend the next meetings of the coordination committees for the coordinated airports of Frankfurt, Düsseldorf, Munich, Stuttgart, and the three Berlin airports.

September 2003

All committee meetings are scheduled to take place in September; Since some meetings probably do not take long, several of these can be held on the same day. For the meetings of other airports, namely Düsseldorf or Frankfurt, more time is needed and therefore separate meeting dates are envisaged. All these committee meetings will be attended by representatives of the airport in question and by ATC as capacity providers, by airlines and their representations as capacity users, by the regional authorities ('Länder') as regulators and by the Airport Coordinator. All committees are chaired by a representative of the Federal Ministry of Transport.

While the complete range of the coordination committee's tasks may be found in Article 5 of the EU Slot Regulation, the focus of the planned September meetings undoubtedly lies on the discussions – or debates – of the coordination parameters which the airport coordinator is supposed to use for the coordination of the summer season 2004.

End September 2003

Meanwhile the airport coordinator has informed air carriers worldwide that they have to submit their schedule requests for the summer season 2004 to reach his office at the set line which lies about four weeks prior to the first day of the schedules conferences, organized by IATA. Normally, this deadline is by mid-October. At an earlier date, which again is part of the worldwide slot allocation procedures, the airport coordinator has informed all airlines about their schedules, which he considers historic. The historical priority or the Grandfather Right is another vital part of the decision making process when it comes to the initial slot allocation for the coming season. Grandfather Rights for the next season (summer 2004) can be obtained if the respective flight was planned and operated, as allocated, for at least 80 per cent of the previous season (the so-called use-it-or-lose-it rule). The information about what the coordinator considers historics will assist the airlines in planning their applications for the next season 2004.

Mid-October 2003

In our example, we meanwhile have reached the deadline for the initial submissions. The Federal Ministry of Transport has informed the coordinator about the coordination parameters, valid for the upcoming summer season 2004, as requested by Article 6 of the EU Slot Allocation Regulation. The IT system used by the airport coordinator has taken the coordination parameters on board. The airlines have submitted their schedule request for this period in time. It can be assumed, that the latter has been done almost without exception because meeting the deadline is in the airline's very own interest: Only requests which would reach the coordinator's offices in time would be subject to any priority consideration by the coordinator. In other words, even schedule requests for which Grandfather Rights normally would be granted might not be treated that way if the slot request did not reach the coordinator's office in time.

The coordinator and his team by now will have an extremely busy time: Within less than three weeks, all the incoming schedule requests, coming up to some 1.3 million flight operations for 17 airports subject to coordination or slot allocation, would have to be played against the coordination parameters, the various environmental and administrative restrictions and the set of priority rules. All this is aimed at generating schedules which are manageable by all the 250 airlines intending to fly into Germany. Although one can register a great deal of stability in the airline scheduling, particularly when looking at the highly congested airports such as Frankfurt, Düsseldorf, Munich or Tegel, there are a large number of cases where an airline might only get slots, which are far away from their original intentions. Furthermore, a great number of slot requests could not be accepted at all and had to be put into a wait list system, which is maintained for all airports subject to coordination.

Approximately Four Days Before The Schedules Conference

After the three weeks time the coordinators would submit whatever they were able to grant to the respective air carriers worldwide. And they also would inform the air carriers about the cases where no slots at all have been found.

This, after all, would give the affected airlines at least a few days to prepare themselves before the IATA Schedule Conference will start ....

... In the Second Full Week in November

The schedules conferences, which were mentioned already several times, are an integral part of the 'worldwide IATA system' of airport coordination.

These conferences take place every half year, namely for the initial coordination of the summer season – as in our example for summer 2004 – or for a winter season. They always take place about five months prior to the beginning of the season which is dealt with during the conferences and we will learn that there are good reasons for it.

The meetings are called 'IATA Schedules Conferences', however, IATA primarily just acts as the organizer of the events. The conferences are open to all entities being active in the field of airline scheduling, airport coordination, and airport slot allocation. The only prerequisite is the proof that the persons attending belong to an airline or to an officially appointed airport coordinator1 or 'schedule facilitator1'. We even have cases where officially authorized agents may represent some of the smaller airlines which might shun away from the costs of attending such a conference. Altogether, the conferences are attended by virtually all airlines doing international air traffic from a certain extent, onwards, and by airport coordinators and schedule facilitators for approximately 220 airports worldwide. The number of people attending might be well over 1,000.

Since a number of years, other industry partners such as airports or suppliers of IT programmes or equipment might attend as observers. However, they are explicitly not allowed to interfere into the discussions between coordinators and their counterparts, the airlines.

It may be of interest to learn that the European Slot Allocation Regulation stipulates that ... 'it is the Airport Coordinator who allocates the slots' (and only the Airport Coordinator).

The purposes of these conferences are to enable discussions between airport coordinators and airlines with the aim of agreeing on arrival and departure slots and officially allocating them. Keeping this in mind, it is obvious that the organization of the conferences is a bit different from other events of such a size: Not only that smaller delegations will use one large meeting room, but larger delegations of airlines as well as of airport coordinators would need their individual working rooms with all the necessary equipment for being able to discuss and allocate arrival and departure times. For instance, airport coordinators like the ones for Germany, for France or for the U.K. would normally install a full local area network with a server plus several workstations in order to do the job properly.

The big advantage of these conferences lies in the fact that airlines can talk to all the coordinators in person by just changing from 'Frankfurt' to 'London' or 'Madrid' by walking into the individual working room of the Spanish or British Coordinator who may have their offices next door. Experience has shown that even in the age of electronic communication personal meetings are extremely useful for achieving solutions, and this forum is also being used for discussions among airlines themselves, eventually to exchange ('swap') slots, or for discussions among different airport coordinators in order to jointly find solutions for international flights.

Mid-November 2003

After the official end of the conference everyone travels home again and now it is the time for airlines to create their schedules on the basis of the airport slots they got from the various coordinators. This is particular important in cases where the original requests to land or depart at certain times could not be fulfilled by the coordinator. Airlines would then not only need to adjust their schedules planning but also their aircraft rotation, crew rotation, seat reservation systems, marketing or traffic rights for international flights. Coordinators, on the other hand, would keep their coordinated arrival and departure times updated constantly and also would keep trying to improve the slots according to the airline's needs.

End November 2003 and Onwards

It cannot not be over-estimated that all airport coordinators have the obligation to make to best use of the resource pool, which is at their disposal. Some of the methods to do so are:

* keeping wait lists;
* considering the SRD;
* organizing regular optimization meetings (SOM);
* doing slot monitoring;
* considering overbooking profiles.

This is the time to briefly explain these four methods:

Wait lists At the end of each Schedule Conference a wait list is set up by the airport coordinator.

The wait lists are updated permanently and continuously.

Available slots will be offered to carriers having schedule requests on the wait lists.

Carriers having schedule requests on wait lists are asked in regular intervals whether they want to keep their request on that list.

The coordinator will cancel the schedule requests from the wait list unless a confirmation to keep them on that list is given by the air carrier.

31 January 2004

Considering SRD The European Slot Allocation Regulation has introduced 'Slot Return Deadlines' (SRD) which are set for 31 January for a summer season and for 31 August for a winter season. It is only from the SRD onwards that the requirements for a use-it-or-lose-it rule would occur, respectively that any infringements or other measures might be taken in cases where slots are not properly used. IATA's WSG has adopted the slot return deadline for its worldwide recommendations.

By introducing the SRD, regulators, coordinators and the rest of the industry acknowledged the necessity for airlines to reconsider and reorganize their schedule intentions after the end of the IATA schedules conferences. On the other hand, the SRD lies some two months before the beginning of the respective scheduling season. By then any airline is supposed to know whether or not all the slots will be needed which had been allocated by the airport coordinators during the conference. If not there is still time enough to re-allocate such slots to someone else who might be in desperate need for them. The introduction of the Slot Return Deadline has proved to be extremely successful and is more and more widely observed properly.

Early February 2004

Schedule optimization meetings (SOM) In order to assist the airline industry even further, the EUACA has introduced the European Schedule Optimization Meetings (SOM) which are placed always shortly after the Slot Return Deadline for the summer season, hence early in February. The SOMs are considered a 'mini-conference', valid for Europe only and they have been endorsed by the European Commission. Of course, there are some differences to the IATA Schedules Conferences. For instance, no Grandfather Rights, New Entrants or any other priorities will be considered. So, an air carrier happy with the slots received, may stay away from any of these SOMs. Also, there is no obligation for any coordinator to attend the meetings. Rather, attendance is up to the coordinator's discretion, but most of them attend. New or additional slots may primarily be allocated only in accordance with the wait list priorities.

The SOMs are conducted since five years and have proved to be a substantial benefit to airports as capacity providers and air carriers as capacity users.

Monitoring and Enforcement of Slot Allocations

Slot monitoring The principle of slot monitoring is very simple:

1) Have flights been operated without an allocated slot, where obtaining airport slots is mandatory?
2) Flights for which the coordinator gave the slot away – Were they properly operated?
3) Has the airline given proof that the respective flight was intended to be operated as allocated by the coordinator?
4) Did the actual operation of that flight show that the airline used the slot as given by the coordinator?

Of course, the reality is somewhat more complex. For instance, a certain amount of irregularities is accepted by the air transport industry, keeping in mind the complexity of flight scheduling with all its possible imponderables. The question is from what points onwards unpunctualities may not be acceptable anymore. One also has to differentiate between 'normal' flight irregularities and those cases where an airline is not able to or even does not want to stick to the arrival/departure time allocated. In such a case the industry would talk about an off slot operation that is 'intentional, significant, and regular'. There is an agreement throughout the industry that such a behaviour is totally intolerable and must be stopped, if necessary by starting infringement processes which may end up in the airline paying considerable fines or by taking up other suitable measures. A similar focus would lie on cases where slots allocated are not properly used. It is evident that slots can be distributed only once and not using them would be a waste of precious capacity. While we have seen that Grandfather Rights can only be obtained for flights, which were operated as allocated for at least 80 per cent of the previous season, the recent development sharpened the requirements for proper slot usage. The updated EU Regulation allows or even requests the coordinator to take away slots not used as allocated and also has imposed some obligation onto the EU member state to develop adequate procedures to prevent airlines from doing so. Spain, Portugal, Germany have already developed infringement processes which ultimately may impose very high fines onto airlines.

The British administration quite recently has started an initiative to introduce an adequate process for the U.K. Even the IATA WSG, after all being primarily an airline publication, strongly encourages airlines to give back slots not used to in order to avoid wasting precious capacities at congested airports.

The European Union Airport Coordinator Association – EUACA has developed some Recommended Practices for all European Airport Coordinators, based on the amended EU Regulation for slot allocation.

By and large, the entire air transport industry would consider slot monitoring an integral tool to improve the usage of the given resources. This view, by the way, is definitely shared by airports and their associations. The recent development shows that no show rates at least at major EU airports became smaller and obviously the air transport community has started to develop adequate tools to prevent capacity wastages.

Overbooking profiles Different to the other four methods to improve the usage of the given resource pools, overbookings must be handled with great care.

While all coordinators supposedly have their experience when, where, and to what extent slots might be cancelled on short notice or just not be performed there is always the danger that the flights allocated shall really be operated.

Depending on the capacity situation and the nature of the airport's constraints, having too many flights in the system may result in a disaster. Therefore, overbooking profiles may be used very rarely if at all.

Conclusions and Final Remarks

In the present environment we are facing mounting capacity shortages at major airports together with more and more market driven scheduling procedures, resulting in fierce competition among carriers. Experience has shown that avoiding congested times or even airports just fail. So would any voluntary mediation process. In the present, admittedly regulated environment, official slot allocation, while being far from the ideal, would bring a great deal of advantages. Advantages to capacity providers, mainly that the given resources are not overused, and to capacity users, namely that a smooth operation of the approved schedules, including the guarantee for the highly important Grandfather Rights, are secured.

All in all one can list a number of advantages linked to the present 'IATA' slot allocation system:

- It is done on a worldwide basis and we all can agree that air transport is a worldwide business. Limiting one of the key elements of scheduling, and capacity distribution, to just one or two areas (e.g. EU and/or US) would lead to nowhere.
- Although, the system is set in a rather regulated environment, it seems to be enough flexibility to cater for special situations.
- We learned that the system has clear priority rules, valid for everybody.
- The entire industry being active in this field accepts identical or at least similar procedures worldwide. This must not be underestimated.
- About 250 airports worldwide, out of which some 150 airports are located in Europe, are subject to airport coordination or slot allocation. With the exception of the US, the system seems to be very attractive to virtually all parts in the world.
- The system is inexpensive. In Germany for instance, the direct costs linked to slot allocation are less than €2.50 for each coordinated/operated flight movement. The costs in other countries lie in the same range. Compared to airport charges, arrival fees, charges for security, etc. these amounts are marginal. Any commercial system for slot allocation would be much more expensive. Furthermore, it has to be anticipated that slots might be considered an asset and only one institution might welcome this: the tax authorities.
- The system may have a built in early-warning-function: the airport coordinator would know best whether or not the given set of parameters at an airport will be sufficient for the next season or whether an overflow in demand has to be

anticipated. In the latter case there might be even time enough to adjust some of the capacity elements.

• The system is democratic for capacity providers and capacity users: We have learned that capacity elements, environmental and administrative aspects and local rules, together with the respective allocations priorities are widely discussed at the coordination committees and that these committees are open to all parties concerned. Arguments, which might be brought forward by whomever, have the chance to find entry into the set of rules, provided they are reasonable.

The present 'IATA Allocation System' is certainly far from ideal and there is always room for improvement. It has to be kept in mind that particularly in the recent past the system has undergone a number of adjustments, most if not all of them aimed at improving the use of the given capacities and avoiding their wastage. The best solution to solve any of the slot problems would of course be having ample capacities everywhere and at all times. We all know that reality is different and it remains to be questioned whether or not the functioning system which is applied worldwide and which has undergone a great number of changes throughout the time of existence should just be thrown away.

Chapter 3

Economic Perspectives on the Problem of Slot Allocation

Batool Menaz and Bryan Matthews

Introduction

Worldwide airline passenger numbers have more than doubled in the last 20 years, whilst air cargo volume has increased by some 300 per cent. In the EU, growth has been similarly remarkable, with passenger numbers increasing by more than 250 per cent from 74,000 million passenger kms in 1980 to 280,000 million passenger kms in 2002 (European Commission, 2004). This growth in demand has, necessarily, been accompanied by a range of supply-side changes, most notably the increased use of larger aircraft and, with the growth in the number of services offered and the development of 'hubbing', increases in the number of aircraft movements. In fact, the number of take-offs and landings have increased from 6,811,000 in 1998 to 7,444,000 in 2003 in total for the top 30 EU airports (European Commission, 2004). Table 3.1 illustrates the EU airports with the fastest growth in aircraft movements over recent years. It can be seen, for example, that Milan Malpensa airport encountered major growth during this period as aircraft movements increased by 182.1 per cent.

The increase in the demand for air travel has, therefore, led to increased demands being placed on capacity at airports. The top ten EU airports, in terms of passengers, experienced 6.8 per cent growth in passenger numbers and 4.6 per cent growth in the number of aircraft movements over the period 2000–2004, (European Commission, 2005). Over that same period, growth in the volume of cargo and mail loaded and unloaded at the top 10 EU airports, in terms of cargo, was even greater – at 13.4 per cent – though the number of aircraft movements grew by only 0.7 per cent (European Commission, 2005). It would seem, therefore, that growth in passenger demand is the main driver of the growth in aircraft movements and, hence, of the increased pressure on airport capacity.

The amount of capacity needed for an aircraft to take off or land is referred to as a 'slot', defined under UK and EU law as the scheduled time of arrival or departure available or allocated to an aircraft on a specific date at an airport. The supply of airport slots is limited by the availability of a bundle of airport facilities associated with take-offs and landings, such as terminals, stands and runways, but the most frequent cause of airlines not being able to schedule flights at desired times is the lack of available runway capacity.

Table 3.1   Total movements (take-off and landing) of aircraft (passenger and cargo) in the top five fastest growing EU airports in the period 1998–2003 for which complete data is available

Airport	1998	1999	2000	2001	2002	2003	% growth 1998–2003
Milan Malpensa (Italy)	76,900	216,146	249,727	237,029	236,409	216,910	182.1
London Stansted (UK)	126,600	155,098	165,776	169,578	170,544	186,477	47.3
Madrid Barajas (Spain)	269,300	306,672	358,487	375,558	368,064	383,804	42.5
Barcelona (Spain)	217,500	233,600	256,905	273,118	271,020	282,021	29.7
Munich (Germany)	278,500	298,969	319,009	337,653	344,405	335,602	20.5

Source: European Commission (2004), *EU Energy and Transport in Figures: Statistical Pocketbook 2004*, Luxembourg: Office for Official Publications of the European Communities.

Many airports are already experiencing excess demand from airlines relative to available airport capacity. Mostly, this is confined to peak periods, though in a number of cases airports experience excess demand throughout most of the day. For the summer 2000 season at Heathrow and Gatwick airports, for example, total demand for slots exceeded supply by over 15 per cent, as illustrated in Table 3.2.

Some have forecast that air travel demand could double over the next ten to fifteen years (Khan, 2001; IATA, 2003) whilst even relatively conservative estimates predict average annual growth rates of approximately 3 per cent over the coming decade. For UK airports, IPPR (2003) predict for 2030 that passenger numbers will increase by nearly triple the 2002 numbers. It therefore seems somewhat inevitable that existing airport capacity problems will be exacerbated and that problems will become more widespread over the coming years (Debbage, 2002; Abeyratne, 2000; Castles, 2000). In light of both the existing situation and this future growth, there would appear to be an urgent need to devise a solution to deal effectively with these capacity problems.

Clearly one means of alleviating capacity problems is to expand capacity and make more slots available. Essentially, this may be achieved either by extracting more capacity from the existing infrastructure or physically expanding the infrastructure. Both of these avenues have been pursued, but both are associated with problems.

Table 3.2   Demand for slots in summer 2000 at London Heathrow and Gatwick airports

	Heathrow	Gatwick
Total demand for slots	335,578	207,910
Total slots allocated	283,681	173,908
Excess demand	51,897	34,002

Source: Adapted from DotEcon (2002) *Auctioning Airport Slots*, report for the HM Treasury and the Department of the Environment, Transport and the Regions.

More efficient runway utilization, resulting from improved technology for example, has enabled airports to achieve increases in capacity. In the 1970s, the capacity of Heathrow and Gatwick airports combined was estimated to be 440,000 annual movements, however in 1997, Heathrow alone handled 426,000 (Starkie, 1998). Even in the period from 1991 to 1999, maximum hourly runway capacity in the summer seasons at Heathrow increased from 74 to 84 and at Gatwick from 41 to 48 slots (DotEcon, 2002). However, at high rates of capacity utilization, service reliability problems and questions of how to ensure an efficient allocation of capacity amongst competing airlines emerge. Therefore, the scope for continuing to squeeze more capacity out of the existing infrastructure, whilst maintaining a reliable service and an efficient allocation of services, must be limited.

Expansion of airport infrastructure has also taken place, in terms of numbers of airports, terminals and runways, but it has failed to keep pace with the demand for flights. Only one major new airport was built in the USA (Denver) and one in Europe (Munich) during the period 1978–1998 (Caves and Gosling, 1999). In 2001 approval was given for a fifth terminal at Heathrow, but only following the longest public inquiry in British planning (nearly four years) and, even now, it is some six years from being fully operational (BAA Heathrow, 2005). Furthermore, whilst the UK government acknowledges that the economic case for an additional runway at Heathrow is persuasive, it has indicated that it would be unlikely to gain approval within the next decade (DfT, 2003). There are many reasons why expansion of the infrastructure is so difficult. Starkie outlined that 'Many airports are still publicly controlled utilities subject to political whims and often tight budgets, expansion may be restricted by environmental limitations and other controls, and building new runways or extending existing ones is not an easy task' (Starkie, 1998).

So whilst some further expansion of airport capacity is likely, it will certainly be insufficient to both alleviate existing capacity problems and keep pace with projected growth in demand. One is left with a problem fundamental to the study of economics – how best to allocate a scarce resource amongst competing parties. Economists should, one might reasonably think, be in an excellent position to provide the appropriate solution to this problem. However, as often is the case, the specifics of the market and the differing perspectives of different economists result in not one but several alternative solutions being proposed. Hence, even within a series of papers by like-minded economists published by the Institute of Economic Affairs (Boyfield, 2003), unfair competition, economic inefficiencies, ad hoc pricing, unnecessary regulation, excess runway congestion and increasing environmental damage are all highlighted as market-specific issues bound up with slot allocation, leading to different authors recommending different solutions.

This chapter seeks to provide an overview of the different economic perspectives on airport slot allocation. We begin by describing the nature of the problem and assessing the current European system in allocating slots, before going on to examine various other types of allocation methods and the views of different authors in the literature. In particular, we examine the relative merits of pricing, auctioning, secondary trading and administrative arrangements. In undertaking our review and in drawing our conclusions, we also draw on our knowledge of the literature and experience of the allocation of scarce capacity within the rail sector, where similar problems arise.

Defining the Problem

NERA (2002) highlight three types of inefficiencies that may occur when prices for airport capacity fail to reflect the marginal social cost of using that capacity, as illustrated in Figure 3.1. Firstly, if price exceeds marginal cost, as illustrated by Demand D1 where the number of slots used (Q1) is less than the efficient number (E1), there is excess capacity. This is inefficient as the potential benefits of flights for which airlines would have been willing to pay a price equal to marginal cost are foregone. Secondly, where prices are less than marginal cost but all slot requests are satisfied, as illustrated by demand D2 where the number of slots used (Q2) exceeds efficient levels (E2), congestion arises. The third type of inefficiency is most common at major EU airports at peak times of the day and is illustrated using demand D3. This inefficiency arises when demand for slots (Q3) exceeds the declared capacity (E2) given the existing prices, therefore there is excess demand for airport use. In such cases, the optimal price (assuming no capacity expansion) is EP3 where slot demand equals declared capacity.

It is useful to make a clear distinction between the two problems arising from shortages in capacity – congestion, represented by D2, and scarcity, represented by D3. Congestion at airports represents the expected delays resulting from the transmission of delays from one aircraft to another. The use of an additional slot at an airport reduces the airport manager's ability to recover from an incident and increases the probability of delays. This becomes worse at high levels of capacity utilization, since there is a lack of spare capacity to recover from any delays. Congestion costs are the costs associated with these expected delays. In this way, the consumption of additional capacity and the resulting congestion at the airport imposes delay costs on airlines and, ultimately, passengers.

Scarcity at airports, on the other hand, represents the inability of an airline to obtain the slot they want in order to operate a particular service. The inability of

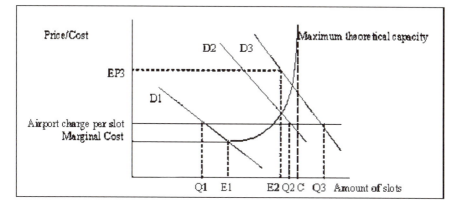

Figure 3.1   Three types of inefficiency when prices deviate from marginal costs (NERA, 2004)

the airline to provide the service it estimates will best meet its customer's demands represents a cost to society equal to the social value of that service, where social value is comprised of profit to the airline, consumer surplus to the passengers and net benefits to third parties – so-called 'externalities' – which may be positive or negative and which, for instance, include changes in pollution levels.

We are concerned here with the problem of scarcity – how to balance the demands of competing airlines in order to make best use of the existing fixed airport capacity. In principle, slots are efficiently allocated when used by those carriers that can generate the greatest overall social value from them. The problem then becomes how to identify, or how to get carriers to identify, the social value associated with a slot.

Firstly, there is a problem as to whether the value an airline places on a slot equates with, or at least approximates to, the social value of the slot. As was mentioned above, social value is comprised of profit, consumer surplus and the value of any externalities. Airlines will, almost by definition, take account of any profit they are able to make, and it is reasonable to assume that they are good at accounting for consumer surplus, particularly given the sophisticated price discrimination strategy most engage in today. However, they will not – again by definition – take account of externalities. So, if externalities exist in aviation – which most observers agree they do – airlines' valuations of slots will not equate with social value. Hence, any system of allocating slots based on the value airlines place upon them must either involve taking steps to internalize the externality (for example, by imposing a charge equal to the marginal external cost or providing a subsidy equal to the marginal external benefit) or assuming that their valuation of a slot approximates to the social value sufficiently well for the purposes of prioritization.

Secondly, there is a problem of whether or not it is possible for airlines to communicate their value within the allocation mechanism. That is, even if the value an airline places on a slot equates with, or at least approximates to, the social value of the slot, does the method of allocating slots allow or encourage them to reveal their valuation? Some mechanisms clearly do not allow this, and whilst others clearly do allow it, it is not always clear whether they achieve it.

The Existing Slot Allocation Mechanism

Internationally, slot allocation is based on a process developed under the auspices of the International Air Transport Association (IATA), centred around their bi-annual scheduling conferences. These conferences provide an opportunity for airlines and coordinators from airports around the world to meet and agree slot allocations for the coming seasons. Prior to the timetabling conferences, airline's (confidential) requests for slots are compiled, according to a set of criteria, into a draft proposal for the number and timing of slots allocated; this is then sent back to each operator. During the conference, the initial proposals are discussed between the operators and an agreed allocation is established.

IATA states that it promotes a slot allocation system that is:

*   globally compatible;
*   market driven and aimed solely at the maximum effective use of airport capacity;
*   transparent, fair and non-discriminatory;
*   simple, practical and economically sustainable (IATA, 2003).

At airports in the European Union (EU) slot allocation is governed by a set of conditions which sit within the framework of the IATA model. These conditions were first set out in EC Council Regulation 95/93 on the common rules for the allocation of slots at Community airports (since replaced in 2004 by regulation 793/2004). The objective of the EU, in setting additional conditions, was said to be aimed to encourage the efficient use of airport capacity through the optimal allocation of slots (DotEcon, 2002). Its principles were based on EU policy (EEC No. 2408/92) to 'facilitate competition and to encourage entrance into the market.'

However, both internationally and within the EU, the process is founded on the highly controversial principle of 'Grandfather Rights'. This principle of historic precedence bestows rights upon incumbent airlines to be granted slots in the next timetable on the basis of them having had those slots in the previous period. It was designed in an era when there were few if any capacity constraints on airports but has survived, with only relatively minor modification, to the present day. There are widely differing views on the appropriateness of grandfather rights. New entrants claim that grandfather rights deny them opportunities to enter the market and compete against the major carriers. On the other hand, incumbent carriers – those with the grandfather rights – argue that it maintains stability and continuity in scheduling which facilitates long-term planning (Pagliari, 2001).

The main modification to 'Grandfather Rights' was the introduction of the 'use-it-or-lose-it' rule, first adopted within the EU regulation and then incorporated by IATA into their overarching process. It means that if the historic right to the slot is to be protected for the next period, the slot has to be used at least 80 per cent of the time in the current period. The rule was introduced as an attempt to prevent incumbents from hoarding slots with the sole intention of restricting competition, though there is still thought to be powerful incentives for airlines to hang on to slots, even if they are not making optimal use out of them. That is, the incumbent carrier may still try to retain a slot, even if this involves operating loss-making flights because of the possibility of them generating 'scarcity rents' defined as additional revenues that accrue because competition is limited by capacity constraints.

Once the grandfather rights are confirmed, the remaining slots are allocated to a 'slot pool' which also contains any newly created slots through increases in hourly schedule limits, slots returned voluntarily and slots otherwise unclaimed by anyone. In an effort to encourage competition and new entry, under the EU Regulation, up to 50 per cent of the slot pool is set aside for new entrant airlines, defined as those carriers with less than 3 per cent of slots at a fully coordinated airport. Slots in the slot pool are allocated free of charge on a twice yearly basis.

However, a large percentage of slots in the slot pool are argued to be of limited commercial value and the number of slots available may not be sufficient to secure the scheduling of a new route. The size of the pool may only permit new entry from very small carriers offering low frequency services, which are unlikely to pose significant competitive challenges to high frequency services offered by established carriers. Hence, it is believed that the turnover of slots is typically small. For example, in summer 2000, 97 per cent and 89 per cent of slots at Heathrow and Gatwick respectively were grandfathered (DotEcon, 2002).

A final point worth noting here relates to the scope for slot exchange and trade within the existing allocation mechanism. Once an allocation of slots has been established at the timetabling conference, further ongoing negotiations are permitted between airlines in order for them to seek mutually beneficial exchanges of slots, so long as this is on a one-for-one basis and involves no monetary payment. This process of slot exchange has been described as 'the lubricating oil of international air transport services' (Smith, 2004). However, the exchange of slots accompanied by monetary side-payments, or any more straight-forward buying and selling of slots, is illegal throughout the EU. However in the UK, in a High Court ruling in March 1999 – the Guernsey Case – a judge ruled slot exchanges accompanied by monetary side payments to be lawful; setting a precedent for further cases of slot trading at UK airports (DotEcon, 2002).

In the face of continued criticism, the process in the EU underwent another reform in 2004, when regulation 95/93 was amended by the introduction of regulation 793/2004. However, changes have been described as being 'largely housekeeping in nature, with some tightening on language, roles and requirements' and that 'slot entitlement is still retained by usage' (Smith, 2004).

Many believe that the current slot allocation procedures perform poorly and do not allocate slots with the objective of generating the greatest benefit for consumers and the economy at large. Allocating slots principally on the basis of historic precedence may be seen as inconsistent with obtaining the greatest possible benefit from available airport capacity as a slot is allocated to the current user even though it is possible that a different user could generate much greater social benefit from it (DotEcon, 2002).

Differential Pricing

For many economists, the natural response to the question of how to ration the use of a scarce resource would be via pricing. In theory, if the price for each slot were derived so as to reflect the marginal social costs of usage, in particular scarcity or opportunity cost, an airline's slot requests and usage would take direct account of such costs, including the benefits that could have been generated if the slot was used by another airline. The result would be that each slot would be used by the airline who attached the highest economic value to it, as expressed through their willingness to pay. Put alternatively, Leveque (1998) argues that those carriers who are able to maximize the efficiency of the slots, i.e., to gain the maximum revenues per slot, should be willing and able to pay a higher price for the slots. As mentioned above, however, whether this maximizes social efficiency depends on how well airline revenues approximate to social value.

The question arises then of how to go about calculating the marginal social costs of usage, in particular the opportunity cost of a slot. One option would be for the airport or aviation regulator to attempt to calculate directly the costs involved. For instance, if an airline has to operate one of its services at a different time from that desired as a result of not securing the desired take-off slot, it would be possible to estimate the value people place on departure time shifts in order then to estimate the value to the airline's customers of the cost involved. Similarly, the costs of a longer journey time resulting from an airline not securing their desired landing slot may be estimated from passengers' values of time. This kind of approach to calculating scarcity costs was attempted for a rail corridor in Britain – the Trans-Pennine route between Leeds and Manchester. This found that 'the value of different types of service may vary enormously according to circumstances, and may also be substantial' (Nash et al., 2003). The method employed in this rail case is complex, and it is not entirely clear how one would translate it for use in the aviation context. In principle, the calculations involved are the same as those needed to identify the optimal use of capacity; so if that exercise has been undertaken then the necessary data should exist. However, Quinet argues – again in the context of rail – that it is only possible to calculate the optimal scarcity prices once an optimal allocation of slots has been identified, by which time the actual need to calculate optimal scarcity prices no longer exists (Quinet, 2003).

A more straight forward approach to using the price mechanism would be to simply differentiate the price of airport slots between different periods, as discussed by Doganis (1992) and NERA (2004). There are often significant and readily recognisable variations in the demand for slots at different times of the day, on different days of the week and during different months of the year. This can result in airports operating at full capacity at peak periods, whilst many slots may remain unused at off-peak periods. At an airport that experiences capacity shortages at peak hours and in peak months, despite having significant excess supply for most of the time, it may be possible to accommodate all requests for slots, but not necessarily at the times requested (DotEcon, 2002). However, if the price of a peak period slot was higher than the price of an off-peak slot, the demand for slots may be rescheduled; a high peak price may induce at least part of the demand to use airport capacity at off-peak periods. The extent to which this occurs will depend on the willingness of airlines to switch between slots at different time periods, which in turn depends on factors such as:

- the time sensitivity of their target customers (business travellers are typically very time sensitive and can be charged premium prices, whereas leisure travellers are typically more price sensitive);
- the need to make efficient use of their assets by optimizing aircraft usage;
- their ability to coordinate landing and departure times with the other airports on the routes they serve; and
- the need to coordinate a particular flight with others running on the same route.

In order to implement this sort of differential peak pricing at all accurately, the price to be charged would rely on the use of advanced forecast demand for slots. If, as one might reasonably expect, the higher charges for slots in the peak periods are reflected in passenger fares then the pattern of passenger demand would alter, perhaps significantly. For example, passengers who are not willing to pay the new peak fares may shift to flights at different times, use different modes of transport or not travel at all. This may lead to airlines that find themselves operating flights that carry insufficient passengers, rescheduling or reducing these flights. Forecasting in light of these sorts of potentially major changes will always involve a degree of uncertainty, particularly during the early stages of any transition between current charges and market clearing levels. In any event, it is postulated that airports would interpret any forecasts with caution and would seek to avoid setting slot prices at a value higher than that airlines would be willing to pay, for fear of suppressing demand too much and leaving slots unsold. Therefore it seems likely that slot prices in the peak would increase but that the increase would tend to fall short of the market clearing level and excess demand would remain (NERA, 2004).

Alternatively, differentiation could be introduced via an iterative process, whereby higher prices for peak and lower for off-peak slots are introduced in phase one and the airport authority could then assess whether or not excess demand has been eliminated. If not, the peak-price for the subsequent time period could be raised, where as if the price is too high and capacity is not being fully utilized, the price could be reduced. However, Nilsson (2002) highlights that this process of iteration could take several periods to arrive at an equilibrium solution and meanwhile, administrative allocation decisions would still have to be taken for the excess demand situations, leading to disruption of schedules and leaving open the possibilities for misallocation and even non-allocation of slots.

The issues of who receives any additional revenues generated by such differential pricing (or, as we shall note below, from auctions) and the use to which such revenues are put are important and contentious. Airports tend to argue that the slots form part of their assets and, therefore, that they should receive the revenues arising from them. Indeed, the funds generated could provide a natural and economically efficient way of recovering the costs of investment in enhanced airport capacity, e.g., by financing the building of more runways or the establishment of satellite airports. However, an airport operator who has market power in the provision of services and who can benefit from additional revenues arising out of differential pricing may actually have an incentive to restrict capacity, to drive up the price of slots and generate excess profits. As Carlsson (2002) suggested, 'Since the revenues from the user charges can be higher when capacity is too small, the incentives from infrastructure investments might be distorted'. Furthermore, it is thought that if runway capacity were limited by planning and environmental concerns, it would be inappropriate for the airport operator to enjoy scarcity rents resulting from these limitations.

Government may also see itself as having a call on these revenues, perhaps as a means of offsetting airport operators' monopolistic incentives. However, many airlines contend that slots for which they have Grandfather Rights actually form part of their assets. They are, therefore, strongly opposed to differential pricing (or auctioning) and any resulting transfers from the relatively less profitable airline sector to the government or the relatively more profitable airports sector.

There is some experience of differential peak pricing of slots having been used in practice. For example, Doganis (1992, p. 106) notes that British Airport Authority's introduction of peak period charges at Heathrow and Gatwick airports 'had only a limited effect of shifting demand out of the peak period.' Though somewhat anecdotal, this serves to throw further doubt on the effectiveness with which this approach could be employed.

The Role of Auctioning

Rather than airport operators or regulators having to estimate and test different price levels, a number of economists argue that it would be more efficient to introduce an auction mechanism, where operators are asked to quote prices for the slots they want, (Nilsson, 2002; DotEcon, 2002). Nilsson (2002) contends that

'The main argument behind the proposal comes from within the industry. In order to keep the airline industry reasonably competitive, it is of strategic importance to provide access to scarce airport capacity on equal grounds for all airlines, large and small, incumbents and entrants'. DotEcon (2002) also argue in favour of auctions, stating that 'A transition to a market-based allocation system with appropriate safeguards against concentration of slots would improve efficiency, encourage competition and yield significant benefits for consumers'.

Auctions come in many different formats, requiring a choice to be made as to the most appropriate format for slot allocation. The main types of auction are described elsewhere in this volume (Button, ch. 16) It has, for example, been suggested that the primary market for airport slots be organized as a 'sealed-bid, one price auction' (Grether et al., 1989). Each bidder submits a sealed bid for each unit desired indicating the maximum price they are willing to pay. They do not have knowledge of the behaviour of other bidders. The bids are then arranged from highest to lowest. If x units are auctioned, then the highest x bids are accepted. The price paid by each of the winning bidders is the value of the lowest accepted bid.

Nilsson (2002) provides an example of how the mechanism might work. Suppose bids are submitted for a specified part of a day at an airport. It is assumed that carrier A submits a single bid for 450; carrier B submits three bids of 3000, 700 and 400; carrier C submits two bids of 550 and 425; and carrier D submits three bids of 1500, 500 and 350. Suppose that six slots are available for this period. The slot that was last accepted would then go to bidder A, while bidders B, C and D would each get one slot less than they want. All six allotted slots would go at price 450, as it is the lowest accepted bid, even though carrier B submitted the bid 3000.

It is argued that this mechanism is able to achieve the optimal bidding strategy. Each carrier would bid close to the maximum they are willing to pay, in order for their bid to be accepted. This value is directly related to the profits the flight will generate. As a result, slots may be allocated to those airlines that value them the highest. As the highest bids do not determine price, profits from carrier's most profitable flights are protected. Price is determined by the lowest accepted bid and therefore by the least profitable flight in the market. Carriers may be tempted to inflate their bids in order to increase the chance of their bid being accepted or to deflate bids in order to increase their profit. A carrier inflating its bid is a relatively risky strategy as theirs could be the lowest accepted bid, requiring them to pay their bid price and leaving them with little or no profit. A carrier deflating its bid may actually be a less risky strategy, as while it may diminish the chance of the carrier winning the auction, it will increase profits if they do.

Whilst slot auctions appear attractive as bidders have an incentive to reveal their valuation for the capacity if the auction is properly designed, there are some powerful practical arguments that make auction design difficult. The first, and perhaps most difficult to overcome, relates to the complementary nature of slots at different airports on a route or within a network. That is, any operator bidding for a take-off slot would certainly need a matching landing slot, and would be likely to seek networks of slots; multiple slots at a node and complementary slots at other nodes. Thus, slots would have to be auctioned in packages. However, there are many different ways in which they might be packaged so as to reflect different origin-destination pairings and different network configurations, some of which may favour certain types of carrier over others. Hence, the auction designs for these tend to end up being highly complex as they seek to reflect different potential packaging options (Pels et al., 2002). So whilst it seems essential that auctions be designed to enable bids for packages of slots, the complexity that this packaging is likely to entail means that it becomes less clear that a particular bid from a particular airline would accurately reflect their valuation of the slots in question.

It is widely acknowledged that certain types of aircraft use more runway and/ or terminal capacity than others and that an increase in the use of these types of aircraft would lead to a loss of airport capacity. Nilsson suggests that a way to handle this would be to make operations that placed disproportional demand on capacity, more costly to bid for. This would, however, make the assessment of bids somewhat more complex as the items being bid for are non-uniform, perhaps varying in terms of both their runway and terminal capacity requirements. One way of normalizing the process might be to simply require carriers wishing to use larger aircraft to bid for two slots (Nilsson, 2002), but this would not necessarily accommodate differential requirements for runway as compared with terminal capacity.

Higher density routes with greater revenue potential would be expected to bid up the market price of slots, hence squeezing out smaller-scale services. In general, this is entirely in keeping with promoting the efficient use of airport capacity. However, where smaller scale services are deemed to have some social value not properly reflected in potential revenues, this may be a problem. This problem may be eliminated by reserving a certain number of slots to be put to the use of serving small communities. The EEC Regulation 95/93 tried to handle this problem by allowing special provisions for regional services under strict conditions. Capacity constrained airports could reserve slots for domestic flights conditional that the route was considered to be vital to the development of a particular region, the slots were being used for that route at the time of implementation of the regulation and no other carrier was operating that route and there was no other mode of transport offering an adequate service. The reservation of these slots ends when another carrier operates an equally frequent service. Slots could also be reserved on routes where public service obligations (PSOs) had been imposed under EU legislation. However DETR (1998) did not believe it was possible for the UK to reserve slots where a PSO was not imposed. Considerable opposition was expected from within the industry to this setting aside of slots for these regional or smaller scale routes. BAA felt that 'ring-fencing' in this way could only be done at the expense of other services that provide greater benefit to the UK economy. However they were likely to feel that way due to the fact that regional services are typically less profitable and use smaller aircraft carrying a small number of passengers. (DotEcon, 2002)

As with pricing, larger carriers are likely to be in a much stronger financial position to acquire the slots. Carriers with large market share are likely to have much deeper pockets and have a greater chance of winning slot auctions than new entrants and carriers with small market share. Some commentators have, therefore, raised concerns regarding the extent to which auctions would actually encourage competition and the potential they might give rise to for anti-competitive practices. For example, Abeyratne (2000) argued that 'the sale of slots ... is considered by many as unduly oligopolistic and favouring only a few powerful air carriers of the world'. However, Nilsson (2002) argues that auction mechanisms can be designed so as to maximize the potential for competition and minimize the threat of anti-competitive practices. For example, if auction rules were designed so that neither winners nor bids were announced and slot trading rules could be designed to conceal the identities of the other carriers involved then restriction of competition via collusion would be made difficult. Furthermore, it could be difficult for a carrier to utilize an auction process to monopolize an airport by, for example, driving up slot prices to prevent competition. Firstly, in doing so they would be using up their presumed monopoly profits. Secondly, any monopolistic tendencies could be further undermined if auction revenues were to be earmarked for airport capacity expansion. If monopolistic tendencies occurred, a monopoly would only be effective if it could withhold supply, involving the monopolist paying for slots that might either go unused or be used for operations involving small numbers of passengers that would not generate sufficient revenue to cover the price paid for them.

As with differential pricing, who receives the revenues and the use to which they are put are important and contentious issues. Revenues from auctions might again be an effective means of funding capacity expansion but again there are potential incentives to restrict capacity, to drive up bid prices and generate excess profits. In the case where the funds cannot be set to expand capacity, e.g., for planning or environmental reasons, Nilsson (2002) suggests that they should be used to hold a negative auction, of the sealed-bid, one price type, in order to encourage off-peak traffic. Carriers bid negatively, indicating the amount of subsidy it would take to persuade them to provide off-peak services. Each carrier operating at a subsidized hour would receive a subsidy equal to the lowest accepted (negative) bid. In the end, however, airlines have tended to be strongly opposed to the use of auctions, as with differential pricing (Grether et al., 1989).

A Secondary Market for Slot Allocation

So far, we have concentrated on primary allocation mechanisms – the distribution of runway slots from the regulator or airport to airlines, e.g., via Grandfathering, pricing or auctioning – as opposed to secondary market mechanisms. Secondary allocation refers to the redistribution of slots among airlines. This may be through slot exchange (barter trade) and monetary trading via for example, auctions or face-to-face transactions. It is believed that slot trading may bring advantages in terms of efficiency, as the monetary payments would create gains from trade if an airline currently holding a slot is not the most efficient user.

The CAA (2001) asserts that legitimizing a monetary secondary market in slots offers scope to improve the efficient allocation of scarce airport capacity by confronting the users of grandfather slots with the true opportunity cost of slots held. Such a market for airport slots has existed for a number of airports in the USA since the mid-1980s, though a secondary market involving monetary payments remains illegal in the EU, outside the UK.

From April 1986, the four major US airports (Kennedy, LaGuardia, O'Hare and National) have used a secondary market based mechanism known as the 'buy-sell rule'. Capacity was still allocated based on grandfather rights but in addition, the new rule allowed any carrier to purchase, sell, trade or lease slots. It was possible to not only buy or sell slots on a permanent basis but also on a temporary basis. The market was restricted to slots used for domestic services, which were divided into two groups: air carrier slots and commuter slots. However the market was overseen by the Federal Aviation Administration (FAA), which stated that air carrier slots could be traded without restriction while commuter slots could not be bought by the large carriers and all other slots, including those for international flights, were excluded from the trading market. Carriers had to 'use-or-lose' their slots. If slots were not used in a predetermined minimum time in a period of two months, the slots had to be returned to the FAA. Slots that had become available were assigned to a pool and reallocated using a lottery with 25 per cent initially offered to new entrants.

Starkie (1998) found that there was an initial surge of activity as airlines acquired the slots that they believed they could use best and disposed those that could be sold profitably. After this sorting out process, the number of outright sales fell but there was an increase in short-term leases, reflecting the fact that some carriers require the use of a slot at only limited times of the year.

Starkie asserts that in the absence of a pricing mechanism, evidence from O'Hare airport in Chicago supports the case for having a secondary market in slots, as it is likely to increase the efficient use of slots at congested airports. But it must also be considered that the emergence of a backdoor secondary market without formal regulation could have many problems. There is the risk of market power concerns as slots could become concentrated amongst relatively few carriers thereby giving them strong market positions. The European Commission in 2001 criticized slot trading for not easing market entry but reinforcing the dominant position of incumbent carriers at congested European airports. There may be a lack of transparency, as an airline that is planning to sell a slot would need to find a buyer. Without a formal market, it may be difficult to identify potential buyers, especially potential new entrants. Established airlines may therefore be the natural parties for a seller to approach. There is also the risk that this secondary market may not promote public interest and may eliminate particular routes that had been safeguarded.

Czerny and Tegner (2002) argued that the current regulatory framework for the allocation of runway slots in the EU neglected the positive effects of market mechanisms. The trading of runway slots was rejected even though it led to an increase in overall efficiency in the US market. DotEcon believes that there is a strong case on efficiency grounds for the development of a secondary market in slots, but if it is not regulated properly, it could allow airlines to gain market power. Therefore to address these concerns, slot trading could be implemented with a strict application of competition law.

Conclusion

The current system for allocating slots at most airports is based mainly on historic precedence and administrative procedures and is, in the context of increasingly scarce capacity at airports, widely viewed to be economically inefficient. With existing users being given priority to use the same slots in the next season, it does not include an explicit mechanism for ensuring that slots are allocated to those who attach the highest value to them and, hence, does not adequately reflect the scarce nature of airport slots. Historic precedence does nothing to encourage competition from new entrants and, even though a slot pool is created for this purpose, it is argued that the slot pool contains slots of a low commercial value, and insufficient slots to allow a new entrant to create a service or route that could compete with incumbent airlines. It is, therefore, highly unlikely that the system results in slots being held by the most efficient user or those users who place the greatest value on the slots. It has, however, survived many years of concerted criticism. Proponents point to its success in maintaining stability within a complex and turbulent industry, whilst airline competition and new market entry has been provided for via competing airports.

A number of alternative systems that could be used to allocate slots based on a reflection of their value as a scarce resource have been proposed. Slots could be priced to better reflect their value and opportunity cost, auctioned in a way that allocates them to the optimum bidder who can best utilize the resource, or traded on a secondary market following either an administrative or market-based primary allocation. Alternatively, some combination of administrative and pricing mechanisms might be used. Yet few of these alternatives have been tested in real world conditions, despite research and investigation into the subject dating back over 30 years (Carlin and Park, 1970; Yance, 1970).

Differentiation of slot prices may be a good way to reallocate the demand for slots from peak to off-peak periods, hence reducing scarcity at peak periods. It would, in all likelihood, lead to some prioritization amongst different air services, but an optimal allocation of slots would require prices to be estimated with a high degree of accuracy. In actuality, it would be difficult to calculate what prices to charge. One proposal is for the airport authority to attempt to calculate directly the opportunity cost involved, though the scale and complexity of this calculation would be overwhelming. Alternatively, a process of iteration towards equilibrating prices may be a pragmatic means of arriving at appropriate price levels, but the time it might take to arrive at these prices, and the implied distortions along the way, are causes for serious concern.

There is a sizeable body of literature relating to the use of auctions for allocating slots (e.g., Nilsson, DotEcon, NERA, Grether et al.). There are many different types of auctions to choose amongst. Nilsson argues that the sealed bid, one price auction – a sealed process where the highest bid is accepted at the lowest bid price – is the most efficient method of allocation as it allowed bidders to reveal their true valuation of the slots, and did not destroy their profits as the lowest bid determined the price. However, there is considerable concern regarding the practicability of designing an auction for airport slots, given the different complexities that most agree would have to be allowed for. Probably as a result of this concern, there is no actual experience of auctions in this setting.

The use to which the revenue generated from differential pricing and/or auctions would be put to, is a very important issue. This depends on who receives the funds, which in turn depends on who is judged to 'own' the slots – governments, airports or airlines. Using the price mechanism as a means of rationing demand would involve the transfer of economic rent from the airlines to the airports (or, in the case of a tax, to the government). In an environment where many airlines are already less profitable than airports, a further transfer in this direction is likely to be strongly resisted.

There is a strong case for a secondary market in slot allocation. After the primary allocation of slots via administrative forms such as grandfathering, slots could then be redistributed among airlines in a secondary market through barter trade or monetary trading, as has been the case at four of the busiest US airports. This may ensure that all slots are efficiently allocated to the appropriate users, however the market needs to be regulated to ensure against market concentration concerns.

There are fears that market-based allocation mechanisms may have a negative impact on airline competition, in that the major airlines are likely to be able to better afford the higher prices for slots than smaller operators or new entrants. In a policy environment where competition is viewed as being 'good' for the market, these competitive concerns are likely to be another barrier to implementation. However, there is perhaps a question mark surrounding the efficacy of encouraging new entry at already capacity constrained airports. It may be that scale economies imply that there is an optimal number of airlines that might operate out of one airport, and that that number might be quite small. Encouragement of competition from new entrants may lead to significant instability; in which case it may be better to encourage competition via some form of franchising.

Our view is that economics definitely provides the scope to improve upon the existing system. However, there is no conclusive evidence, and hence a lack of consensus, regarding which economic approach would be most appropriate to the specific conditions of the aviation market. From the point of view of implementation, it appears clear that wholesale reform in one particular direction, involving the abandonment of the existing system, is not feasible. Slot trading is, therefore, attractive as it offers the potential to gradually increase the efficiency of slot allocation without any major disruption to current practices. It would, however, probably require regulation to ensure fair trading prevailed. The potential for auctions could also be tested, again without abandoning the existing system, by allowing for experimental auctioning of new capacity and/or slots from the slot pool. These moves would provide for a gradual improvement of slot allocation; any more comprehensive changes would depend on the success of these moves and/or further research to demonstrate the degree to which the existing system fails to provide for an efficient outcome.

References

Abeyratne, R.I.R. (2000), 'Management of Airport Congestion through Slot Allocation', *Journal of Air Transport Management*, 6, pp. 29–41.

ACI Europe (2004), 'ACI Europe Comments on the NERA Study to Assess the Effects of Different Slot Allocation Schemes', Airport Council International, Brussels: http://www.aci-europe.org.

Adler, N., Nash, C. and Niskanen, E. (2003), 'Towards Cost-Based Pricing of Rail, Air and Water Transport Infrastructure in Europe', paper presented at the IMPRINT seminar, May, Leuven.

BAA Heathrow (2005), *Heathrow Airport Interim Master Plan*, June.

Barrett, S.D. (1992), 'Barriers to Contestability in the deregulated European Aviation Market', *Transportation Research*, A 26 (2), pp. 159–65.

Bass, T. (1994), 'Infrastructure Constraints and the EC', *Journal of Air Transport Management*, 1 (3), pp. 145–50.

Boyfield, K. (ed.), Starkie, D., Bass, T. and Humphreys, B. (2003), *A Market in Airport Slots*, London: Institute of Economic Affairs.

Bruzelelius, N. (1997), *The Airport Problem: An Economic Analysis of Scarce Runway Capacity*, KFB-report 1997:6.

Carlin, A. and Park, R.E. (1970), 'Marginal Cost Pricing of Airport Runway Capacity', *American Economic Review*, 60, pp. 310–19.

Carlsson, F. (2002), 'Airport Marginal Cost Pricing: Discussion and an Application to Swedish Airports', Working Papers in *Economics*, No. 85, December, Department of Economics, Goteborg University.

Castles, C. (2000), 'Development of Airport Slot Allocation Regulation in the European Community', in Bradshaw, B., and Smith, H.L. (eds) (2000), *Privatization and deregulation of transport*, Basingstoke: Macmillan.

Caves, D. and Gosling, G. (eds) (1999), *Strategic Airport Planning*, Oxford: Pergamon.

Civil Aviation Authority (1995), *Slot Allocation: A Proposal for Europe's Airports*, CAP 644, London: UK CAA.

Civil Aviation Authority (2001), *The Implementation of Secondary Slot Trading*, London: UK CAA.

Civil Aviation Authority (2004), *Secondary Trading in Airport Slots: A Consultation Paper*, London: UK CAA..

Czerny, A.I. and Tegner, H. (2002), *Secondary Markets for Runway Capacity*, Berlin University for Technology, IMPRINT.

Debbage, K.G. (2002), 'Airport Runway Slots: Limits to Growth', *Annuals of Tourism Research*, 29 (4), pp. 933–51.

Dempsey, P.S. (2001), 'Airport Landing Slots: Barriers to Entry and Impediments to competition', *Air and Space Law*, Vol. XXVI/1.

DETR (1998), *Select Committee on Environment*, Transport and Regional Affairs, Eighth Report, 28 July.

DfT (2003), *The Future of Air Transport*, White Paper, London: Department for Transport.

Doganis, R. (1992), *The Airport Business*, London: Routledge.

DotEcon (2002), 'Auctioning Airport Slots', report for the HM Treasury and the Department of the Environment, Transport and the Regions: http://www.dotecon.com.

European Commission (2004), *EU Energy and Transport in Figures: Statistical Pocketbook 2004*, Luxembourg: Office for Official Publications of the European Communities. Available at: http://europa.eu.int/comm/dgs/energy_transport/figures/pocketbook/index_en.htm.

European Commission (2005), *EU Energy and Transport in figures: Statistical Pocketbook 2005*, Luxembourg: Office for Official Publications of the European Communities. Available at: http://europa.eu.int/comm/dgs/energy_transport/figures/pocketbook/index_en.htm.

Forsyth, P. (1997), *Price Regulation of Airports: Principles with Australian Applications*, Transportation Research-E, Vol. 33, No. 4, pp. 297–309.

Grether, D., Isaac, M. and Plott, C. (1989), *The Allocation of Scarce Resources: Experimental Economics and the Problem of Allocating Airport Slots*, Boulder, CO and London: Underground Classics in Economics, Westview Press.

IATA (2003), 'Policy Paper: Ensuring an Effective and Globally Compatible Slot Allocation System', International Air Transport Association. Available at: www.iata.org.

Institute of Air Transport (1997), *Airport Charges in Europe*, Paris: ITA.

Jones, I., Viehoff, I. and Marks, P. (1993), 'The Economics of Airport Slots', *Fiscal Studies*, 14 (4), pp. 37–57.

Khan, A.M. (2001), 'Airport Slot Controls', in Button, K.J. and Hensher, D.A.(eds), *Handbook of Transport Systems and Traffic Control*, London: Elsevier Science.

Kleit, A. and Kobayashi, B. (1996), 'Market Failure or Market Efficiency? Evidence on Airport Slot Usage', in B. McMullen (ed.), *Research in Transportation Economics*, Greenwich, CT: JAI Press, pp. 1–32.

Leveque, F. (1998), 'Insights from Micro-economics into the Monetary Trading of Slots and Alternative Solutions to Cope with Congestion at EU Airports', Working Paper, Paris: CERNA.

Maldoom, D. (2003), 'Auctioning Capacity at Airport', *Utilities Policies*, 11, pp. 47–51.

Majumdar, A. (1994), 'Barriers to the Liberation of European Aviation – 'Just How Open Are the Skies?', 22nd European Transport Forum, PTRC (Planning and Transport, Research and Computation), 12–16 September, UK.

Nash, C.A., Coulthard, S. and Matthews B. (2003), 'Rail Track Charges in Great Britain: The Issue of Charging for Capacity', *Journal of Transport Policy*, 11 (4), pp. 315–28.

National Economic Research Associates (1998), *An Examination of Rail Infrastructure Charges*, London: NERA.

National Economic Research Associates (2004), *Study to Assess the Effects of Different Slot Allocation Schemes*, London: NERA.

Nilsson, J. (2003), *Marginal Cost Pricing of Airport Use: The Case for Using Market Mechanisms for Slot Pricing*, Swedish National Road and Transport Reasearch Institute, VTI notat 2A-2003.

Oum, T. and Zhang, Y. (1990), 'Airport Pricing: Congestion Tolls, Lumpy Investment and Cost Recovery', *Journal of Public Economics*, 43, pp. 353–74.

Pagliari, R. (2001), 'Selling Grandfather: An Analysis of the Latest EU Proposals on Slot Trading', *Air and Space Europe*, 3 (1/2), pp. 33–5.

Pels, E. and Verhoef, E.T. (2002), *Airport Pricing*, Tinburgen Institute Discussion Paper, TI 2002-078/03.

Productivity Commission (2002), *Price Regulation of Airport Services*, Report No. 19, Canberra: AusInfo.

Quinet, E. (2003), 'Short Term Adjustments in Rail Activity – The Limited Role of Infrastructure Charges', *Transport Policy*, 10 (1), pp. 73–9.

Rassenti, S., Smith, V. and Bulfrin, R. (1982), 'A Combinational Auction for Airport Time Slot Allocation', *Bell Journal of Economics*, 13 (2), Autumn, pp. 402–17.

Smith, C. (2004), 'Killing the Golden Goose: Assessing the Benefits and Pitfalls of Airport Slot Auctions, and the Consequences for Hub Development in Europe', presentation for the 11th Global Airport Development conference, 25 November, Prague.

Starkie, D. (1994), 'Developments in Transport Policy: The US Market in Airport Slots', *Journal of Transport Economics and Policy*, 28 (3) September, pp. 325–9.

Starkie, D. (1998), 'Allocating Airport Slots: A Role for the Market?', *Journal of Air Transport Management*, 4, pp. 111–16.

Starkie, D. (2001), 'Reforming UK Airport Regulation', *Journal of Transport Economics and Policy*, 35, Part 1, January, pp. 119–35.

Yance, J.V. (1970), *The Theory of Air Carrier Demand for Slots*, Secretary of Transportation, USA, 1970/01.

Chapter 4

Airport Slots: A Primer

David Gillen

Introduction

The first rules and regulations governing access at airports became official with the creation of the High Density Rule (HDR) in 1969 in the United States. However, the idea of a slot vis-à-vis airport access appeared three years earlier, in 1966, when airlines agreed to limit their flights at National airport to 40 per hour. The enactment of the HDR measure was seen at the time, as a temporary solution to congestion and delays and thus the idea of a slot in this context was considered provisional. It should also be noted that these decisions were made in the era of economic regulation of airlines in the US. Europe at this time did not face the same degree of airport congestion due to the tight economic regulation of aviation, the need for bilaterals and the dominance of home markets by national carriers.

Forty years later we now have airline deregulation but limited infrastructure liberalization.[1] The EU has also now equaled the US in the number of severely congested airports. However, not only has the issue of congestion exacerbated but, in practice, little can be said on the variety of alternative approaches of regulators to the congestion problem; it is the use of slots, albeit with differences between the US and EU, that is the mainstay of dealing with congestion and excess demand generally.

The purpose of this chapter is to provide an introduction to and basic coverage of airport slots; what they are – definitions, how they are allocated and the various types of slot transactions. We present a brief describing of the major factors that have contributed to the status quo. We also provide a comparison between the US and EU markets as well as differences that exist within the EU as between the UK and the rest of the EU.

What is a Slot?

A slot is most commonly known as a landing or take-off right at airports during a specified period of time. In the United States Code, Sec. 41714, a slot

---

1   The point being that deregulating infrastructure is not the panacea that some might think as local planning and zoning regulations present formidable barriers to expansion.

is defined as 'a reservation for an instrument flight rule takeoff or landing by an air carrier of an aircraft in air transportation'.[2] This definition implies a slot is associated only with the use of runway and no other infrastructure. There is also no implication of a property right to the slot being given to the user, only the use of infrastructure; legally the FAA claims to be the 'owner' of all slots at all Federally funded US airports.

The definition of a slot is quite different in the EU. Under EU rule 793/2004 'slot' is given a definition, 'slot shall mean the *entitlement* of an air carrier to use the full range of airport infrastructure necessary to operate an air service at a coordinated airport on a specific date and time for the purposes of a landing and takeoff'. Note the use of the word 'entitlement' implies lack of property rights since the carrier is given the right it does possess the right. Two other important differences are the slot refers to a range of airport infrastructure not just runways and second, the slot is confined to 'coordinated airports'.

How are Slots Exchanged?

Slots can be '*swapped*' between two or more carriers. This is fairly commonplace among many carriers for scheduling and logistic reasons. Second slots can be '*leased*' which is more attractive than selling as the holder retains control. They are also useful for short-term agreements with early termination clauses. Slots can be *sold* but there are relatively few outright sales. There are a number of reasons for this; first, airlines do not hold the property rights to slots only the right to use them and therefore slots are treated as quasi-permanent assets with a sizable amount of risk of loss attached. Second, incumbent slot holders engage in strategic behavior based on the potential network opportunity costs and knowledge of whom current and potential competitors are. This creates an incentive for hoarding and babysitting. Third, the value of slots is higher as a package than individually – incumbents if selling will want the full package value of each slot, but potential buyers may not be willing to pay full package value for a single slot.

Slots can also be reallocated due to bankruptcy proceedings, as part of a route transfer between carriers, to redeploy slots within an alliance group and to baby-sit surplus slots for current owner.

IATA Scheduling Process

The IATA scheduling process is a well recognized means of allocating slots and managing demand at slot coordinated airports around the world; approximately 213 and all international airlines take part in the scheduling meetings which are held twice per year before the winter and summer schedule is announced. Each slot coordinated airport has a slot coordinator (termed ASC – airport slot

---

2   United States Code Online, Sec. 41714.

coordinator) whose appointment varies from jurisdiction to jurisdiction (see Table 4.1 for differences between US, Canada and EU).

The allocation and trading of slots follows the following process:

- Each airline submits its desired schedule to the ASCs about six months before the start of the season, and the ASCs then allocate slots according procedures.
- The ASCs' decisions are formally announced at the start of the relevant international conference and this is also when airlines first see the planned schedules of their competitors.
- Trading starts, with airlines who did not receive their desired slots seeking to improve their allocations and also ensuring that they have consistent sets of departure and arrival times.
- ASCs provide information showing slot mismatches and who 'owns' various slots (brokerage role).
- Airlines may trade slots at the same or different airports alter the type of aircraft flown and the destination (origin) of the flight, subject to the approval of the ASCs.
- All trades must be authorized by the relevant ASC to ensure that there is sufficient terminal capacity and parking space to accommodate any changes.
- Trading can be complex, involving many parties in simultaneous swaps of slots.

Not all airports are slot coordinated. There are of three types of airports worldwide, namely level 1, level 2, and level 3.[3] Level 1, also called *non-coordinated airport*, describes a situation where capacity adequately meets demand and there is no need for slots. Level 2, or otherwise kown as *schedule facilitated airports*, requires a formal intervention in order to avoid a situation where demand exceeds capacity. This intervention consists of schedules being submitted to a schedule facilitator who relies on cooperation and voluntary schedule changes from air carriers so as to avoid congestion. In this case the rules and regulations of slot allocation based on Grandfather Rights do not apply. With Level 2 airports the property rights for slots would seem to go to the airlines since airports cannot reject their requests unless capacity constraints exist. Note such capacity constraints would differ under US and EU slot definitions.

Level 3 or *fully coordinated airports* describe situations where demand exceeds capacity and the shortage of resources cannot be resolved simply through voluntary cooperation between airlines. In this scenario the allocation of slots is based on historic precedence and supervised by a slot coordinator. As reported in the 2005 IATA Scheduling Guidelines document, there are 96 level 3 and 52 level 2 airports in North America and Europe combined (see Table 4.3 in the Appendix for more details). However, it should also be pointed out the validity of the aggregate number of airports labeled as either level 2 or 3 is subject to the

---

3   IATA, Worldwide Scheduling Guidelines, 12th Edition: December, 2005.

basis of how an airport is deemed as a Level 3 airport. Property rights for slot would for Level 3 airports seem to rest with the airport as they can deny access to an airline.

Worldwide, the formal allocation of slots relies on varying and different criteria including Grandfather, use-or-lose provisions, the services the use of such a slot would bring and the size of the marker the slot would serve. Grandfather or what is typically referred to as the use-it-or-lose it provision. The IATA based Grandfather ordinance states an air carrier is entitled to use the same slots in the following season if the slots in the current season are utilized eighty percent of the time. In United States, the grandfather rule applies to slots at four airports designated as High Density. These airports are categorized as either Level 2 or 3, where, as stated above, demand either approaches capacity or exceed it.

The use-or-lose-it provision was introduced to reduce inefficiency, yet there are different types of inefficiency that could easily be confused. A task force report, published on 7 December 2005, on slot allocation procedures defines inefficiency to mean 'operational practices that lead to the inefficient use of slots allocated under the current capacity regime'. It also refers to seasonality. It does not include, for example, economic inefficiency or the percentage of slots not usable (for example for purely traffic structure reasons) and therefore not available as real capacity'.[4] It is thus important to differentiate between technical and economic efficiency/ inefficiency. Technical efficiency is only a branch of overall economic efficiency that generally measures the 'maximum amount of an output produced for a given set of inputs'. As applied to airport access, technical efficiency could be measured by calculating the total number of slots used as a percentage of the total allocated slots. Whereas economic efficiency measures the total welfare created as a sum of the effects on air carriers, airports, and consumers.

Factors Leading to the Status Quo

There are different events that have helped define the present situation in United States. For an overview see Table 4.4. Hoever, six major events were:

High Density Rule 1969: The traffic at the High Density Airports is regulated under Subpart K of 14 CFR Ch. 1, Section 93.123. In 1969, the Administration designated five airports under the HDR, however in the 1970s, restrictions at Newark airport were lifted. The five HDR airports and their respective slot controls per hour in brackets are Kennedy International (80), Chicago O'Hare (135), La Guardia (60), Newark (60), and Washington National (60). These slot restrictions have changed over time.[5]

Buy-and-sell Rule 1 April 1986: The provisions outlining the transfer of slots are explicitly stated in the Code of Federal Regulations (title 14), Subpart S, Section

---

4   Task Force, Outcome of Study on Slot Allocation Procedures, European Civil Aviation Conference: December 7, 2005, DGCA/124-DP/6.

5   [Doc. No. 9974, 35 FR 16592, Oct. 24, 1970, as amended by Amdt. 93-27, 38 FR 29464, Oct. 25, 1973].

Table 4.1   A comparison of slot allocation, rules and definitions

	US	EU	Canada
Trading allowed	Yes, bought/sold	Not allowed but de facto slots are bought and sold	Not an issue
Swaps allowed	Yes, for operational purposes	Yes	Yes
Secondary trading	Yes	Yes but differences in UK and rest of EU	Not an issue
Part of IATA	No, anti-trust issues. Level 3 airports are: LAX, JFK, ORD, SFO, EWR	Yes, most primary and secondary airports in EU	Yes YVR and YYZ Level 3
Designation	FAA sets rule	Member states can designate airport as Level 2 or 3	Airport provides argument as Level 2 or 3
Coordinator	Airport	Government body appointment	Airline or airline/ airport
Impact on congestion	Low	High	Low
Impact on capacity	Effectively none	Effective reduction	negligible

93.221. As a preamble to this section the law states 'slots may be bought, sold or leased for any consideration and any time period and they may be traded in any combination for slots at the same airport or any other high density traffic airport'. Section 93.221 of the CFR also includes the necessary conditions required for the transfer, lease or sale of a slot.

Use-it-or-Lose-it provision: The purpose of the use or lose provision, introduced in 1985, is to encourage the efficient use of slots and to avoid slot hoarding by incumbent airlines. If the slot is not used 80 per cent of the time in the current season, the airline loses right to use the same slot in the following season and the slot will be recalled by the FAA. Initially, air carriers were required to use the allocated slots only 65 per cent of the time over a two month period but it was later raised to 80 per cent. A more detailed outline of the use or lose provision can be found in 14 CFR, Section 93.227.

Slot exemptions: As early as 1994 the United States Congress enacted the Federal Aviation Administration Authorization Act which, inter alia, 'authorized the Department to grant exemptions from the High Density Rule[6] for the provision of basic Essential Air Services (EAS) at eligible communities, for international

---

6   In this chapter the High Density Rule will be used interchangeably with Slot Rule.

air service, and for the service by 'new entrant' carriers'.[7] The Act represented a step toward a complete revocation of the slot rule in hopes it would help promote competition in the airline industry. The Act indeed promoted 42 slot exemptions for air service to LaGuardia of which 30 were allocated to new entrants and 12 were reserved for international flights.[8]

Wendell H. Ford Aviation Investment and Reform Act [AIR-21] 5 April 2000: The 1994 Act also invited the FAA to conduct a study in May 1995 so as to determine whether it is beneficial to abolish slots.[9] The study mistakenly assumed the elimination of slot controls would not substantially increase daily flights (i.e., at LaGuardia airport the study estimated an increase in daily flights of 6.9 per cent) while the increase in service would benefit consumers, air carriers as well as airports. Based on this study the FAA submitted a proposal that helped build the AIR-21 Act. The legislation eliminated slot controls at O'Hare Airport in 2002 and extended the expiration date for Kennedy and LaGuardia Airports until 2007. However, the AIR-21 law also directed the Administration to grant slot exemptions for two groups, namely new entrants and low volume flights (i.e., less than 71 seats) between a small hub or non-hub airport and a high density airport. Naturally, the slot exemptions granted by AIR-21 increased delays to an average of 48 minutes per flight at LaGuardia airport and in September 2000 9,000 flight delays were recorded at LGA.

Lottery 2001: The inability to accommodate the increase in flights, on account of the slot exemption provision in AIR-21 Act, lead to a moratorium on additional flights at LaGuardia airport as declared by the Port Authority of New York and New Jersey.[10] The new limit placed on flights at LGA by the Administration, beginning on January 31, 2001, of 75 commercial airline flights per hour would accommodate only 159 of the slot exemptions granted post AIR-21 Act. Thus, as a temporary solution the FAA introduced a lottery mechanism for the distribution of the 159 slot exemptions.

Status Quo

United States:[11] the last FAA amendment to the 1969 High Density Rule, published on 1 October 1999, addresses important features that still dictate the slot allocation process today and are key in understanding the present terms and conditions navigating airport access. The second document is a final rule on a reservation system for unscheduled arrivals at Chicago O'Hare International

---

7   Department of Transportation, Order Granting and Denying applications for slot exemptions at Chicago O'Hare Airport, Docket OST-97-2771-8, 24 October 1997.

8   The Subcommittee on Aviation Hearing on The Slot Lottery at LaGuardia Airport.

9   Ibid.

10   Ibid.

11   This section will draw its information from two main documents published by the FAA on the slot allocation procedures at certain High Density Traffic Airports.

Airport. This section will therefore expand on the following points: i) Domestic vs. International Air Carriers and ii) Scheduled vs. Unscheduled Flights.

i)   Domestic vs. International Air Carriers

There are three important characteristics underlining the slot allocation process in United States:

*   international air carriers are given priority at the slot constrained airports;
*   slots may be withdrawn from air carriers;
*   international slots may not be bought, sold, leased, or otherwise transferred.

The set of regulations for the allocation and transfer of air carrier and commuter slots in United States differentiates between international and domestic slots. In order to ensure the promulgation of the 1985 buy/sell rule would not impede access to slot constrained airports by foreign air carriers, the Department of Transportation granted power to the FAA to determine the allocation of new slots to international carriers at Kennedy, Chicago O'Hare, and LaGuardia Airports.

The international aviation policy gives priority to foreign carriers under the Code of the Federal Regulations, title 14, section 93.2 17 (a) (6) and charges the FAA to allocate slots for international operations at Chicago O'Hare upon request. The request must be fulfilled even if it means a domestic carrier's operations will suffer; in other words if a slot is not made available within 60 minutes of the requested time then the FAA will withdraw a slot from a domestic carrier. In Subpart S of the CFR (title 14), section 93.223 (a) the law specifically states 'slots do not represent a property right but represent an operating privilege subject to absolute FAA control. Slots may be withdrawn at any time to fulfill the Department's operational needs, such as providing slots for international or essential air service operations or eliminating slots'.[12]

At LaGuardia Airport, Section 93.2 17 (a) (7) of the CFR (title 14) provides power to the FAA to allocate slots to air carriers for international operations. However these slots will be imparted only if required by bilateral agreement. While at Kennedy airport, Section 93.2 17 (a) (8) dictates that domestic slots will be withdrawn for international operations only if required by international obligations.

The aftermarket in slots, or otherwise known as the secondary trading market regulated under the buy/sell rule, is restricted only to domestic landing and take-off rights. While domestic slots may be used for either international or domestic services, international slots may only be employed as originally intended.[13] Furthermore, international slots are not to be allocated through a lottery mechanism. Prior to the 1999 HDR amendment, FAA regulations

---

12   Government Printing Office, Code of Federal Regulations Online, http://www.gpoaccess.gov/cfr/index.html.

13   See ch. 12.

restricted the use of lotteries only for domestic slots and only US carriers were permitted to participate. The 1999 HDR amendment, inter alia, permits Canadian carriers to participate in any lotteries of domestic slots as a result of the US-Canada Open Skies agreement. Other relevant changes include: reclassification of certain international slots as domestic and a clause that enables the FAA to 'reduce the international allocation for air carriers that hold and operate more than 100 permanent slots at Chicago O'Hare by the number of international slots reclassified as domestic slots.'

A look at the number of domestic slots held by incumbent airlines in 1986 and 1999 shows a dramatic increase and entrenchment of incumbent presence at the four most congested airports in US which might be indicative of slot hoarding. For example statistics show an increase of 18 per cent in the slot shares of the two largest operators at ORD from 66 per cent in 1986 to 84 per cent in 1999. However, the more striking boost in market share of incumbents has occurred at LGA, JFK and DCA.

ii)   Scheduled vs. Unscheduled Flights

On January 21, 2004 the FAA issued an order limiting the number of scheduled operations at Chicago O'Hare Airport during peak period hours that would take effect in March 2004. Given the two largest operators at O'Hare airport were American and United Airlines, accounting for 88 percent of all scheduled services, the order applied only to these carriers and their affiliates.[14] The renewal of the limit on air traffic activity, first eliminated in July 2002, occurred as a result of record levels in delays. The FAA Aviation System Performance Metrics have shown 'from November 1 through December 31, 2003, 39 percent of O'Hare arrivals were delayed, with an average of 492 delays per day and an average of 57 minutes delay per delayed aircraft'.[15] There has been a significant decline in the number of flights arriving on time, from a high of 79 per cent in January 2003, to a low of 65 in January 2006.

The following restrictions were introduced in 2004 and are currently open to discussion as the FAA has invited all interested parties to show cause as to why the FAA should not extend the August 2004 order through to 28 October 2006:

- 'American Airlines shall not conduct more than 505 scheduled operations at O'Hare during the hours of 1:00 p.m. through 7:59 p.m. (local time) daily; and
- United shall not conduct more than 655 scheduled operations at O'Hare during the hours of 1:00 p.m. through 7:59 p.m. (local time) daily'.[16]

---

14   Operating Limitations at Chicago O'Hare International Airport, Docket FAA-2004-16944.

15   Ibid.

16   Ibid.

The benefits achieved by limiting the number of scheduled operations during the peak operating hours between 7:00 a.m. through to 8:59 p.m. at Chicago O'Hare International Airport in August 2004 would have dissipated had it not been for the development of a new regulation that would officially cap the number of unscheduled flights as well. The latter regulation came into effect on 8 July 2005 with the publication of SFAR No. 105 'Reservation of System for Unscheduled Arrivals at Chicago's O'Hare International Airport' [70 FR 39610]. This notice was extended on 31 March 2006 until 28 October 2006 for unscheduled arrivals during peak hours only. In the preamble of the SFAR 105 document it is stated the expiration date of the unscheduled arrivals cap should coincide with the final rule limiting scheduled operations. However, since the FAA has not yet set an expiration date for scheduled arrivals cap the FAA has extended SFAR until 28 October 2006.[17]

European Community

> One of the main difficulties of the current system of slot allocation has been to find the right balance between the interests of incumbent air carriers and new entrants at congested airports so as to take due account of the fact that incumbent air carriers have already built up their position at an airport and have an interest to expand it further, while new entrants or air carriers with relatively small operations need to be able to expand their services and establish a competitive network.[18]

A look at the European Community legislation on slot allocation procedures draws ones' attention to the similar events that helped define the status quo in United States, but with a substantial lag; for an overview of events that influenced the defining of the status quo of slot allocation procedures in the European Community, see Table 4.5. In Europe the Grandfather Rights provision was introduced in the EC legislature as late as 1993, when the Council Regulation (EEC) No. 95/93 on common rules for the allocation of slots at Community Airports became effective – a lag of eight years. Furthermore, the EU has relied a hundred percent on administrative measures and refrained from introducing market mechanisms in the slot allocation procedures, which, as exemplified by

---

17   For more detailed information regarding the limitations placed on both scheduled and unscheduled flights at Chicago O'Hare International Airport we recommend to consult the following primary FAA rulemaking documents:
- Operating Limitations at Chicago O'Hare International Airport, Docket No. FAA-2004-16944, Order Limiting Scheduled Operations [August 2004];
- Reservation System for Unscheduled Arrivals at Chicago's O'Hare International Airport, Vol. 70, No. 130 [8 July 2005];
- Reservation System for Unscheduled Arrivals at Chicago's O'Hare International Airport, [Docket No. Faa-2005-19411; SFAR No. 105], Vol. 71, No. 62, 31 March 2006.

18   European Commission, Activities of the European Union: Summaries of Legislation.

the secondary market in slots, are growing strong roots in the United States. The 1993 EU Council Regulation underwent only two significant amendments, the last one taking place on 30 April 2004. These changes however, were limited to definitions of certain terms and did not produce significant changes to the slot allocation process.

Two of the major discrepancies in the EC Regulations between 1993 and 2004 pertain to the interpretation of the words slot and new entrant. In Europe, the word slot means 'the entitlement established under this Regulation, of an air carrier to use the full range of airport infrastructure necessary to operate an air service at a coordinated airport on a specific data and time for the purpose of landing and take-off as allocated by a coordinator in accordance with this Regulation'.[19] Notice in contrast to the FAA definition, it is evident the EU definition does not limit a slot to mean a period of time set aside for the use of the runway space but it also considers the infrastructure needed to complete an arrival or a take-off at an airport. Another radical change in content relates to the word 'new entrant'. The 793/2004 regulation expanded the definition to include the number of slots held by the air carrier as a percentage of the total number of slot available on the day in question as a determining factor of whether the party should be labeled as a new entrant.

Air carriers operating in the EU depend on two provisions in their choice to service new routes at congested airports, namely acquisitions of slots from transfers and/or slot pools. After the primary allocation of slots, airlines are allowed to transfer slots between themselves but only under the supervision of a slot coordinator and under the provisions outlined in Article 8a of the 793/04. Airlines also have access to the slots retrieved by slot coordinators and placed in the slot pool. The rules governing access to the landing and take-off rights placed in the slot pool are outlined in Article 8 of the 793/2004. However, some scholars have argued these slots do not possess a high commercial value. As a result, in the UK a compensatory grey market has developed whereby monetary slot transactions are disguised as slot swaps. This situation determined the EC to conduct a staff working document on commercial slot allocation mechanisms at Community airports. Other studies that critically assess the current administrative slot allocation policy and potential benefits of market mechanisms include DotEcon (2001), Nera (2004), Task Force (2005) and Madas and Zografos (2005).

In defense of their policy of allowing slot 'trading', the UK argued it improved the economic efficiency of the use of this resource. In the table below, it is clearly evident that the use of secondary markets leads to a substantial increase in operational and economic efficiency. The numbers refer to changes between 2001 and 2006; along with these changes long-haul flights increased by near 18 per cent.

There has also been a significant shift in slot holdings among carriers between 2001 and 2006. The direct welfare gains would be shown as more efficient use while the indirect or downstream affects would be more competitive airline markets.

---

19   Regulation (EC) No. 793/2004 of the European Parliament.

Table 4.2   Changes in average seats, sectors and ASK with slot trading

	Before	**After**	**Difference**
Av. seats	135	255	+ 90%
Av. sector	575 km	6800 km	× 12
ASK/slot	77,625	1,734,00	× 22

Notes

1 Av. seats – average number of seats per aircraft.
2 Av. sector – average length of the sector flown.
3 ASK – available seat kilometers.

Source: James Cole, paper presented at EU Slot Coordinators Conference, Amsterdam, June 2006.

Concluding Remarks

Over a time span of almost four decades the institutional bodies authorized to manage airspace traffic, have found this task to be challenging: regulators were unable to lift the initial slot controls imposed on the five HDR airports in United States, and the 2002 attempt at Chicago O'Hare Airport resulted in significant delays, indirectly affecting other parts of the NAS. The Administration is currently soliciting interesting parties to show cause as to why the slot controls on scheduled operations at ORD should not be extended until the end of 2006. Plus, restrictions on unscheduled operations at ORD have already been prolonged until 28 October 2006 as embodied in the latest SFAR rulemaking document.

So, will slot controls be lifted as planned by the FAA for LGA and JFK airports in 2007? The FAA forecast for 2006 displays a reduction of 0.7 per cent in domestic capacity but an expansion of 5.9 per cent in international markets.[20] Overall, the market is estimated to increase only by 0.9 per cent in 2006, owing in part to the expected cutback in flights serviced by legacy carriers. However, the buy/sell rule enacted in 1986, in conjunction with the grandfather rule, did not eliminate the anti-competitive effects of slot hoarding. Evidence shows incumbent air carriers have established a powerful presence at the HDR airports over the last twenty years. Moreover, the course taken by the relatively recent published works on airport access, as directed by both the FAA and European Commission, illustrates a shift toward market mechanisms. As a replacement of the grandfather rule, some of these measures might completely redefine the initial allocation of slots through the creation of an ongoing and anonymous primary market, such as slot auctions. More importantly, however, the flight movement forecast over the next ten years indicates the suspension of slot controls without an impartial allocation apparatus is an unlikely scenario.

---

20   Federal Aviation Administration, FAA *Aerospace Forecast Fiscal Years 2006-2017*.

Slot allocation will not go away even with airport expansion; there will always be a need to have some basis of allocation. Issues of equity, efficiency and competition are all intertwined in such methods of allocation. The US seems to favour more market based measures although there are still those who prefer administrative rules. The EU seems split between the UK and EU. The EU clearly favours administrative rules and despite a growing grey market in the EU argues slot trading is not appropriate for improving the system. This approach would seem to protect incumbent, particularly current or former, national carriers. The UK is using secondary trading of slots principally at LHR to improve economic efficiency and have over the period 2001 to 2006 realized significant gains. The EU slot coordinators have stated they feel the current system is working and the issue is more one of fine tuning and adding capacity than a wholesale change in the IATA approach; the IATA process protects status quo, entrenches incumbents, is anti-competitive, and is generally blocking effective entry.

Table 4.3   Airport slot designation

Country	City (airport)	Airport code	SCR Level 3	SMA Level 2
Canada	Montreal	YUL		Yes
	Toronto	YYZ	Yes	
	Vancouver	YVR	Yes	
United States	Chicago-O'Hare	ORD		Yes
	Los Angeles	LAX		Yes
	Newark	EWR		Yes
	New York-J.F. Kennedy	JFK	Yes	
	San Francisco	SFO		Yes
	Orlando	MCO		Yes
Austria	Graz	GRZ		Yes
	Innsbruck	INN		Yes
	Klagenfurt	KLU		Yes
	Linz	LNZ		Yes
	Salzburg	SZG		Yes
	Vienna	VIE	Yes	
Belgium	Brussels	BRU		
Bulgaria	Sofia	SOF		Yes
Cyprus	Larnaca	LCA		Yes
Czech Republic	Prague	PRG	Yes	
Denmark	Copenhagen	CPH	Yes	

Country	City (airport)	Airport code	SCR Level 3	SMA Level 2
Finland	Helsinki	HEL	Yes	
France	Lyon-Satolas	LYS	Yes	
	Nice	NCE		Yes
	Paris-Ch. De Gaulle	CDG	Yes	
	Paris-Orly	ORY	Yes	
Germany	Berlin-Schoenefeld	SXF	Yes	
	Berlin-Tegel	TXL	Yes	
	Berlin-Tempelhof	THF	Yes	
	Bremen	BRE		Yes
	Cologne	CGN		Yes
	Dresden	DRS		Yes
	Duesseldorf	DUS	Yes	
	Erfut	ERF		Yes
	Frankfrt	FRA	Yes	
	Hamburg	HAM		Yes
	Hannover	HAJ		Yes
	Leipzig	LEJ		Yes
	Munich	MUC	Yes	
	Muenster	FMO		Yes
	Nuremberg	NUE		Yes
	Saarbruecken	SCN		Yes
	Stuttgart	STR	Yes	
Greece	Athens	ATH		Yes
	Chania	CHQ	Yes	
	Chlos	JKH	Yes	
	Corfu	CFU	Yes	
	Heraklion	HER	Yes	
	Kalamata	KLX	Yes	
	Karpathos	AOK	Yes	
	Kavala	KVA	Yes	
	Kefallinia	EFL	Yes	
	Kos	KGS	Yes	
	Lemnos	LXS	Yes	
	Mikonos	JMK	Yes	
	Mytilene	MJT	Yes	
	Paros	PAS	Yes	

Country	City (airport)	Airport code	SCR Level 3	SMA Level 2
	Patras	GPA	Yes	
Greece (cont'd)	Preveza/Lefkas	PVK	Yes	
	Rhodes	RHO	Yes	
	Samos	SMI	Yes	
	Skiathos	JSI	Yes	
	Skiros	SKU	Yes	
	Thessalonika	SKG	Yes	
	Thira	JTR	Yes	
	Zakinthos	ZTH	Yes	
Hungary	Budapest	BUD		Yes
Iceland	Reykjavik	KEF	Yes	
Ireland	Dublin	DUB	Yes	
Italy	Bologna	BLQ	Yes	
	Cagliari	CAG	Yes	
	Catania	CTA	Yes	
	Florence	FLR	Yes	
	Lampedusa	LMP	Yes	
	Milan-Linate	LIN	Yes	
	Milan-Malpensa	MXP	Yes	
	Milan-Orio al Serio	BGY	Yes	
	Naples	NAP	Yes	
	Palermo	PMO	Yes	
	Pantelleria	PNL	Yes	
	Pisa	PSA		Yes
	Rome-Ciampino	CIA	Yes	
	Rome-Fiumicino	FCO	Yes	
	Turin	TRN	Yes	
	Venice	VCE	Yes	
Kosovo	Pristina	PRN	Yes	
Luxembourg	Luxembourg	LUX		Yes
Netherlands	Amsterdam	AMS	Yes	
	Eindhoven	EIN	Yes	
	Rotterdam	RTM	Yes	
New Zealand	Auckland	AKL	Yes	
	Christchurch	CHC	Yes	

Country	City (airport)	Airport code	SCR Level 3	SMA Level 2
	Wellington	WLG	Yes	
Norway	Bergen	BGO		Yes
	Oslo-Gardemoen	OSL	Yes	
	Stavanger	SVG	Yes	
Poland	Gdansk	GDN		Yes
	Katowice	KTW		Yes
	Krakow	KRK		Yes
	Poznan	POZ		Yes
	Rzeszow	RZE		Yes
	Szczecin	SZZ		Yes
	Warsaw	WAW		Yes
	Wroclaw	WRO		Yes
Portugal	Faro	FAO	Yes	
	Funchal	FNC	Yes	
	Lisbon	LIS	Yes	
	Ponta Delgada	PDL		Yes
	Porto	OPO	Yes	
Slovakia	Bratislava	BTS		Yes
Slovenia	Ljubljana	LJU		Yes
Spain and Canary Island	Alicante	ALC	Yes	
	Almeria	LEI	Yes	
	Barcelona	BCN	Yes	
	Bilbao	BIO	Yes	
	Fuerteventura	FUE	Yes	
	Gerona	GRO	Yes	
	Gran Canaria	LPA	Yes	
	Ibiza	IBZ	Yes	
	La Coruna	LCG		Yes
	Lanzarote	ACE	Yes	
	Madrid	MAD	Yes	
	Malaga	AGP	Yes	
	Menorca	MAH	Yes	
	Palma Mallorca	PMI	Yes	
	Reus	REU	Yes	
	Santiago de Compostela	SCQ		Yes
	Seville	SVQ	Yes	

Country	City (airport)	Airport code	SCR Level 3	SMA Level 2
	Tenerife-Reina Sofia	TFS	Yes	
Spain and Canary Island	Tenerife-Norte	TFN	Yes	
(cont'd)	Valencia	VLC	Yes	
	Vitoria	VIT		Yes
	Zaragoza	ZAZ		Yes
Sweden	Gothenburg	GOT		Yes
	Stockholm-Arlanda	ARN	Yes	
	Stockholm-Bromma	BMA	Yes	
Switzerland	Basel/Mulhouse	BSL		Yes
	Geneva	GVA	Yes	
	Zurich	ZRH	Yes	
Ukraine	Kyiv	KBP	Yes	
United Kingdom	Aberdeen	ABZ		Yes
	Birmingham	BHZ		Yes
	Edinburgh	EDI		Yes
	Glasgow	GLA		Yes
	London-Gatwick	LGW	Yes	
	London-Heathrow	LHR	Yes	
	Lodon-City	LCY		Yes
	Manchester	MAN	Yes	
	Newcastle	NCL		Yes
	Stansted	STN	Yes	
Worldwide			96,00	52,00
North America			3,00	6,00
Europe			93,00	46,00

Source: http://www.iata.org/NR/ContentConnector/CS2000/SiteInterface/sites/
whatwedo/scheduling/file/fdc/WSG-12thEd.pdf. See appendix, 'Characteristics
of Slot Coordinated Airports, 2003–2004', for details on other airports.

Table 4.4   United States slot regulatory highlights 1966–2006

1   **1 September 1966**: A voluntary agreement effective this date limited operations at Washington National Airport to a maximum of 60 Instrument Flight Rules (IFR) operations per hour.

2   **1 June 1969**: The High Density Rule at five of the nation's busiest airports, between 6 a.m. and midnight. The airports are Kennedy International (80), Chicago O'Hare (135), La Guardia (60), Newark (60), Washington National (60).

3   **3 November 1980**: FAA published a special rule allocation reservation, or 'slots', for take-offs and landings under instrument flight rules at Washington National Airport.

4   **3 August 1981**: Nearly 12,300 members of the 15,000 member Professional Air Traffic Controllers Organizations (PATCO) went on strike

5   **6 December 1981**: A new Metropolitan Washington Airports Policy became effective consisting of a maximum of 60 landing slots per hour.

6   **18 February 1982**: A special rule issued this date amended the Interim Operations Plan for air traffic control (see 3 August 1981). The new rule provided procedures to be used in scheduling and in allocating airport landing reservations at the 22 airports at which operations were limited due to the PATCO strike.

7   **10 May 1982**: FAA began an experimental program of allowing airlines to buy, sell, and transfer airport landing 'slots' among themselves.

8   **6 July 1982**: FAA announced it was suspending the buy-sell policy, but would continue to allow the exchange or trade of slots; on Aug. 5 it announced that it was easing certain restrictions on this slot trading.

9   **20 December 1985**: DOT published a new rule on allocation of take-off and landing reservations at the four high density airports. Beginning on **1 April 1986**, any person might buy, sell, trade or lease air carriers or commuter slots (with the exception of international and essential air service slots, which were subject to certain transfer restrictions). A lottery procedure was provided for the allocation of new slots.

10 **1 October 1999**: High Density Airports: Allocation of Slots (Final Rule); This document amends the regulations governing takeoff and landing slots and slot allocation procedures at certain High Density Traffic airports.

11 **April 2000**: The FAA enacted the 'Wendell H. Ford Aviation Investment and Reform Act for the Twenty-first Century', known as the AIR 21. A four-year reauthorization bill, covering fiscal years 2000 through to 2003. http://ostpxweb.dot.gov/aviation/Data/air21summary.pdf.

12 **1 July 2002**: Slot restrictions at Chicago O'Hare airport expired.

13 **1 April 2004**: FAA issued two orders granting Beyond-Perimeter and Within Perimeter Slot exemptions at Ronald Reagan Washington National Airport. http://ostpxweb.dot.gov/aviation/domesticaffairs.htm#slots.

14 **1 November 2004**: FAA limited arrivals at Chicago O'Hare airport to 88 per hour during peak hours and effectively prohibited entry at O'Hare without FAA permission.

Table 4.4 (cont'd)

---

15 **8 August 2005**: Special Federal Aviation Regulation (SFAR) No. 105. The FAA set a limit on unscheduled operations at Chicago O'Hare International Airport. www.fly.faa.gov/ecvrs/circulars/ORD_SFAR_105.pd.f

16 **31 March 2006**: The FAA issued an order to show cause, soliciting written views on extending through 28 October 2006, the August 2004 order limiting scheduled operations at O'Hare International Airport.

Source: Federal Aviation Administration, FAA Historical Chronology 1926–1996, http:// www.faa.gov/about/media/b-chron.pdf.

Table 4.5   European Commission slot regulatory highlights: 1993–2006

---

1   **18 January 1993**: Council Regulation (EEC) No. 95/93 on common rules for the allocation of slots at Community Airports.

2   **25 June 1993**: Commission Regulation (EEC) No. 1617/93.

3   **24 July 1996**: Commission Regulation (EC) No. 1523/96.

4   **5 September 2003**: Regulation (EEC) No. 1554/2003 amending Council Regulation (EEC) No. 95/93.

5   **30 April 2004**: Regulation (EC) No. 793/2004 of the European Parliament and of the Council of 21 April 2004 amending Regulation (EEC) No. 95/93.

References

Commission of the European Communities (2004), 'Commercial Slot Allocation Mechanisms in the Context of a Further Revision of Council Regulation (EEC) 95/93 on Common Rules for the Allocation of Slots at Community Airports,' Commission Staff Working Document, 17 September.

Council of the European Communities (1993), Council Regulation (EEC) No. 95/93 on Common Rules for the Allocation of Slots at Community Airports, 18 January: http://europa.eu.int/eur-lex/lex/LexUriServ/LexUriServ.do?uri= CELEX:31993R0095:EN:HTML.

Council of the European Communities (2003), Regulation (EC) No. 1554/2003 of the European Parliament and of the Council, *Official Journal of the European Union*, 22 July: http://europa.eu.int/eur-lex/pri/en/oj/dat/2003/l_221/l_22120030904en00010001. pdf.

Council of the European Communities (2004), Regulation (EC) No. 793/2004 of the European Parliament and of the Council, *Official Journal of the European Union*, 21 April: http://europa.eu.int/eur-lex/pri/en/oj/dat/2004/l_138/l_13820040430en00500060. pdf.

Department of Transportation (2003–2006), *Air Travel Consumer Reports*, Aviation Consumer Protection Division: http://airconsumer.ost.dot.gov/reports/.

Department of Transportation (1997), *Order Granting and Denying Applications for Slot Exemptions at Chicago O'Hare Airport*, Docket OST-97-2771-8, 24 October.

DotEcon (2001), 'Auctioning Airport Slots', report for HM Treasury and the Department of the Environment, Transport and the Regions.

European Commission, Activities of the European Union: Summaries of Legislation: http://europa.eu.int/scadplus/leg/en/lvb/l24085.htm.

Federal Aviation Administration (1999), *High Density Airports; Allocation of Slots*, Docket No. FAA-I 9994971-11, Amendment No. 93–78, 1 October.

Federal Aviation Administration (2004), *Operating Limitations at Chicago O'Hare International Airprt*, Docket No. FAA-2004-16944-1, 21 January.

Federal Aviation Administration (2005), *Reservation System for Unscheduled Arrivals at Chicago's O'Hare International Airport*, 70 (130), 8 July.

Federal Aviation Administration, *FAA Aerospace Forecast Fiscal Years 2006–2017*: http://www.faa.gov/data_statistics/aviation/aerospace_forecasts/2006-2017/media/FAA%20Aerospace%20Forecasts%202006-17.pdf.

Federal Aviation Administration (2006), *Reservation System for Unscheduled Arrivals at Chicago's O'Hare International Airport*, Docket No. Faa-2005-19411; SFAR No. 105, 71 (62), 31 March.

IATA, Worldwide Scheduling Guidelines (2005), 12th edn, December: http://www.iata.org/NR/ContentConnector/CS2000/SiteInterface/sites/whatwedo/scheduling/file/fdc/WSG-12thEd.pdf.

National Economic Research Associates (2004), 'Study to Assess the Effects of Different Slot Allocation Schemes: A Final Report for the European Commission', January: http://europa.eu.int/comm/transport/air/rules/doc/2004_01_24_nera_slot_study.pdf.

Times (London) (2004), 'Special Report: Soaring Cost of Touching Down', 22 February.

Task Force (2005), *Outcome of Study on Slot Allocation Procedures, European Civil Aviation Conference*, 7 December, DGCA/124-DP/6.

The Subcommittee on Aviation Hearing on The Slot Lottery at LaGuardia Airport: http://www.house.gov/transportation/aviation/hearing/12-05-00/12-05-00memo.html.

US Government Printing Office, *United States Code*: http://www.gpoaccess.gov/uscode/index.html.

PART B
Congestion, Slots, and Prices

Chapter 5

Setting the Slot Limits at Congested Airports

Peter Forsyth and Hans-Martin Niemeier

Introduction

Airports facing demand in excess of capacity are now common around the world, especially in North America and Europe. In North America, the predominant means of rationing demand to capacity is through queues and congestion. In Europe and elsewhere, the preferred method of constraining demand is through a slot capacity system. A maximum capacity of the airport is declared, and slots matching this capacity are allocated to airlines. To use the airport at a specific time, an airline must have a slot. The usual objective of this system is to reduce congestion, in the form of delays to aircraft and passengers, to an acceptable level. However, this system is not supposed to eliminate delays entirely, since this would unduly limit the throughput of the airport. Slot limits may also be imposed to reduce environmental externalities, such as noise.

The slot system has attracted considerable attention in both theoretical and policy oriented literature. Most of this attention has, however, been given to the issue of whether slots are efficiently allocated between airlines. Slot allocation is usually on an arbitrary basis (such as grandfathering) and slot trading is often limited or prohibited.

In this chapter, a different issue is explored, namely that of choice of the slot capacity of an airport to be allocated to airlines. We do not discuss the efficiency of the slot allocation system per se, though recognize that how efficient it is can have a bearing on the choice of slot capacity. The slot capacity choice is a trade off between the costs of additional delays, and possibly, environmental costs, and the benefits which result from additional throughput. With slot pairs for London Heathrow airport trading for as much as £10m, the benefits from additional throughput are large. Even if they add perceptibly to delays, would increases in slot capacity made available to be used be worthwhile?

We also discuss an additional aspect of slot systems. While the impact of slot constraints on congestion, due to the reduction of through reducing throughput, is an obvious one, a slot system may also reduce congestion at any given level of throughput. It can do this by lessening the clustering and randomness of arrivals and departures, and thus lessening queues. Thus the congestion cost function for a slot constrained airport may lie below that for a non slot controlled airport.

This could be an important feature of slot systems, and it will have implications for the choice of slot capacities.

We begin with a review of the way slot capacity systems typically work, especially in Europe. Next we explore the theory of setting slot capacities, and how slot prices should, ideally, be related to the marginal and average costs of delay. The effects of slot systems on reducing congestion, and the implications this has on the slot constraint choice are examined. Then we examine the empirical evidence on the issue. This evidence is patchy, but it is consistent with the theory. The levels of slot prices (such as are available) and congestion costs, at least in London, suggest that slot capacities could be about right. However the big divergence between delays at slot controlled US airports and European airports raises the question of whether both have made correct decisions on the slot capacity. Finally, we note some of the main results of the chapter and indicate the gaps which call for further research.

Slot Constraints at Airports – A Background

'In recent years it has become increasingly apparent that the rapidly growing demand for air travel has been outstripping the supply of infrastructure both in the air and on the ground.' This quote is not from a recent publication, but goes back over a dozen years. Doganis (1992) notes that the late 1980s were 'uncomfortable years for airport users' (p. 339). It seems to be a permanent feature of the airport industry that certain airports face excess demand, even though the airline industry has been lobbying for new airport capacity for decades, and regional authorities have seen airport expansion as a tool for boosting growth and creating new jobs. Airlines and airports have developed certain ways to live with the problem of congestion and queues.

In the US queues operate on a first come first serve basis without ex-ante coordination. This results in a relatively high level of congestion. The exceptions are the so-called high density airports such as New York's La Guardia and Kennedy, and until recently Chicago's O'Hare airport. Slots are allocated and subject to secondary trading, but the trading is restricted to national slots, not slots for international carriers. The preferred solution outside the US is for a restricted number of slots to be made available and flights are coordinated according to the IATA-system of slot allocation (for details of this system, see Gillen and Ulrich in this volume, and IATA, 2005).

The current discussion on reforming the slot allocation rules has so far focused nearly entirely on the question how to allocate the given amount of slots. The IATA system is criticized for being inefficient, leading to a decrease in welfare and to less competition within the airline market (for a discussion, see NERA, 2004). Market based mechanisms for allocating slots, such as secondary trading and auctions are preferred but so far, with the exception of the US airports and some trading of slots at UK airports, have rarely been tried out. The current focus of the debate tends to overlook how the given amount of slots is defined by the relevant body, which may be a government agency or a committee of stakeholders.

Are the slot constraints set at the optimal level or are they set too high, with too much congestion, or too low with not enough congestion? These questions need to be answered independently from any preferred slot allocation mechanism as they are part of the problem of optimizing a given capacity.

Little research has been directed so far to these questions, even though this aspect of the problem seems crucial as an efficient allocation of a given amount of slots is only a partial solution if the level is not set optimally. One of the criticisms made of the slot system, as compared to a price allocation system, is that there is no guarantee that slot constraints will be set efficiently. So far it is not clear if the system has set the slot constraint at the optimal level of congestion. Doubts are quite legitimate, as cost benefit reasoning seems to often play no role in determining the level of capacity. Furthermore, the question to what extent the system has reduced congestion relative to the first come first served system has also not been answered yet.

This chapter is based on the presumption that, *ceteris paribus*, congestion and delays rise with a increase of movements. We assume a simple standard congestion model in which delays stay low for low capacity utilization, but begin to rise when capacity becomes fully utilized (see, for example, Mohring, 1976). The relationship between movements and delays is subject to a good deal of noise. Factors other than movements often affect delays. These include:

- weather and seasonal changes like vacation times;
- strikes, accidents;
- night curfews;
- starting problems – in the morning times there might be more delays;
- delays caused in other parts of the system like ATC, airline delays, delays at other airports.

These factors vary from airport to airport. In spite of these, there is an underlying relationship between movements (relative to capacity) and delays. Additional throughput can only be achieved if a higher level of delays is accepted.

In this chapter we focus on runway slot constraints. It should be noted that some airports have capacity problems at terminals, and have implemented slot constraints to handle these. Generally, the runway slot problem is more important, since it is usually much easier to expand terminal than runway capacity. Similar principles apply to terminal slot constraint choices as do to runway slot constraints.

Slot coordination systems do not only impose a limit on the number of movements that can be handled within a period such as, for example, an hour. Typically, the coordination process also involves the allocation of specific times to specific movements within this period. This is done to even out the flow of movements, and to lessen congestion that might arise from too many movements clustering around a popular time. Many airports operate slot coordination systems not because they are subject to excess demand, but simply to even out the flow of movements.

Slot Constraints and Efficiency

In this section, we set out the basic theory of determining slot constraints, and explore the efficiency properties of different approaches to the airport capacity allocation problem. We only consider short run aspects – we do not consider investment issues (see Knieps, 1992; Turvey, 2000). In particular, we consider several stages of the problem in sequence:

- determination of the slot capacity constraint, and integration with slot allocation under certainty;
- allowance for heterogeneous users;
- allowance for the possibility that some congestion costs are internalized by users;
- environmentally determined slot constraints;
- recognition of the implications of uncertainty;
- analysis of the role of price regulation or profit constraints on the problem; and
- identifying the empirical relationships which might be expected to hold.

Before considering the possibilities, it is useful to note the different aspects of allocative efficiency which are relevant to the airport capacity allocation problem. There are at least two distinct aspects: the first of these concerns the choice of the level of utilization for the airport – this is a matter of optimising the extent of congestion. This will be a problem even if all users are homogeneous. The second aspect concerns the allocation of this capacity to different users. It deals with questions such as how the capacity of the airport is rationed – by prices, slot auctions, congestion or by other means? Different discussions of airport efficiency focus on different aspects – the discussion of slot trading focuses exclusively on the second of these, for example. In this chapter, both aspects of efficiency will be drawn upon.

Homogeneous Users under Certainty

We begin with the simplest case, where users are homogeneous, except to the extent that they have a different willingness to pay for the use of the airport. In all other relevant aspects they are the same. This may mean, for example, that the cost of a particular level of delay is the same for all users. The users of the airport consist of a large number of independent airlines, and there is no internalisation of the cost of congestion by the users. The model is a single period model- there is no distinction between peak and off peak, and there is no build up of traffic over the day. Rather, demand stays constant for a constrained facility. For simplicity, operating costs of the airport are assumed to be zero, and there are no charges for the utilization of the airport other than congestion prices which may be introduced.

The above-mentioned situation is depicted in Figure 5.1. The airport has a fixed capacity for flights of K. As the number of flights increases, average delays and average delay costs increase as well – this is shown by the curve AC. This curve rises sharply as output tends towards capacity. There is a marginal cost curve, MC which corresponds to this average cost curve. Demand is shown by curves D_1 and D_2. In keeping within the traditional congestion theory framework, it is assumed that the atomistic users of the facility face an average cost of using it of AC, and do not internalize the additional costs they impose on other users, which is measured by the difference between marginal and average cost at an output. Thus with demand of D_1 they use the airport up to the point where the demand curve crosses the average cost curve – i.e. at E_1 in Figure 5.1, with an output of Q_1.

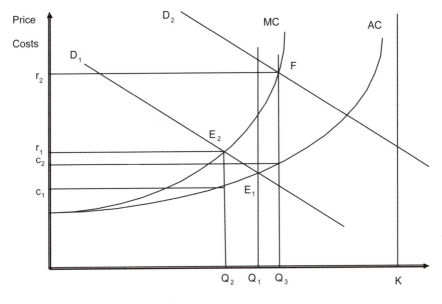

Figure 5.1   Slot constraints for homogeneous users under certainty

Because congestion costs have not been internalized, use of the airport is excessive. Optimality would be achieved if the use of the airport is restricted to Q_2, where the demand curve intersects the marginal cost curve at E_2. This could be achieved by setting a congestion toll of r_1-c_1. An alternative means of achieving this would be to set a slot constraint at an output of Q_2. Demand for slots would ensure that slots had a positive price. Suppose there is free slot trading – the price of slots would be bid up to r_1-c_1, and the slot price would equal the price which optimizes use. Alternatively, if the slots were auctioned, a price of r_1-c_1 would be achieved at the auction.

In this simple case, all three methods of allocating capacity would be equally efficient. The solution addresses both aspects of efficiency – choice of output/congestion level, and allocation of scarce capacity. It provides a rule for the efficient choice of a slot restraint. The number of slots allowed should be such that the slot auction or trading price is be equal to the difference between marginal and average cost at that output, or, in other words, the marginal external cost.

Imposition of a slot constraint improves efficiency, assuming that the slots which are created are efficiently allocated. The welfare gain can be shown as E_1E_2F. If it is simply a matter of choosing between a slot constraint or a congestion solution, an efficient slot constraint will be preferable.

The outcome would change if demand were to change. For example, suppose demand increases from D_1 to D_2. The new optimal slot constraint would be set at Q_3. Comparing the two solutions, it can be seen that the new optimum has a greater output, but also a higher level of congestion and a higher slot price r_2-c_2. Thus if slot constraints are being set efficiently across a range of airports facing different degrees of excess demand, we should expect to see slot prices and the levels of average congestion to be correlated. For busier airports, it is worth allowing a higher output, even though this leads to an increase in congestion levels. The gain in benefits from additional users being able to use the scarce facility would exceed the additional costs which arise from allowing the airport to become more congested.

A further observation is that it is quite possible for congestion levels at an optimally slot constrained airport to exceed congestion levels at another, less busy, airport where slot controls are not imposed. In Figure 5.1, output Q_3 exceeds output Q_1, and the average level of congestion at Q_3 exceeds that at Q_1.

Heterogeneous Users/Imperfect Capacity Allocation

In the case above, it was assumed that users were identical except in their willingness to pay for the use the airport. When users differ, there is a potential allocation problem – how can it be guaranteed that scarce capacity or slots go to those who value them most? This is a particular problem for slot control systems where slots are not auctioned or traded freely. Compare a slot allocation system to the option of setting market clearing prices. In Figure 5.1, if a price of r_1-c_1 is set, there will be an efficient allocation of users to capacity – only those users who are willing to pay at least this price will be able to use the facility. Alternatively, if a slot constraint is imposed but slots are auctioned, the auction price achieved will be r_1-c_1, and this price will allocate slots efficiently. However, if slots are simply allocated on some other basis, such as by grandfather rights, there is no guarantee that they will be allocated efficiently. If trading can take place, it has the potential to achieve efficiency, though as the literature stresses, there can be several distortions to the market. Thus there is a distinct possibility that a slot constraint system will result in an efficient choice of output/congestion levels, but an inefficient allocation of the scarce slots. This poses the question of whether it would be any superior to a system under which capacity is rationed by congestion only. In the scenario depicted in Figure 5.1, is it possible that the gains from a more efficient output level choice could be more than cancelled out by an inefficient allocation of slots? If this is the case, a congestion rationing system, however imperfect, may be preferable on efficiency grounds to a slot constraint system.

This, however, relies on there being only one source of inefficiency when allocating capacity. The allocation of capacity which comes about when congestion is the rationing device will not be efficient when users have different delay costs. Rationing by congestion gives an advantage to those who have low delay costs, such as small aircraft. The users who are willing to queue are not necessarily those who have the highest willingness to pay for capacity. Thus the congestion solution not only results in an excessive use of the airport, but it also results in an inefficient allocation of capacity.

A way of illustrating this is as follows. Suppose it were possible for places in the queue to be traded. If this were so, larger aircraft with a high willingness to pay but also a high delay cost, would be prepared to purchase places from smaller aircraft, with low delay costs but also a low willingness to pay. In other words, since such trading improves allocation, delay allocates capacity inefficiently. Thus it is inappropriate to compare imperfectly allocated capacity under a slot constraint system with efficiently allocated capacity under congestion, since under the latter, an efficient allocation will not come about. While the exact level of inefficiency of capacity allocation under the two different systems is very difficult to judge, if they are comparable, a slot constraint system will be preferable to the congestion solution because it ensures a more efficient choice of use of the airport.

Internalization of Congestion Externalities

At some airports, there are airlines which account for a significant proportion of total traffic. This is typically true of congested US airports and slot constrained European hubs. There is the possibility that these airlines will internalize the congestion costs that one flight causes for other of their own flights, though not for flights of other airlines (Brueckner, 2002). If this occurs, the congestion solution will be less inefficient. The total number of flights scheduled for the congested period will be reduced as compared to the atomistic competition case, and the average and marginal congestion costs will be reduced. The result will be closer to the optimum, though it will only coincide with the optimum if there is a single user of the airport which internalizes the externalities that its flights cause on itself. To this extent, the gains from implementing a slot constraint system will be reduced.

The effect noticed by Brueckner will cease to exist under a slot constraint system. No airline can have an effect on reducing congestion by reducing the number of its own flights because the total number of flights is fixed by the slot constraint. If an airline reduces the number of its own flights, other airlines will take up the slack. However there will be an analogous effect which comes into play in a slot constrained system.

If there are users who are obtaining a large share of the slots, they have an incentive to allocate them efficiently between their own flights. Thus, even if the willingness to pay for slots differs from airline to airline, at least there is some chance that slots will be allocated efficiently within airlines (though it is an empirical question exactly how well slots are allocated within airlines – for a discussion, see Bauer in this volume). Furthermore, it is likely that an airline which is a major user of slots would have a good idea of what the slots are worth – in particular, it would be willing to try to gain slots from others when it perceives that they are undervaluing them, and it would be keen to sell slots when it notices that other airlines are willing to pay high prices (though it is possible that such an airline may be hoarding slots for the future, or to limit competition). The overall result is that slots would be more efficiently allocated than under the assumption of atomistic competition with inefficient slot trading.

The allocation of capacity amongst possible flights under the congestion solution would also be improved in the case where there are airlines with large shares at a given airport. As noted, these airlines recognize the externality which one of their flights causes on their other flights. In scheduling flights they take this into account, and one way in which they might be conceived of as doing this would be to work out the externality cost, or shadow price of a flight, and insist that the net revenue generated by a flight be sufficient to cover this congestion externality cost or shadow price. Thus they will be taking into account their willingness to pay to use the airport, not just the delay cost of a single flight. In the extreme case of the airline being the single airport user, it would not only schedule the profit maximizing number of flights, but would, in addition, allocate flights to the airport according to the net revenue they generated, or their 'willingness to pay'.

The net result is that if there are airlines with large shares of traffic at the airport, the allocation of flights to the airport will be improved, relative to the case of atomistic competition, whether there is a slot constraint, or whether congestion is used to ration. The gain (from a more efficient output choice) from implementing a slot constraint, when slots are inefficiently traded, will be less than where there is atomistic competition.

If, however, a slot constraint system is put in place and slots are efficiently traded, the constraint system will be superior to the congestion solution for two distinct reasons. Firstly, the output choice will be better, and secondly, the allocation of flights to the scarce capacity will be improved, since delay costs will have no role in the allocation process.

Environmentally Determined Slot Constraints

Slot constraints may be imposed to lessen congestion. However, in some cases, they are set primarily to limit environmental externalities, such as noise or emissions. Thus the slot constraint at Düsseldorf is set well below the practical capacity of the airport so as to limit the noise associated with the airport. When considering environmental motivations for slot constraints, there is still the problem of determining at which level to set the constraint. In this case, there is the added complication of integrating environmental externality considerations with congestion considerations.

The problem is illustrated in Figure 5.2. The demand and congestion cost curves are as before. To these is added a new curve which shows the marginal externality cost MEC. For simplicity, this is shown as a straight horizontal line – while in practice the curve could also take on other shapes; these will not affect the points being made here. To optimize airport utilization, the slot constraint should be set where the demand curve intersects the total marginal cost curve, MTC, which is the vertical summation of the marginal congestion cost curve, MC, and the marginal externality cost curve, MEC. The slot premium now equals the sum of the marginal externality cost plus the excess of the marginal congestion cost over the average congestion cost.

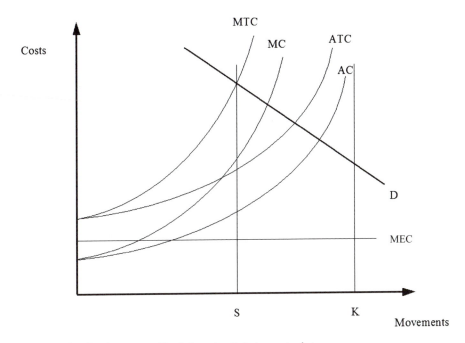

Figure 5.2 Environmentally determined slot constraints

Thus the theory of environmentally set slot constraints is straightforward, but its implementation may not be. While it may be reasonably easy to derive approximate measures for congestion cost externalities, this is unlikely to be so for environmental externalities. These are likely to be quite difficult to evaluate, although progress has been made by recent studies (Dings et al., 2003), any values which are produced are likely to be strongly debated. Thus accuracy in the setting of slot constraints when mitigating environmental externalities is a key objective which is more difficult to achieve.

The analysis here supposes that environmental costs are linked to airport movements. In reality, matters are more complex, though, as noted above, sometimes aircraft movement slot constraints are imposed to limit environmental costs. Different aircraft can impose different noise costs, and generate different greenhouse or toxic gas emissions, and thus aircraft movement limits are a crude means of handling environmental costs. A preferable approach would be to set prices refecting environmental costs, or noise and emissions limits with tradeable permits. These issues are beyond the scope of this chapter.

Slot Constraints under Uncertainty

It is recognized that the presence of uncertainty affects the choice between price versus quantity (slot constraint) solutions (Forsyth, 1976; Czerny in this volume). In the case of certainty, as modelled in Figure 5.1, price and quantity solutions are equivalent (assuming that slots are allocated efficiently). When uncertainty is present, the price which is set can be non-optimal, and be either too high or too low. Likewise, the slot limit which is set can be too tight or too loose. In either case, there is a loss of efficiency. For example, when the price is set too low granted the demand which actually comes about, prices will not fully ration demand – congestion will be too high. Alternatively, there can be an efficiency cost when the slot limit is set too tightly, and slot prices are above the marginal external cost of using the airport. The nature of uncertainty (whether it shifts demand or cost curves, or both) and the demand and supply elasticities, will determine which solution is preferable (see Weitzman, 1974). For example, if demand is highly variable, but the cost curve is stable, and steep near capacity, the efficient output can be determined moderately accurately, though the efficient price would be highly variable. A slot limit set in advance would result in a tolerably efficient solution, though a price set in advance could be far too high or too low. In this situation, a slot limit would be more efficient than a price. In other situations, the price solution would be more efficient.

For present purposes, however, the choice is between slot limits and congestion solutions. Suppose, initially, that slots are allocated efficiently, either through an auction or a well functioning trading system. Will the gain from implementing a slot limit system be less than it would be under uncertainty?

In Figure 5.3, it is supposed that all uncertainty comes from variations in demand. Demand could be as shown by D_1 or D_2. Average and marginal congestion costs are known, and shown by AC and MC. A slot limit is set at Q_3 – in fact the setting of the slot limit would ideally be done so as to maximize expected gains from restricting flights and lowering congestion. If demand is D_2, the slot limit restricts output too much – the slot price would be higher than the marginal externality cost. However, if demand is D_1, the slot limit is too low – in fact, it is ineffective as the congestion solution, E_1, with output of Q_1, is below the slot limit. In this outcome, there is no gain from implementing the slot limit. If E_1 and E_2 correctly represent the congestion rationing solutions, then the gain from implementing slot limits is less than under certainty, because it is not possible to set slot limits correctly.

This said, one may question whether congestion works as effectively as a rationing device as is supposed in the figure. In essence it is supposed that airlines take actual congestion levels into account when determining how many flights to schedule. In practice, they may not be able to do this – they may have to make estimates of expected delays and use these when scheduling. Granted uncertainty, actual delays will diverge from expected delays. Thus delays may not work as well as might be supposed from Figure 5.3, and actual output may be too high or too low (equilibria such as E_1 and E_2 are not really feasible under uncertainty). The

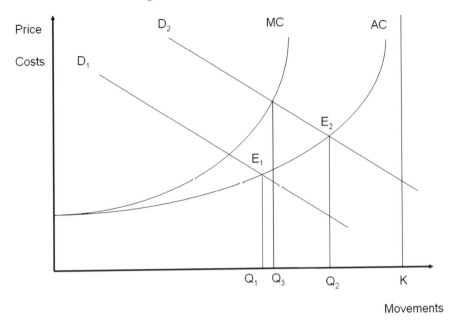

Figure 5.3   Slot constraints under uncertainty

presence of uncertainty will make both slot limit and congestion solutions less efficient, and the advantages of the slot limit will remain.

The analysis above supposes that slots are allocated efficiently. Clearly they may not be. There may be no slot auction, and trading in slots may be quite imperfect. If this is so, the advantages of a slot limit system are less – though it needs to be recalled that the allocation of capacity by congestion costs is not efficient either. There will still be a gain from moving from a congestion solution to a slot limit system, since there will be an expected improvement in the output and congestion cost level choice.

This said, however, it may be that slot allocation systems have the potential to perform well, and possibly better than either prices or congestion costs, in allocating users to capacity. This is because trading can take place up to the time the slot is used. While the slot limit may have been set too high or too low, given the outcome of demand, the available slots can be continuously traded, and actual prices for the slots can reflect the actual demand supply situation. With preset prices, prices can be too high or too low, and there is not much that can be done to adjust them. If they have been set too low, capacity will, to an extent, be rationed by congestion, though as noted above, it will be expected rather than actual congestion costs which do the rationing. This will also be the case under a pure congestion cost solution. It may be feasible for airlines to respond, to an extent, to actual congestion costs (though these cannot be known exactly until the time of use), but they will find responding to actual slot prices, as they change in response to changing demand and supply conditions, more straightforward.

In summary, slot limits perform well as compared to the congestion solution under uncertainty. Even if slots are not particularly well allocated, it must be remembered that congestion is not a very efficient allocator of capacity either. Slot limits will reduce the expected use of the airport, and will lead to expected gains from lower congestion and more efficient output levels. Slots might be efficiently allocated, if auctions are present or trading is effective. Furthermore, the slot trading approach has distinct advantages under uncertainty, because trading can take place up to the time of use, which makes it possible to the available slots to be allocated as efficiently as feasible under conditions of uncertainty.

Price Regulation, Profit Constraints and Slots

There are several ways in which excess demand at busy airports can be rationed. The most prominent are: congestion costs; slot limits where slots are given to airlines, and perhaps traded; slot limits where slots are auctioned to airlines, and pre set rationing prices.

While the first two are common, the second two, in spite of their good efficiency properties, are rarely, if ever, observed.

Busy airports are very often subjected to price or profit constraints. Privately owned airports, such as those in London, are subjected to price regulation. This regulation is in place to limit their ability to exploit their monopoly position. Typically, even though (some) regulators recognize the incentive problems, they seek to keep revenues at a level around costs – highly profitable monopolies are difficult to justify to the public or to governments. The problem is that, with busy airports, price regulation serves to keep prices below efficient levels. Where there is excess demand, efficient rationing prices would generate large profits – if, for example, Heathrow capacity were rationed by price, the airport would achieve extremely high profits. The problem is that price regulation rules out using prices to ration scarce capacity.

Prices could be used to ration capacity when an airport is only moderately busy. Efficient prices may still be below those allowed under the price cap. Many airports are busy for only part of the day, and it would be feasible for prices to be set at peak times which were sufficiently high to ration capacity, but which did not result in price caps being exceeded. Significantly, however, few airports, be they publicly owned, private but price regulated, or owned by not for profit firms, actually implement peak pricing. For very busy airports, peak pricing may be feasible and desirable on efficiency grounds, but efficiently set prices would still exceed the price cap.

Price regulation is likely to be implemented for privately owned airports. Publicly owned, and not-for-profit airports are often not subjected to formal price regulation. However, they are often required to consult with users and are likely to be subjected to formal or informal profit constraints. Busy airports facing excess demand are expected to not to make large profits out of their situation. Again, prices cannot be used to ration demand.

It might be feasible to use slot limits to constrain demand, but to auction the slots. However with this option, there is a real problem of how to use the auction earnings. Slot auctions would generate large sums of money at the expense of the airlines. These could go the government or the airport – this would be an efficient solution but one which would also generate large distributional transfers. It might be feasible to distribute the proceeds back to the airlines, but there are real problems in devising a way of doing this which does not interfere with the auction process and its efficiency properties.

Granted this, price or auction solutions to the airport excess demand problem are rare if they exist at all. Airports rely on either congestion or slot limits to ration demand to capacity. There are some airports which have peak charges, but these are not sufficient to ration all excess demand (CAA, 2001; Hague Consulting Group, 2000).

Thus the advantage of slot limit systems may not be that they are preferable to price solutions. Rather, they are a means of achieving an efficient level of utilization of the airport, and if the slots are traded effectively, an efficient allocation of capacity to users. This in return is consistent with price or profit constraints which will invariably be imposed on busy airports. Price and auction solutions are not consistent with these constraints. In practice, the effective choice may well be one between congestion rationing and slot limits.

Slot Systems and Congestion Costs at Less Busy Airports

The basic model of congestion costs, as set out above, has assumed that congestion costs are the same regardless of whether capacity is rationed by congestion, prices or slots. This would be appropriate if the slot system worked in such a way that a movement obtained a slot to the use the airport in a specific hour or half hour, and it chose to use the slot any time during that period. Movements could either arrive/ depart on a random basis, or they could be clustered around preferred times, or both of these.

However, this is not the way slot systems normally work (see chapters by Gillen and Ulrich in this volume).Within the slot coordination process, specific movements are given a specific time during the period. When there are too many movements requesting a popular time, such as on the hour, movements will be shifted to even out the flow. While there is some flexibility needed to allow for uncertainty, this is limited, and airlines are expected to keep to time. Sanctions for non performance are not likely, but they could be very severe – loss of a slot pair worth £10m is a strong incentive to operate according to allocated times. Given the slot rents which airlines enjoy, they have a strong interest in ensuring that the system works well.

This process has the effect of reducing delays. The problem of clustering is lessened or eliminated, and arrivals are not random. Short term queues within the slot period have less opportunity to form. This solution has many parallels in other service industries. Doctor's clinics often operate on an appointment basis. By doing this they can operate with a very high utilization of the doctor's time, yet waiting times can be very low. Other clinics operate on the basis of first come first served, and at these, waiting times can be quite long.

Thus the slot system has two distinct functions. Firstly, as generally recognized, it limits throughput of the facility (runway or terminal) so that congestion is reduced. Secondly, it evens out traffic flows, and lessens the congestion at any given level of throughput. This second function is unique to slot management systems. It would not be feasible to even out flows under a first come first served queuing system, or a pricing solution (unless there was a very detailed structure of distinct prices for different short blocks of time, such as intervals of five minutes).

The system is not costless. It does reduce delay costs. However it imposes costs in that users cannot always obtain their preferred time to use the airport. Further, since it is an administrative system, the allocation of movements to times may not be very efficient (though there could be scope for airlines to trade times within slot periods). Airlines will face costs in rearranging schedules to fit slot times. Delays may be imposed at secondary airports, as movements will need to depart at times which enable them to arrive at slot controlled airports at their allocated slot time.

This said, it must be remembered that all systems of rationing of capacity, congestion, pricing and slots, impose similar costs. An airport which can handle 60 movements in an hour can handle only five movements in a five minute period. Under a congestion rationing system, an airline which strongly preferred a particular time can queue up so that it achieves it. Under a slot system with secondary trading an airline which strongly prefers a particular time can trade with other users.

This smoothing effect can also be achieved in a situation of first come first served rationing when an airline accounts for a large proportion of movements at an airport. Just as it is recognized that the airline will take into account its own impact on congestion (Brueckner, 2002), it will also internalize its impacts on congestion through scheduling its flights so that they are evenly spread, and do not cause congestion through clustering. Thus American Airlines would not seek to schedule all its departures from Dallas on the hour. In fact it may be easier for an airline with a high share of flights at an airport to internalize this aspect of congestion rather than to reduce congestion by reducing its own flights (because it may be more concerned that its competitors may take up the slack if it reduces its own flights).

The function of slots in reducing congestion is illustrated in Figure 5.4. The basic congestion cost model, as discussed in the section above, is given, with average congestion costs of AC_1 and marginal cost of MC_1. With no change in the congestion cost, the optimal number of slots to allocate in the high demand situation would be Q_3, and the throughput would fall from the original equilibrium. With slots coordination performing a smoothing function, the congestion cost would fall to AC_1 and MC_2. The new optimum number of slots would be Q_4. It is quite possible that this would be larger than the original – congestion solution – throughput. Thus, as compared with a congestion solution, a slot system will not only reduce congestion, but it may also enable a higher throughput.

Slot coordination may have an important role to play even when the airport has ample capacity. This possibility is illustrated in Figure 5.4. If the demand

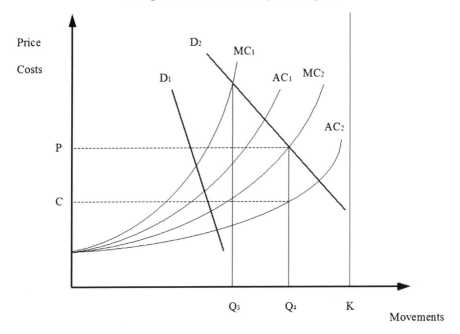

Figure 5.4   Slot systems and congestion costs at less busy airports

curve is D_1, there will be some congestion, but the gains from reducing throughput to reduce congestion, though positive, could be quite small. However, the slot coordination process will also lower the congestion cost curve – the gain from this could be considerable. Many airports which have no problem of inadequate capacity impose slot coordination for this reason.

The slot coordination process may also have implications for the shape of the congestion cost curve. Coordination keeps delays quite low even when throughput gets close to the maximum theoretical capacity. However, when throughput is very close to capacity, congestion will increase sharply (throughput cannot exceed capacity for an extended time). Thus the delay function could be close to a reverse 'L' shape.

The congestion cost curve is not likely to be completely flat, however. This is because average delays depend not only on throughput but on other factors, such as weather. Poor weather will increase delays. It will do this by reducing the effective capacity of the airport. An airport which can handle 60 movements in an hour may only be able to handle 50 movements in poor weather. If more than 50 movements are scheduled during an hour, delays will build up, and carry over to subsequent hours. The greater the scheduled throughput is, the greater the probability will be that it exceeds capacity during poor weather, and the greater the probability of significant delays. Thus expected delays will increase as the scheduled throughput of the airport approaches and exceeds the effective capacity during poor weather. Therefore the average cost of congestion increases the larger the throughput.

This said, it is still probable that the congestion cost curve will be less gradually rising than in the basic, non-coordinated, model. Average and marginal congestion costs will be fairly flat, up to the point where scheduled throughput is close to maximum capacity, and beyond this they will rise sharply.

This has practical implications for the choice of the slot constraint level. The optimum slot constraint will be close to the maximum capacity of the airport – this will be so regardless of the level of demand or the level of congestion costs. It becomes possible to choose the number of slots to allow without detailed cost benefit analysis of the delay costs and the benefits from allowing extra costs. While a simple choice will not achieve the optimum exactly, it will not be far off it.

Empirical Evidence

There are a number of empirical questions which arise out of the analysis above. These include:

- Do slot limits reduce congestion delay costs and improve the efficient utilization of airports?
- How large are the gains from implementing slot limits likely to be?
- Does slot coordination reduce delay costs at any given throughput? and
- Are slot limits being set efficiently?

There are other empirical questions which are associated with slot allocation systems – the most important of these concerns how efficiently the slots are allocated to airlines. We are not concerned with this question in this chapter, except to the extent to which the efficiency of allocation affects the answers to the other questions above. We explore these questions in the next section.

While the questions which we have identified are essentially empirical ones, there is not a wealth of empirical study of them. While studies of some aspects do exist, the questions have not always been posed in the terms which we have used. Data do exist, on delays at individual airports, on use of airports at different times, on values of time and costs of delay to airlines. However, these data have not been summarized in a way useful for our purposes. On other issues, critical for our purposes, data are very difficult to come by – information on the prices or shadow prices for slots is very difficult to obtain.

The upshot of this is that rather than providing a comprehensive study, the best we can hope to do at this stage is gather together the diverse pieces of evidence, and examine how consistent they are with expectations, and what they might imply for choices of slot capacity.

There is good evidence that slot systems do work in terms of their primary objective, namely that of reducing delays. When European airports, which are mainly slot constrained, are compared with US airports, which are mainly not slot constrained, it is clear that delays are significantly lower at European airports (though care must be taken in making such comparisons, as delay measures may differ). Average delays, and the average delays experienced by flights which are delayed, in Europe are around about half those in the US (see chapter by Janic in this volume). Interestingly enough, however, delays at the slot constrained airports such as La Guardia do not appear any lower than those at non-slot constrained US airports (see also Brueckner, 2002). As noted in section 3 above, this is not inconsistent with the theory of slots reducing congestion. However, it is also possible that slot limits at US constrained airports have been set too high, and are not constraining delays by as much as they could.

Another point which emerges from these data is that delays at slot controlled European airports are well below those of US airports even when utilization is not high relative to capacity. This is not a result which would be expected if the sole purpose of slots were to reduce congestion through taking pressure off capacity – for non busy airports, the effects of reducing throughput on congestion would be small. This suggests that the smoothing function of slots coordination is effective in reducing congestion, and that its effect could be very significant.

Table 5.1   Airport utilization and delays

Airport type	Demand/ capacity indicator	Average delays – departures (min.)	Average delays – arrivals (min)
Almost continuous excess demand	5.2	3.44	3.32
Excess demand at peaks	3.17	2.99	2.82
Occasional excess demand	1.93	2.91	2.16
Non slot constrained	–	2.63	1.90

Sources: NERA (2004); Eurocontrol/ECAC (2002).

In the section on 'Slot Constraints and Efficiency' it was suggested that, even when slot constraints are set optimally, we should expect that there would be higher delays at airports which are very busy and subject to considerable excess demand. Some evidence for this is presented in Table 5.1. This table reports an indicator of demand relative to capacity, taken from the NERA Report (NERA, 2004). This Report divided airports into categories according to how busy they are – whether they are busy all day, at peaks only etc. Data on average delays for arrivals and departures for these, and other, non busy airports, were taken from Eurocontrol (2002). One thing to note is that the airports which are subject to excess demand do achieve higher utilization – this need not happen if slot constraints are set too tight. As can be seen, there is a clear relationship between delays and capacity utilization, amongst airports which are slot controlled. More is being squeezed out of the busier airports, at a cost in terms of average delay. Given the greater shadow or market price of a slot at these airports, this is warranted.

It is also apparent from these data that average delays do not increase substantially as airports become busier. This could be because of the high degree of aggregation of these figures – more disaggregated data might show a delay curve which rises more sharply as utilization rises compared to capacity. However, it would also be consistent with the suggestion, in the previous section, that the delay or congestion cost curves could be relatively flat until utilization comes very close to capacity.

One possible experiment would be to consider whether it would be worthwhile to alter slot constraints so that an airport shifted from one category to another. London Heathrow is in the top category of busiest airports. Suppose that slot availability was reduced so that it achieved a utilization equal to that of the next category of airports. Doing this would reduce delays per movement by one half of a minute. It would require a large reduction in total movements at Heathrow – about 40 per cent – to achieve the lower utilization factor. Suppose this were done, and slot constraints were tightened to remove some 182,000 movements per annum. Using the cost of delay of $US79.1 per movement, as was estimated in a recent ITA (ITA, 2000) study, the delay cost per minute per movement would be about £43.5 (though delays at Heathrow could be a little higher because of larger aircraft and passenger loads). The value of the saving in delays would be about £6.2m per annum. On the other hand, the value of the lost slots would be much greater. Taking the most recent price of £10m for a daily slot pair in perpetuity, this would be equivalent to £1 m per annum at a 10 per cent interest rate. This is equivalent to £1370 per slot (about £10 per passenger). The value of the lost slots would be £249.3m per annum. This figure is much larger than the value of the delay reduction. It should be recognized that the average delays as reported in Table 1 are not points on a delay curve, and that a reduction in throughput of the order of magnitude supposed would produce a much larger impact on delay costs. However, this suggests that more, rather than less, delays could be appropriate for Heathrow.

To test whether an airport slot constraint is optimal, information about average and marginal delay costs are needed along with the value of a slot. For Heathrow, we can find the value of a slot (£1370) and the average delay (4.5 minutes). From this, it is possible to make an estimate of the average delay cost – at £43.5 per minute per movement, this would be £195.75. We do not know the marginal delay cost, or the difference between it and the average delay cost. Nor do we know the slope of the average delay cost function – this would enable an estimate of the marginal delay cost.

However, could these figures be broadly consistent, in the sense that the marginal delay cost was £1565.75 (i.e. the slot price of £1370, equal to the marginal external delay cost, plus the average delay cost of £195.75 faced by each user)? This is about eight times the average cost. This could be the case if the slope of the average delay cost curve was high, and marginal was well above average cost. Given the nature of the problem, this is quite possible. Close to maximum capacity, the average and marginal delay cost curves will be sharply rising, and a high to very high ratio of marginal to average cost is to be expected. In short, while a good deal more evidence on marginal delay costs are needed to be conclusive, it does look as if the slot constraints at Heathrow are set at very roughly the right level. If the slope of the average delay cost is not very high at current delay levels, and if the ratio of marginal to average delay cost is somewhat below eight, then some small increase in slots could be worthwhile. It also seems very unlikely that too many slots have been made available.

It is interesting to contrast the results for Europe, and especially London Heathrow, with those for the US. Average delays for the slot constrained airports in the US are much higher. The choice of the trade off between slot availability and delay in the two regions appears quite different – and this suggests that they cannot both be right. It is possible that the value of a slot at a US airport may be very much higher than that of a slot at Heathrow, but this does not seem all that likely. It is also possible that values of time and aircraft delay costs are much lower in the US than Europe, but with higher per capita real incomes in the US, the reverse should be the case. On the face of it, it seems as if too many slots are being made available at slot controlled US airports if the European choice is about right, and that delays are being allowed to be too large. This suggests that there are efficiency gains to be made, especially at US airports, through obtaining better empirical estimates of delay costs and slot prices, and optimizing slot constraint levels.

This could be a reflection of the balance between the diverging interests in the choice of slots to be made available. Airports may wish to have high slot availability to maximize throughput, as might national aviation authorities. Competition authorities are often sceptical about slots limits, for fear of creating market power. Airlines may wish to keep slots more tightly constrained, to lower delays for themselves, and to prop up the value of slots. In Europe, the single dominant airlines at the busy hubs may be more effective in convincing the slot decision makers to limit slots, and congestion, than the airlines in the US, where competition concerns are more to the fore.

Setting Slot Limits – Theory and Practice

In this chapter, we have set out the factors which, ideally, should be taken into account when slot limits are set. These include congestion costs, other external costs, and the value of additional output. In reality, slot limits are set administratively. If congestion is the only factor, managers or bureaucrats with experience in airport operations will set the slot limits with a view to keeping delays at some arbitrarily determined 'acceptable' level. Where noise or emissions are a consideration, limits will tend to be set more arbitrarily, perhaps as a local political compromise. The choice of slot limits will be affected by pressure from the interested parties, such as airlines. For very busy airports, there may be pressure from the airlines to set limits so that utilization and delays are higher than at less busy airports. It is rare for a formal cost benefit assessment, estimating delay and other external costs, and comparing them to the benefits from additional output, to be carried out.

It is desirable that the setting of slot limits be subject to an explicit cost benefit assessment. The rule of thumb which administrators have been using might be getting the slot limits about right – but they might not. It is possible that delays could be reduced for little reduction in output at some airports, and that highly valuable additional output could be achieved at a small cost in terms of additional delay at some busy airports.

This said, it is possible that gains from an efficient choice of slot limits are positive, though not large. This is the case if the shape of the delay curve is that of a reverse 'L', as was suggested as a possibility in the section on 'Slot Systems and Congestion Costs at Less Busy Airports' above. If so, the range of utilization rates over which the optimum will fall is quite narrow, and this range is close to full utilization. Thus if the rule of thumb used by the administrators results in a slot limit which is a little below maximum theoretical capacity, at a level of utilization at which delays start to rise sharply, it will tend to be about right. It is an empirical question whether this is the case – rules set for a particular airport could be quite inappropriate, especially if they are taken from other airports operating under different circumstances.

Thus the choice of slot limits is likely to be an efficiency issue of medium level importance. There could be efficiency gains from more rigorous determination of the efficient level of slots to be permitted at an airport, though it is quite possible that administrators have been using a rule of thumb which gets the choice approximately right. The gains from more efficient slot limit choice may not be as large as those from better allocation of slots or better pricing. However, this is an empirical issue which needs to be investigated.

Conclusions and Future Directions

While much attention is now given to slot systems to limit congestion at airports, most of this is given to the problem of allocating slots efficiently. Comparatively little attention is given to the problem of setting the correct slot limits. The optimal slot limit does not result in the elimination of congestion, but it will reduce congestion as compared with the no control situation. The efficient extent of slot restriction will depend on the demand pressure that an airport is under. Where there is very high demand to use the airport (and the value of slots is very high), it is efficient to make more slots available and accept higher delay levels. In particular, at the efficient solution, there is a relationship between slot values and delay costs. The value of a slot should equal the difference between the marginal and average costs of delay.

In principle, it is possible to test whether this is the case at actual airports. In practice, lack of data limits testing. Data on slot values are very difficult to come by, and data on the shape of the delay function are also not readily available. This said, it is possible to develop some simple tests. The pattern of delays between US and European airports is consistent with the theory. In addition, the patterns of delays at the different European airports, which show differing degrees of excess demand, fit the theory as well. Some simple calculations suggest that slot limits at London Heathrow airport could be about right – if anything, slots availability could be increased a little. US and European authorities do seem to be making very different choices between slot availability and delays, and it does not seem that both are right. Differing bargaining strengths of the key stakeholders, such as airlines, airports and national aviation authorities could possibly explain the different choices.

Addressing the data problem is the top priority. The raw data with which to make better estimates of delay cost functions do exist. It should be possible to gain better measures of the marginal external delay cost at current delay levels at slot controlled airports. On the demand side, better information on the true values of slots is needed. It should not be impossible to come by. While lack of open slot trading means that it is not possible to obtain a single measure of the value of a slot, slots do have values and those who use them know what they are. With better information on these key parameters, it should be possible to assess how efficiently slot limits are being set.

References

Brueckner, J. (2002), 'Airport Congestion when Carriers have Market Power', *American Economic Review*, 92 (December), pp. 1357–75.

Civil Aviation Authority (UK) (2001), *Peak Pricing and Economic Regulation*, Annex to CAA (2001), *Heathrow, Gatwick, Stansted and Manchester Airports Price Caps – 2003–2008: CAA Preliminary Proposals – Consultation Paper*, London: http://www.caa.co.uk.

Dings, J.M.W., Wit, R.C.N., Leurs, B.A. and Davidson, M.D. (2003), *External Costs of Aviation*, Berlin: Federal Environmental Agency.

Doganis, R. (1992), *The Airport Business*, London: Routledge.

Eurocontrol: Central Office for Delay Analysis (2002), *Delays to Air Transport in Europe, Annual Report*.

Forsyth, P. (1976), 'The Theory of Pricing of Airport Facilities with Special Reference to London', DPhil thesis, Oxford.

Hague Consulting Group (2000), *Benchmarking Airport Charges 1999*, The Hague: Directorate General of Civil Aviation.

Institut du Transports Ariens (ITA) (2000), *Cost of Air Transport Delay in Europe, Final Report*, November.

International Air Transport Association (IATA) (2000), 'Peak/Off-peak Charges', Ansconf Working Paper No. 82, Montreal: ICAO.

International Air Transport Association (IATA) (2005), *Worldwide Scheduling Guidelines*, 11th edn, Montreal: IATA.

Knieps, G. (1992), 'Infrastrukturprobleme im europäischen Luftverkehr', *Schweizerische Zeitschrift für Volkswirtschaft und Statistik*, 128, pp.1 643–53.

Mohring, H. (1976), *Transportation Economics*, Cambridge, MA: Ballinger.

National Economic Research Associates (NERA) (2004), *Study to Assess the Effects of Different Slot Allocation Schemes A Final Report for the European Commission, D G Tren*, London, January.

Turvey, R. (2000), 'Infrastructure Access Pricing and Lumpy Investments', *Utilities Policy*, 9, pp. 207–18.

Weitzman, M. (1974), 'Prices vs Quantities', *Review of Economic Studies*, 41 (4), pp. 477–91.

Chapter 6

The Problem of Charging for
Congestion at Airports:
What is the Potential of Modelling?

Milan Janic

Introduction

Congestion and aircraft/flight delays in the air transport systems of both Europe and US have generally increased over the past decade. The most obvious have been growing demand, constrained infrastructure capacity, and unplanned disruptions of airline schedules (Janic, 2003). The congestion and delays due to an imbalance between air transport demand and infrastructure capacity has generally been alleviated by improvements in the utilization of existing capacity, physical expansion of infrastructure, and demand management. The first option has shown to have limited effects. In many cases, implementation of the second option has been difficult or even impossible at least in the short-term due to the various political and environmental constraints in terms of noise, air pollution and land use. The last option, demand management has recently been considered a potentially viable option to relieve the congestion problem (Adler, 2001; DeCota, 2001; FAA, 2001).

In addition to institutional instruments, demand management at airports embraces economic instruments such as charging of congestion and auctions of slots. In general, a central problem of dealing with charging of congestion consists of recognizing the right case (i.e., the type and nature of congestion) and an estimation of the marginal delay costs imposed by an additional flight to all subsequent flights during a given congested period. In such a context, the additional flight has to (theoretically) pay its own delay costs and a charge equivalent to the marginal costs of delays imposed on all subsequent flights. This charge will raise the flights total operating costs and consequently compromise its profitability. Currently, the charging system at European and US airports is mainly based on aircraft weight and has a little in common with the above-mentioned concept of charging congestion (ACI, 2001; Adler, 2002; Doganis, 1992).

This chapter develops a model for charging congestion at an airport. In addition to this introduction, the paper consists of four sections. The second section provides an insight into the problem of congestion at particular European and US airports. The third elaborates on the conditions under which charging of congestion could be implemented. The next section deals with modelling procedures. Next, we provide an application of the proposed model and finally offer conclusions.

Demand, Capacity and Congestion at Airports in Europe and the United States

Dealing with charging for congestion at airports should include an analysis of the relevant parameters such as demand, capacity and congestion. Demand is represented by the flights schedule at an airport carried out by one or more airlines. At many large European and US hub and non-hub airports, one (incumbent) or few (competing) airlines, their subsidiaries and alliance partners carry out the flights. These flights use the available arrival and departure slots, i.e., the airport's declared (practical) capacity, to get service at the airport. In Europe, the maximum number of arrival and departure flights accommodated at an airport during a set period of time (usually one hour) under given conditions determines the airport declared (practical) capacity. This capacity is based on IMC (Instrumental Meteorological Conditions) and IFR (Instrumental Flight Rules). Usually, this capacity is an agreed value between airlines, airports and air traffic control (EUROCONTROL, 2002). In the US, the airport capacity has usually two values: the higher under VMC (Visual Meteorological Conditions) and VFR (Visual Flight Rules), and the lower under IMC and IFR (FAA, 2001).

Demand and Capacity

Many large and congested European and US airports are 'slot-controlled'. This implies that the number of flights/aircraft is balanced with the airport's declared capacity in order to maintain congestion and flight/aircraft delays within prescribed boundaries. At these airports, the initial demand and the available airport capacity are balanced through a multi-stage process of negotiations between airlines, airports and air traffic control. In such context, the demand is generally allowed to reach at most the level of airport capacity over the longer period of time (day), which, under regular operating conditions enables planned congestion and delays. However, due to the system's imperfections and other disrupting factors, the actual demand frequently exceeds the airport's declared capacity and consequently causes unexpectedly longer congestion and delays (ATA, 2002; EUROCONTROL, 2002; FAA, 2001, 2002a; Liang et al., 2000; Janic, 2003).

Figure 6.1a) illustrates an example of two-stage balancing of demand and capacity at New York LaGuardia airport (US) during a peak day, 30 June 2001 (beginning of the Independence Day holiday) (FAA, 2002a). The balancing became necessary when the demand for new flights (slots) grew much above the available airport capacity mostly driven by the legislative document 'Aviation Investments and Reform Act of the 21st Century' (approved in April 2000) (i.e., AIR-21 'Slot Exemption Act'). The document abandoned the slot permit requirements from 1969 for LaGuardia and four other of the most congested

US airports. Particularly at LaGuardia airport the slot-exemption rule enabled free entry of new airlines/flights serving small hub and non-hub airports with smaller aircraft (up to 71 seats). Six months after the introduction of the Act, about 200 new daily flights were added. Eight months latter, 300 new daily flights were additionally planned. Such actual and prospective growth of traffic caused a significant increase in the congestion and delays at LaGuardia. The number of flights delayed by more than 15 minutes had increased from 10 per cent before to 30 per cent after the introduction of the Act. Counter actions aiming at controlling demand followed mostly in terms of capping the number of new slot exemptions, introducing a lottery distribution of these limited exempted slots, and allowing an increase in the number of new slots up to the airport hourly capacity of about 80 operations per hour under VFR conditions. This was carried out periodically, and on daily basis in the specific cases like this. Over time using the above-mentioned measures however proved unsuccessful in deterring new demand. Consequently, the idea of charging congestion fees as an economic measure to better manage demand came under consideration by the FAA and LaGuardia airport authorities.

As can be seen, at the first stage the initial demand consisting of flights of about twenty competing airlines was much higher than the airport's actual capacity (which was lower than the declared capacity). In the second stage, the demand was suppressed below the declared capacity but remained slightly above the actual capacity during the morning and early afternoon hours.

Figure 6.1b illustrates the relationship between the airport demand and capacity at the US Atlanta Hartsfield airport during the same day, 30 June 2001. As can be seen, the pattern of demand was quite different than at LaGuardia airport due to different – hub-and-spoke – operations of the main incumbent Delta Airlines. In the morning and afternoon hours the realized demand was lower than the actual capacity. However, it exceeded this capacity (which also changed during the day) considerably during the late afternoon and evening hours.

In both examples, the relationships between demand and capacity imply congestion. However, this congestion is of a distinctive nature. In the former case, many competing airlines use the airport as origin and destination of their flights. In this case, congestion occurs mainly due to competition. In the later case, an incumbent uses the airport as its hub and intentionally creates congestions through the hub-and-spoke operational pattern. However, in both examples, the level of congestion is negotiated and consequently the access to the airports controlled.

Congestion and Delays

Congestion causes flight delays. In general, delay is defined as the difference between the actual and scheduled time at the 'reference location'. The threshold for either arrival or departure flight delay is any period longer than 15 minutes behind schedule (AEA, 2001; BTS, 2001; EUROCONTROL, 2001; FAA, 2002a,).

Congestion and delays have become common (and inherent) operational characteristic at many European and US airports. Table 6.1 shows some relevant statistics. As can be seen, the proportion of delayed flights has been different in

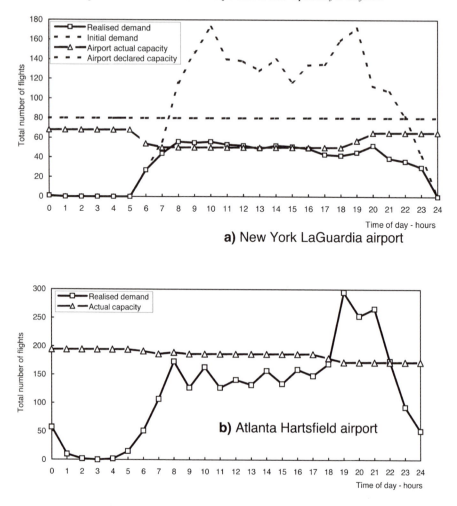

Figure 6.1   Relationship between demand and capacity at two US airports

Source: compiled from FAA, 2002a.

both regions. In Europe, it has varied between 17 per cent and 30 per cent for arrivals and between 8 per cent to 24 per cent for departures. In the US, this proportion has varied between 22 per cent and 40 per cent for arrivals and between 19 per cent to 38 per cent for departures. In general, more frequent delays have taken place at US airports.

Delays at airports are generally expressed as the average time per flight or the average time per delayed flight (the total delay divided by the number of all delayed flights or by the number of only delayed flights per period, respectively) (EUROCONTROL/ECAC, 2002; FAA, 2002a).

In addition, total delays are further classified into arrival and departure delays. Figure 6.2 (a and b) shows both types of delays plotted against the average annual

Table 6.1　Delays at some congested European and US airports

European airports (2001)	(%) of delayed flights		US airports (1999)	(%) of delayed flights	
	Arrivals	*Departures*		*Arrivals*	*Departures*
Paris CDG	24.6	21.8	Chicago-O'Hare	33.6	29.9
London Heathrow	17.4	21.0	Newark	38.4	31.0
Frankfurt	30.8	18.9	Atlanta	30.9	26.8
Amsterdam	25.7	23.2	NY-LaGuardia	40.1	28.9
Madrid/Barajas	19.6	20.0	San Francisco	32.1	21.5
Munich	19.0	19.0	Dallas-Fort Worth	21.7	23.7
Brussels	29.8	27.7	Boston Logan	37.7	29.3
Zurich	23.2	23.8	Philadelphia	40.4	37.9
Rome/Fiumicino	-	12.5	NY-Kennedy	28.0	19.0
Copenhagen/K	17.8	10.3	Phoenix	29.6	30.8
Stockholm/Arlanda	–	8.0	Detroit	24.6	26.3
London/Gatwick	19.6	24.3	Los Angeles	26.1	20.8

Sources: EUROCONTOL/ECAC, 2002; FAA, 2002a.

demand/capacity ratio (i.e., utilization of airport capacity) for 32 US and 17 European of the most congested airports.

As can be seen, the average delay per flight – either departure or arrival – has been generally longer at the US airports. At the US airports, departure delays have generally been longer than arrival delays. Departure delays varied between 10 and 20, while arrival delays were between five and 15 minutes. At European airports, no obvious variances between average arrival and departure delays are observable. Almost all delays were shorter than 10 minutes.

According to the 15-minutes threshold, flights in both samples should not be considered delayed. In both regions, a very slight increasing of delays for both types of operations with increasing of the demand/capacity ratio has been noticeable.

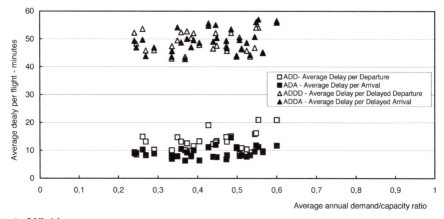

a) US Airports

Source: FAA, 2002a.

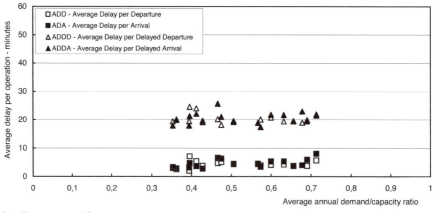

b) European Airports

Figure 6.2   Dependability between the average flight delay and the average utilization of airport capacity

Source: EUROCONTROL/ECAC, 2002.

The picture changes when the average delay per delayed flights is considered. In given sample, this delay appears again longer at US airports, 40-60 minutes compared to 15–25 minutes at the European airports. In both regions, this delay has been similar for both types of operations and seemingly non-influenced by airport utilization (i.e., the demand/capacity ratio). According to the 15-minutes threshold these delays were counted as delayed flights. In the US, on average, bad weather has caused about 70–75 per cent and congestion about 20–30 per cent of these delays (BTS, 2001; FAA, 2001; 2002a). In Europe, on average, severe weather has caused only 1-4 per cent and congestion about 30-40 per cent of these delays (AEA, 2001; EUROCONTROL, 2001).

Figure 6.2 (a and b) also shows that the average utilization of airport capacity has varied between 25 per cent and 65 per cent at US airports and between 35 per cent and 75 per cent at European airports. This indicates that at almost all airports the demand has always been kept below the capacity on an annual scale, as a rule for preventing extreme congestion and delays (Welch and Lloyd, 2001; Odoni and Fan, 2001).

Charging Congestion Prices at Airports

Background

Interaction between demand and capacity, which causes congestion at airports is commonly measured by the ratio between the intensity of demand and capacity (or the capacity utilization ratio). This ratio can generally take the values lower, equal or greater than one. Specifically, if the intensity of demand equalizes with the capacity, this ratio takes the value 1.0 (or 100 per cent) (Newell, 1982). At most European and US congested airports, contrary to the above-mentioned annual averages, the utilization ratio during the short peak periods of a quarter or an hour often reaches or even exceeds the value 1.0 (100 per cent).,This implies significant congestion and delay[1] (FAA, 2001, 2002a). In cases where this ratio cannot be institutionally regulated (i.e., negotiated) the economic instruments of demand management, one of which is the charging congestion prices, could be considered (Vickery, 1969).

Inherent Complexity of Implementation

Currently, despite being conceptionally well developed and recommended by economists and the policy-making literature alike, charging for congestion has still not found much practical application. The main causes could be summarized as 'collision with the overall airport objectives including the lack of real cases', 'complexity of measurement', 'ambiguity of the concept' and 'barriers within the industry'.

3.2.1 Collision with the airport objectives and the lack of real cases Most airports worldwide have traditionally tended to grow under given circumstances due to their internal (economic) as well as wider external (economic and political) regional and national interests. Growth is largely driven by attracting as much traffic as possible. In this context and despite potential short-term social, political

---

1   It is well known that if demand/capacity ratio is lower or close to 1.0 (100 per cent), congestion occurs mainly due to the random variability of the flight inter-arrival and service time. If this ratio is greater than 1.0 (100 per cent), the excessive demand dominates as the cause of congestion (Newell, 1982).

and environmental barriers, physical expansion of infrastructure capacity has always been used as the most feasible long-term solution to relieve congestion. Consequently, airports have very rarely considered charging for congestion as a viable short-term remedy. The revenues from combined aeronautical[2] and non-aeronautical charges have provided coverage of the airport operational costs and partly fund investments.

3.2.2 Complexity of measurement of conditions Typically many simultaneous events cause congestion, which materialize as demand peaks. These demand peaks by and large differ between particular airports in terms of *'frequency and duration'* and often also depend on the type of operation and aircraft involved.

The *'frequency and duration'* implies that the short and frequent peaks have been created by the airline hub-and-spoke operations. The long and infrequent peaks have been created by the large demand exceeding capacity during several hours of the day.

Type of operation and aircraft implies that congestion during peaks affects both arrival and departure flights of the same and/or different airlines and sometimes transfers delays between them.

Extraction of real causes for charging for congestion has shown to be a complex or even impossible task. There has been a lack of criteria for setting up the relevant level of congestion and delays to be charged (or internalized). Since delays up to the 15-minutes threshold have not been counted, only the longer ones (but which?) have thus far been internalized (Airport Council International, 2001, Odoni and Fan, 2001; Janic, 2003) Internalizing congestion within the airline hub-and-spoke operations or due to disruptions of the airport capacity, which might cause delays longer than 15 minutes is questionable. In the former case, the internalisaion could compromise integrity of the airline schedule. In the later case, internalizing of congestion cause by factors out of the control of airlines, airports ad air traffic control coul be difficult to jusify.

3.2.3 Ambiguity of the concept and barriers within the industry Charging for congestion at airports seems to be an ambiguous topic. These charges are meant to be equivalent to the cost of marginal delay, which an additional flight imposes on the succeeding flights during the congestion period. The objective is to deter (i.e., prevent) access of such flights and all other similar flights. This however appears to conflict with the guaranteed freedoms of the unlimited access to the 'slot uncontrolled' airports (Corbett, 2002). In addition, this charge is supposed be effective, which in the case of real market imperfections might not be true. The charge simply may either be too low to be effective, or too high which unwillingly suppresses elastic demand. The relationship between this and other airport externalities such as noise and air pollution, as well as the relationship between this and other existing charging schemes based on the aircraft take-off weight are unclear. Finally problems may arise from the allocation of the collected charges.

---

2   A part of the aeronautical charges has consisted of landing fees based on the aircraft weight (ACI, 2001).

One option could entail the allocation of the monies to those airlines which did not cause the congestion airlines, but bear the additional costs imposed by it. Another option would be to use the funds to expand airport capacity, which ultimately might result in a vanishing of the source of the revenues, which is congestion. Finally, if the threat of imposing charges does work as a deterrent, no additional revenue will be collected, although the benefits will be less delays and less costs associated with the necessary accommodation of flights. All these arguments have contributed to building a case against congestion charging. Adler (2002) has identified the following three groups of barriers:

1) institutional, organizational, political and legal barriers maintained by the monopolistic powerful hub airports (Europe) and powerful airlines (both in Europe and US). This also includes the lack of harmonization of charging conditions across the countries (Europe) and across the airports of different size (both in Europe and US);
2) large airlines and their alliances (lobby groups) are opposed to the concept, due to a lack of similar concepts in most other transport industries (ie rail) in both Europe and the US; and
3) unavailability of the relevant data on the actual causes of congestion at airports including the precise data on the changes of airport capacity (Europe). Relatively useful databases already exist in the US (FAA, 2002a).

A Model of Charging Congestion at an Airport

State of the Art

Economic theory has long noted that the optimal use of a congested transport facility – in this case an airport – cannot be achieved unless each user (flight) is forced to pay the marginal delay cost it imposed on all other subsequent users (flights) during the congestion period. In the nineties, the cost of marginal delay has been considered as an externality to be internalized together with others such as air pollution, noise and air traffic accidents (Adler, 2002; Brueckner, 2002; Daniel, 1995; Daniel and Pahwa, 2000; Daniel, 2001; EC, 1997; ECMT, 1998; Odoni and Fan, 2001; Vickery, 1969). In such context, some researchers proposed charging for marginal delays caused by the hub-and-spoke operations. They used the steady-state and time-dependent, analytically efficient and attractive, queuing models to estimate the costs of congestion and delays to be internalized. Nevertheless, it was quite unclear why alleviation of peaks by charging for congestion was suggested since airports, airlines and passengers already balanced their interests within the given circumstances (Daniel, 1995, 2001). Nevertheless, comparison of different models of charging for airport congestion produced some interesting results on their performances (Daniel and Pahwa, 2000). In addition, some research also tackled the problem of charging the airport congestion when the airlines with different market shares already internalized their congestion costs (Bruckener, 2002).

Despite being theoretically sophisticated, almost all economic models of congestion charging have thus far stayed within the academic domain. With the partial exception of the work by Daniel (1995, 2001) and Brueckner (2001), who made an analogy between congested roads and airports. They showed that the only similarity between the congested roads and congested airports is 'predictable' queues (Hall, 1991). Most recently some research has suggested the non-selective implementation of congestion charges (in addition to other charges for externalities) at almost all European airports. This move however would undoubtedly face strong opposition from both airports and airlines (Adler, 2002).

Assumptions

The proposed model of congestion charges aims at extending previous research and is based on the following assumptions:

- The time–varying demand and capacity profile at the candidate airport are known during typical (representative) day. The demand profile can be obtained from the published airport (and airline) schedule(s). In such context, each flight is considered with respect to the average operational costs and revenues. The capacity profile(s) can be obtained from the airport or air traffic control operator for given conditions (IMC or VMC). This capacity reflects the average service time of particular arriving and departing flights. The runway system is assumed as critical element of congestion.
- Only the congestions during the long peaks in which the demand/capacity ratio is close to or exceeding 1.0 (100 per cent) is considered for internalizing. This assumption can be used for selection of the candidate airports (Adler, 2002).
- The number of flights causing congestion during such long peaks is assumed to be large (at least several dozens), which makes congestion mainly dependent on the predictable variations (and positive differences) between the demand and capacity.[3] This assumption makes the application of the diffusion approximation of queues for estimation of congestion and delays convenient (Hall, 1991; Newell, 1982).

The Model Structure

Estimating the queues at a congested airport Charging congestion at an airport requires estimation of the system marginal delay, which consists of the sum of i) the private cost of delay of an additional flight, and ii) the cost of marginal (additional) delays, which this flight imposes on the succeeding flights during the congested period (Ghali and Smith, 1995; Hall, 1991). Both delays can be

---

3   For example, for the non-stationary Poisson arrival/departure processes, if the number of users-customers during given period is greater, the smaller will be the random variations of this number (Hall, 1991).

estimated by using the various queuing and simulation models (Hall, 1998; Newell, 1982; Odoni et al., 1997). In these models, congestion is usually related to the time-dependent demand/capacity ratio $\rho(t)$. At time (t), $\rho(t) = \lambda(t)/\mu(t)$ where $\lambda(t)$ is the flight arrival rate (i.e., demand for service) and $\mu(t)$ is the flight service rate (i.e., capacity). Different techniques are developed to estimate congestion and related delays in dependence on $\rho(t)$. One of them, a graphical representation of the typical queuing process at the congested facility (an airport) during period T (one day) is shown in Figure 6.3.

The curves $A(t)$ and $D(t)$ represent the cumulative counts of flights requesting service and being served, respectively, by time (t). Since the number of flights in the system is assumed to be large (>> 1.0) both types of counts, actually the step functions of time, can be considered as their continuous (smooth) counterparts. Consequently, $\lambda(t) = dA(t)/dt$ and $\mu(t) = dD(t)/dt$. The functions $A(t)$ and $D(t)$ may either relate only to one realization or be the averages of many daily realizations of serving the flights at the congested airport(s). Dependent on the relationships between two curves, three sub-periods can be identified. In the first one $(0, t_1)$, $A(t)$ lies below $D(t)$ and $\rho(t)$ is less than *1.0*. In this case only the 'random effects' cause congestion. During the second sub-period $(t_1, t_2) \equiv \tau$, $A(t)$ exceeds the curve $D(t)$. The values of $\rho(t)$ fluctuate from being equal, greater, again equal, and finally less than one. In this case, 'deterministic effects' are the main causes of congestion while the previously important 'random effects' are negligible. Finally, during the sub-period (t_2, T), $A(t)$ again drops bellow $D(t)$ and the similar developments as in the first sub-period take place.

Obviously, only congestion during the period (τ) should be internalized since it has the potential produce delays longer than the threshold of 15 minutes (Hall, 1991; Newell, 1982). To estimate these congestion and delays, the period (τ) is divided into K equal increments Δt (i.e., $K*\Delta t \approx \tau$). As compared to (τ), each increment Δt should be sufficiently short[4] in order to register changes of the congestion and delays on the one hand, and sufficiently long to guarantee the independence between the cumulative flight arrival and departure processes and their independence during the successive increments on the other.

Thus, the two processes $A(t)$ and $D(t)$ can be treated as the processes of independent increments or the diffusion processes (Newell, 1982). Under the assumption that the differences between the cumulative flight demand and corresponding airport capacity in (k)th and $(k+1)$st time increment Δt, $A(k+1) - A(k) \equiv A_{k+1} - A_k$ and $D(k+1) - D(k) \equiv D_{k+1} - D_k$, respectively, are considered as the stochastic variables with a normal probability distribution. The difference $Q_{k+1} = A_{k+1} - D_{k+1}$, which represents the queue in $(k+1)$st increment Δt, will also be the stochastic variable with normal probability distribution $(k \in K)$ (Newell, 1982). Consequently, the flight queue in $(k+1)$st interval Δt, can be approximated as follows:

$$Q_{k+1} = Q_k + \bar{Q}_{k+1} + B_{k+1} = Q_k + (\lambda_{k+1} + \mu_{k+1})\Delta t + B_{k+} \text{ for } k = 0, 1, 2, , , , K-1 \quad (1)$$

---

4   For example, if (τ) is the period of several hours during the day, Δt will certainly be quarter, half or an hour.

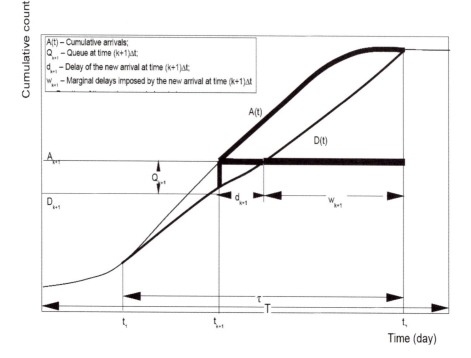

Figure 6.3   A scheme of a typical queuing process at the congested airport – cumulative count of flights

where

Q_k is the queue in $(k)th$ increment Δt;

$\bar{Q}_k$ is the average queue in the $(k+1)st$ increment Δt;

λ_{k+1} is the intensity of flight demand in $(k+1)st$ increment Δt;

μ_{k+1} is the airport capacity (i.e., the flight service rate) in $(k+1)st$ increment Δt;

B_{k+1} is the anticipated deviation of the actual flight queue (i.e., a 'buffer') from its average in $(k+1)st$ increment Δt.

As can be seen, the average flight queue either increases or decreases accordingly as $\lambda_{k+1} > \mu_{k+1}$ or $\lambda_{k+1} < \mu_{k+1}$.

The anticipated deviation B_{k+1} in the expression (1) can be estimated as follows (Newell, 1982):

$$B_{k+1} \cong \sqrt{\Delta t(\sigma^2_{a,k+1} \, /\bar{t}^3_{a,k+1} + \sigma^2_{d,k+1} \, /\bar{t}^3_{d,k+1})} *C \text{ for } k = 0, 1, 2, \ldots, K\text{-}1 \qquad (2)$$

where

$\bar{t}_{a,k+1}; \bar{t}_{d,k+1}$ is the average flight inter-arrival and service time, respectively, in $(k+1)st$ increment Δt;

$\sigma_{a,k+1}; \sigma_{d,k+1}$ is the standard deviation of the flight inter-arrival and service time, respectively, in $(k+1)st$ increment Δt;

C is constant $(C = \Phi^{-1}(1-p)$, where Φ^{-1} is the inverse Laplace's function and p is the probability that the flight queue in $(k+1)st$ increment Δt will spill out of the confidence interval $(\bar{Q}_{k+1} \pm B_{k+1})$.

In the expression (2), the variance of distributions of the flight inter-arrival and service time are assumed to be independent in the successive $(k)th$ and $(k+1)st$ increment Δt (Newell, 1982).

In expression (1) and Figure 6.3, at the beginning of period (τ), the intensity of flight demand becomes equal to the capacity for the first time, and the deterministic queue starts to build up. However, this queue builds upon the already existing queue, due to the previously dominating 'random effects'. The later queue $\bar{Q}_0$ can be approximated as follows (Newell, 1982):

$$\bar{Q}_0 \equiv Q_{0/(\lambda_m = \mu_m)} = \left| \left(\frac{1}{\left[(\sigma_{a,m}/\bar{t}_{a,m})^2 + (\sigma_{d,m}/\bar{t}_{d,m})^2 \right]^2} \right) * (1/\mu_m) * (d\rho_m/dt) \right| \tag{3}$$

where m represents the index of time increment Δt in which the intensity of flight demand becomes equal to the flight service rate (i.e. capacity) $(m \in K)$. All other symbols are analogous to those in the previous expressions.

Determining the system delays and costs From expressions (1)–(3), delay of a flight joining the queue in $(k+1)st$ increment Δt can be approximated as follows:

$$d_{k+1} = Q_{k+1} * (\bar{t}_{d,k+1} + B_{d,k+1}) = Q_{k+1} * \left[\bar{t}_{d,k+1} + \sigma_{d,k+1} * \Phi^{-1}(1-p) \right] \tag{4}$$

where the symbols are analogous to those in the previous expressions.

Expression (4) assumes that the flight service rate (i.e., the airport capacity) does not change while the queue Q_{k+1} is served.

In Figure 6.3, the marginal delay imposed by an additional flight arriving during $(k+1)st$ increment Δt imposes on all subsequent flights until the end of the period (τ) can be determined as:

$$w_{k+1} \cong \tau - \left[(k+1)\Delta t + d_{k+1} \right] \equiv (\bar{t}_{d,k+1} + B_{d,k+1}) * \sum_{l=k+1}^{K} \left[1/(\bar{t}_{a,l} + B_{a,l}) \right] * \Delta t =$$

$$= \left[\bar{t}_{d,k+1} + \sigma_{d,k+1} * \Phi^{-1}(1-p) \right] * \sum_{l=k+1}^{K} \left\{ 1/\left[\bar{t}_{a,l} + \sigma_{a,k} * \Phi^{-1}(1-p) \right] \right\} * \Delta t \tag{5}$$

where all symbols are analogous to those in the previous expressions.

From the expression (5), the marginal delay, which the additional flight imposes on the succeeding flights, is proportional to the product of its service time (i.e., the airport flight service rate – capacity – at the time it takes place) and the number of the succeeding – affected – flights. Diminishing of the airport capacity combined with its increased volatility will certainly increase the marginal delays. Furthermore, if the additional flight is scheduled closer to the beginning of the peak, more succeeding flights will be affected and marginal delays will become longer, and vice versa.

If the additional flight belongs to the group of $N_i(\tau)$ uniformly distributed flights scheduled by airline *(i)* during the peak *(τ)* in addition to the flights of other *M–1* airlines, i.e., $N(\tau) = N_i(\tau) \sum\limits_{i=1}^{M} A(\tau)$, the total cost it imposes to all succeeding flights can be determined as follows:

$$C^i_{m,k+1} = [1 - N_i(\tau)/N(\tau)]*[\bar{t}_{d,k+1} + \sigma_{d,k+1}\Phi^{-1}(1-p)]*\sum\limits_{l=k+1}^{K}c_l*\{1/[\bar{t}_{a,l} + \sigma_{a,l}\Phi^{-1}(1-p)]\}*\Delta t \quad (6)$$

where c_l is the average cost per unit of delay of a flight scheduled in *(l)*th increment Δt (in the monetary units per unit of time). Other symbols are analogous to those in the previous expressions.

The cost per flight c_l may include the aircraft operational and passenger time costs. Expression (6) shows that the total marginal cost imposed by the additional flight of airline *(i)* on succeeding flights will increase as the airport service rate (capacity) decreases and its own volatility increases. In addition, this cost will rise as the number and size (expenses) of flights involved in the peak increases. Holding all other conditions fixed marginal cost will also decrease as the number of flights scheduled by a given airline, which has already internalized its congestion externality, is increased. This implies that congestion charges might favour the economically stronger airlines and disadvantage airlines endeavouring to strengthen their market position by adding more flights and new entrants without any current flights. This could arguably be interpreted as a further protection of the oligopoly position of major carriers.

Estimating profitability of an additional flight The congestion charge should also be able to calculate the expected profitability of additional flights. If $C^i_{m,k+1}$ is the charge and C^i_{k+1} is the average cost per unit of time of an additional flight of capacity *(n)* of airline *(i)* (in the monetary units per unit of time) in *(k+1)*-th increment Δt, the total cost of this flight will be estimated as follows:

$$C^i_{f,k+1} = c^i_{k+1}(n)*[t^i_{f,k+1} + d_{k+1}] + C^i_{m,k+1} \quad (7)$$

where is the duration of the additional flight of airline *(i)* scheduled in *(k+1)* st increment Δt. All other symbols are analogous to those in the previous expressions.

The expected revenues from the additional flight can be estimated as follows:

$$R^i_{f,k+1} = p^i_{k+1}(L)*\lambda^i_{k+1}[p^i_{k+1}(L)]*n^i_{k+1} \quad (8)$$

where

p^i_{k+1} is the average airfare per passenger of the additional flight on a route of length (L) scheduled by airline (i) in $(k+1)$st increment Δt;

$\lambda^i_{k+1}[p^i_{k+1}(L)]$ is the expected load factor of the additional flight carried out by airline (i) in $(k+1)$st increment Δt assumed to be dependent on price;

n^i_{k+1} is the seat capacity of the new flight of airline (i) in $(k+1)$st increment Δt.

Demonstrating the prospective influence of the congestion charge on the price, this new flight will be unprofitable, if the following condition is fulfilled:

$$\Pi^i_{f,k+1} = R^i_{f,k+1} - C^i_{f,k+1} = p^i_{k+1}(L)*\lambda^i_{k+1}[p^i_{k+1}(L)]*n^i_{k+1} - c^i_{k+1}(n)*[t^i_{f,k+1} + d_{k+1}] - C^i_{m,k+1} \leq 0 \ (9)$$

where all symbols are as in the previous expressions.

To achieve the above condition, the charge $C^i_{m,k+1}$ should be slightly greater than the maximum value between the expected profits per flight and the cost of marginal delay, all other factors remaining constant. In this case, the airline will try to compensate the charge by increasing airfares. The congestion charge will be independent of the size of the flight. Thus it will impact more strongly on smaller flights or flights which generate lower revenue. In practice, this means that small regional planes intending to operate at the congested airport(s) in the morning peak(s) will be penalized more heavily. Consequently, if demand is elastic, the airline will find that it is not viable to operate the flight.

An Application of the Model

In the following the proposed model of congestion charges is demonstrated on the case of New York (NY) LaGuardia airport where this measure has been considered in addition to other measures for demand management after the introduction of the AIR-21 Act. The Act institutionally was expected to make the slot exemption, but actually it turned into the lifting of the 1969 slot regulation the High Density Traffic Airports Rules (HDR). LaGuardia is among three biggest airports serving the New York area (US). In terms of the type of traffic, the three largest airports mainly co-operate with each other. LaGuardia airport mainly serves the US domestic short- and medium-haul traffic. About 92 per cent of flights are origin-destination flights carrying about 45–55 per cent business passengers. One of the advantages of LaGuardia attracting business travellers is its proximity to the centre of NY. Manhattan is only about 18km away from the airport. Since the September 11/2001 terrorist attack which triggered an immediate sharp decline in demand for air services, air traffic has gradually recovered at LaGuardia. By the end of 2002, LaGuardia counted 358 thousands flights and an annual passenger number of 22 million. The average number of passengers per flight has remained relatively stable during the past five years (58–62) (PANYNJ, 2003).

At present, 20 airlines operate at the airport. Three have the greatest market share in terms of number of flights and number of passengers, respectively: US Airways (38 per cent; 14.2 per cent), Delta (18 per cent; 17.2 per cent), and American (17 per cent; 18.5 per cent). Two right angle-crossing runways, each 7000ft (2135 m) long, mostly influence the airlines fleet structure in terms of the aircraft size and length of routes-markets they serve. Fleets mostly consist of aircraft categories B737/717, A320 (100–150 seats), and smaller regional jets and turboprops (70–110 seats). The average route length is 1200 km (Backer, 2000; PANYNJ, 2003).

The current runway capacity is about 80 (40/40) flights per hour under VMC (Visual Meteorological Conditions) and 64 (32/32) under IMC (Instrumental Meteorological Conditions) rules. These flights are accommodated at 60 parking stands at the apron.

The hourly and daily demand in terms of the number of flights frequently exceeds the capacity of runway and apron system, which causes severe congestion and delays. Since there is no available land for any further physical expansion, the options for relieving the expected flight congestion and delays under conditions of growth (19 per cent until the year 2010 compared to the year 2002) appear to be very limited. The possible options consist of increasing of the average aircraft size on the one hand and rising of the runway capacity by introducing innovative operational procedures and technologies on the other. The former has already taken place by introducing B767-400ER (about 280 seats) in 2001 (AIRWISE NEWS, 2001). The later still have to take place. It is expected to increase the runway capacity by about 10 per cent under VMC and about 3 per cent under IMC rules (Federal Aviation Administration, 2003a). Nevertheless, both options seem unable to efficiently manage the expected congestion beyond the year 2010. This may again initiate thinking about implementing economic measures of demand management. For example, the auctioning of slots (i.e., 'slottery') implemented in 2000 substantially relieved congestion at that time. For the future, congestion charging might be reconsidered. At present, the airport's landing charge is based on aircraft weight. The unit charge is $6.55 for each 500 kilograms (1,000 lbs) of the aircraft maximum take-off weight. In addition, each operation (flight) between 8:00 a.m. and 9:00 p.m. is charged a fixed amount of US$100 (PANYNJ, 2003a).

Description of Inputs

Three groups of inputs are used in the application of the proposed model: data on demand and capacity to estimate the level of congestion and delays under given circumstances, and the aircraft operating costs and airfares to assess the profitability of particular flights.

Data for estimating congestion and delays The hourly rate of flight demand and corresponding capacity at NY LaGuardia airport for every day in July 2001 were used to estimate congestion and delays. The distributions of the hourly flight demand and their service rate (i.e., the airport capacity) have been derived based on 31 daily realizations (Federal Aviation Administration, 2003, 2003a). Each distribution for each hour has been taken to be normal or nearly normal and independent from each other. Furthermore pairs of these distributions for different hours were also assumed to be independent (Newell, 1982). Table 6.2 gives the main parameters of these hourly distributions.

In addition, in all experiments, the constant *C* was equal to 1.96 implying that the queues have stayed within the given confidence boundaries with the probability of 95 per cent (Newell, 1982).

Aircraft operating costs and airfares The aircraft operating cost per block hour, depending on seat capacity, was estimated by regressing data on US airlines as follows (FAA, 1998):

$$c(S) = 21.97\ S + 11.993\ (R^2 = 0.934;\ N = 45).$$

According to this equation, the average cost of a flight of 100–150 seats (B737/717) operated at NY LaGuardia airport varies between US$ 2,209 and US$3,307 per hour (or US$37 and US$55 per minute). The cost of a flight of 280 seats (for example B767-400ER), is US$ 6162 per hour (or US$ 103 per minute).

The average airfare per passenger at NY LaGuardia airport depending on the non-stop flying distance has been determined by using the regression technique applied to the US data from 1998 modified for changes in the value of US$ for the year 2002 (Mendoza, 2002; Sheng-Chen, 2000) as follows: $p(L) = 9.5605$ $L^{0.3903}$ $(R^2 = 0.941;\ N = 28)$. For NY LaGuardia airport the average length of flight is around 1200 km, wich translates into an average airfar of about US$152 (Mendoza, 2002).

Analysis of the Results

The results from the experiments with the model are shown in Figures 6.4, 6.5, and 6.6. Figure 6.4 (a, b, c) shows the congestion and delays of flights caused by an additional flight.

Figure 6.4a shows how, during an average day, the queue starts early in the morning just after opening the airport (06:00 hours), gradually increasing during the day, and reaching its peak between 19:00 and 20:00 hours. Then, within the next three hours (from 20:00 to 23:00 hours), the queue is being cleared. When the airport has been operated at declared capacity, the average queue has been 35 and the maximum queue 59 aircraft/flights. When the airport's declared capacity has been reduced by 10 per cent, during the second half of the day (from 13:00 hours on, for example, the queue has additionally increased, reaching a maximum of 93 flights between 20:00 and 21:00 hours, and persisting until midnight.

Table 6.2   **The main parameters of distributions of the flight inter-arrival and inter-departure time in given example**

Time of day	Demand Flight inter-arrival time		Capacity Flight service time	
Hour (k)	*Mean* $(t_{a,k})$ *(s/flight)*	*St. dev.* $(\sigma_{a,k})$ *(s/flight)*	*Mean* $(t_{d,k})$ *(s/flight)*	*St. dev.* $(\sigma_{d,k})$ *(s/flight)*
1	–	–	–	–
2	–	–	–	–
3	–	–	–	–
4	–	–	–	–
5	–	–	–	–
6	90.72	9.972	52.56	7.488
7	52.20	3.942	52.92	7.524
8	50.76	3.123	52.2	4.608
9	49.68	4.716	52.56	7.776
10	50.04	4.860	52.20	7.164
11	50.40	1.764	51.12	6.912
12	50.76	2.376	51.12	6.984
13	48.96	3.096	50.76	6.912
14	51.84	3.744	50.76	6.336
15	50.04	3.312	50.40	7.056
16	48.24	2.916	50.40	7.020
17	48.60	5.148	50.04	7.022
18	51.48	8.640	50.04	7.020
19	50.76	5.292	50.40	7.704
20	51.84	7.992	49.68	6.624
21	59.67	5.220	49.32	6.012
22	78.12	16.236	49.32	5.976
23	123.84	36.468	50.40	7.308
24	–	–	–	–

Note: s = seconds.

Source: FAA, 2003.

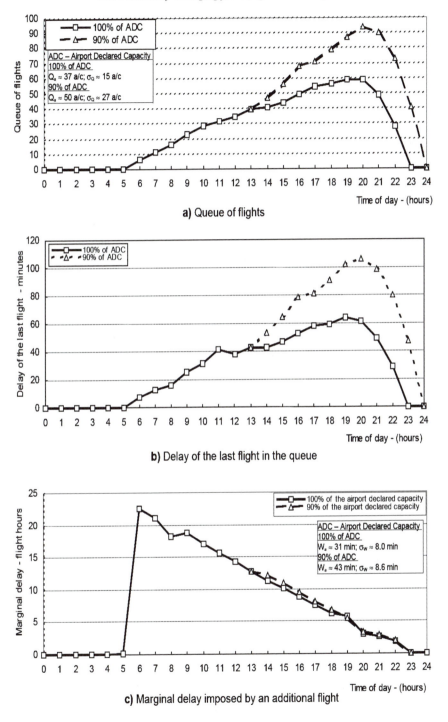

a) Queue of flights

b) Delay of the last flight in the queue

c) Marginal delay imposed by an additional flight

Figure 6.4   The system congestion and delays in a given example

Figure 6.4b illustrates the flight delays as a consequence of the queue. As can be seen, the delay of the last flight in the queue has changed in line with changing the queue length. When the airport has operated at the declared capacity, the average and maximum delay per flight has been 35–40 and 65 minutes per flight, respectively. In the case of deterioration of the airport capacity for by 10 per cent, from 13:00 hours on, the average and maximum delay per flight has increased to about 55-65 and 105 minutes, respectively.

Figure 6.4c illustrates changes of the marginal delay caused by a change of the scheduling time of an additional flight. As can be seen, a flight scheduled early in the morning has imposed longer marginal delays then otherwise. In the given example, one such flight scheduled at 06:00 hours has imposed about 22 additional flight-hours of delay on the succeeding flights scheduled by the end of the congestion period. Scheduling of the new flight later during the day has affected a smaller number of succeeding flights and consequently caused less additional delays, as had been intuitively expected. The average marginal delay imposed by an additional flight at any time during the day has been about 10–12 flight-hours. Deterioration of the airport declared capacity by 10 per cent has increased this and the flight private delay. Figure 6.4b and 6.4c show that the marginal delay imposed on others has been considerably greater than the flight private delay.

Figure 6.5 (a and b) shows the costs of an additional flight in the given example. Figure 6.5a shows the cost of delay of an additional flight in dependence on the time of day and aircraft size. As can be seen, this cost has generally increased with increasing of the aircraft size due to its higher operating costs. For a given aircraft size, this cost increased in proportion with the delays. For example, for an additional flight of capacity of 100, 150 and 280 seats scheduled between 19:00 and 20:00 hours, the cost of delay has been US$2,300, US$3,500 and US$6,500, respectively. Figure 6.5b shows that the costs of marginal delay imposed by an additional flight on the succeeding flights have changed in proportion to the marginal delays. They are highest when the additional flight has been scheduled early in the morning and gradually decrease if this flight has been scheduled later during the day. These costs have been dependent on the aircraft (flight) types succeeding the additional flight, and vice versa. As expected, these costs are higher if larger aircraft have been behind the additional flight. For example, the additional flight scheduled around 06:00 has generated the marginal cost of about US$50,000, US$75,000 and US$150,000, when all flights behind it have been carried out by planes with a capacity of 100, 150 and 280 seats, respectively. Comparison of these marginal delay costs and current landing fees based on aircraft weight have indicated the existence of a large disproportion.

By summing up the costs of delays in Figure 6.5a and 6.5b, the costs of the total system delays caused by an additional flight have been obtained and the profitability of the additional flight estimated.

Figure 6.6 shows how profitability of a flight depends on congestion charges in the given example. The additional two-hour flight has been carried out by an aircraft of 150 seats with an operational cost of US$3,300, an average load factor of 60 per cent and an average airfare of US$152, consequently contributing a

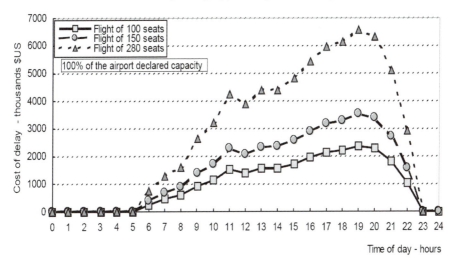

a) Cost of delay of an additional flight

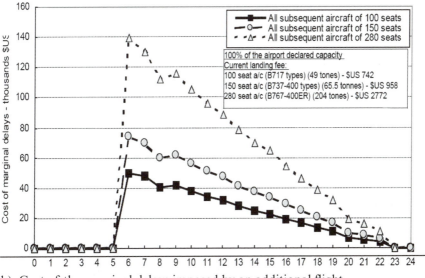

b) Cost of the marginal delays imposed by an additional flight

Figure 6.5 Conditions of profitability of an additional flight in a given example

revenue of US$13680. If this flight had been operated by an airline with minimal market share, and the full congestion charge had been paid, it would have been highly unprofitable during the whole day, except sometimes after 22:30 hours. Obviously, such entry would not be feasible under the condition that all succeeding flights are 100 seaters. However, when the given airline has already a significant market share in terms of the number of flights at the airport (85–90 per cent), the additional flight is profitable independently from the time when it has been scheduled and despite being charged the full congestion fee.

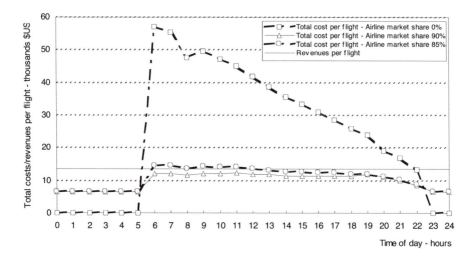

Figure 6.6   Conditions of profitability of an additional flight in a given example

In Figure 6.6, under these circumstances, the airline would have had to have a market share of at least 85 per cent and thus internalizing 85 per cent of the marginal congestion costs it created in order for the additional flight to be profitable. As can be seen, this flight would be at the edge of profitability if being scheduled until early afternoon and absolutely profitable if being scheduled later. Under given circumstances, by increasing the airline's market share above 85 per cent, the additional flight would be profitable independently from scheduling time. This result confirms doubts that congestion charging might disfavour the development of competition at the airport since it may impose unacceptably high congestion charges on flights of the new entrants on the one hand, and very modest charges on flights of airlines already being strongly present at the airport on the other. Congestion charging might stimulate flights to be carried out by larger aircraft if being scheduled before flights carried out by smaller aircraft, and vice versa. Furthermore under given conditions, the charge might discourage new flights during the first half of day, and particularly early in the morning.

Conclusions

This chapter has developed a model for charging congestion at an airport. Currently, congestion charging is not practised at airports despite introducing different charging mechanisms for peak and off-peak periods. In addition to an analysis and comparison of airport delays in Europe and the US, different conditions influencing airport congestion have been examined. It has been emphasized that understanding the nature of congestion and reliable and transparent estimation of relevant parameters are essential for developing models and the system of congestion charging.

The presented model consists of two sub-models: diffusion approximation of queues to quantify the relevant queues and delays during congestion period; a model of marginal delays and their costs imposed by each flight on other flights during the congestion period; and a model of profitability of each flight burdened by the congestion charge. The models have been applied to New York (NY) LaGuardia airport (US) operating under the slot exemption regime (Act AIR-21 introduced in April 2000), which ultimately enabled free access in comparison to the previous slot constraining High Density Traffic Airport Rules (HDR) from 1969. The data was taken from the real airport operations and therefore the results could be considered as a realistic indication of the prospective efficiency of congestion charges in the given case.

The application of the model has indicated that it could efficiently deal with the problem of charging congestion at airports. In particular, the diffusion approximation of queues has enabled quantification of flight queues and delays realistically. Other two sub-models have estimated the congestion charge and profitability of particular flight(s) for the given traffic (congestion) scenario and charging conditions.

The results from the model have shown that the congestion charge at an airport could be used as an economic instrument of demand-management under the following condition:

• congestion as an exclusive consequence of the relationships between demand and airport capacity.

Congestion created by many flights of different competing airlines means that airlines do not fully internalize the costs of the delays they cause, since a single airline with many flights has already internalized costs of marginal delays of these flights. In such a context, airports with many competing airlines performing point-to-point operations are more likely candidates for congestion charges than airports with a few airlines operating hub-and-spoke networks despite the fact that in both cases the congestion can be significant and relevant for charging.

Congestion charges appear to be more effective in preventing access earlier during the congestion period of flights carried out by smaller aircraft before flights carried out by the larger aircraft, and vice versa. This implies that the congestion charge stimulates exclusion of earlier arrivals by smaller-regional aircraft and favours the use of the larger aircraft.

References

ACI (2002), 'Airport Charges in Europe', Report of Airport Council International (Europe), p. 39.

Adler, N. (2002), 'Barriers and Implementation Paths to Marginal Cost Based Pricing: Rail, Air and Water Transport', 3rd MC-ICAM Seminar, 2/3 September, p. 5.

AEA (2001), *Yearbook 2000*, Brussels: Association of European Airlines, p. 80.

AIRWISE NEWS (2001), 'Delta Launches Boeing 767-400ER at LaGuardia': http://news.airwise.com.

ATA (2002), *System Capacity: Part I: Airline Schedule, Airport Capacity, and Weather*, Air Transport Association, Industry Information, USA: http://www.air-transport.org/public/industry.

Backer, C. (2000), 'Airports: Top 1000 Ranking', *Airline Business*, pp. 55–86.

Brueckner, J.K. (2002), 'Internalisation of Airport Congestion', *Journal of Air Transport Management*, 8, pp. 141–7.

BTS (2001), *Airline Service Quality Performance Data*, Washington DC: US Department of Transportation, Bureau of Transportation Statistics, Office of Airline Information.

Carlsson, F. (2002), 'Airport Marginal Cost Pricing: Discussion and an Application to Swedish Airports', Discussion Paper, Department of Economics, Göteborg University, Sweden, p. 20.

Cheng-Shen, A.H. (2000), 'An Analysis of Air Passenger Average Trip Lengths and Fare Levels in US Domestic Markets', Working Paper UCB-ITS-WP-2000-1, NEXTOR Aviation Operations Research, Institute of Transport Studies, University of California, Berkeley, p. 14.

Corbett, J.J. (2002), 'Small Communities Are Concerned about Congestion Pricing', *The Air and Space Lawyer*, 17 (1), Summer, pp. 17–21.

Daniel, J.I. (1995), 'Congestion Pricing and Capacity of Large-Hub Airports: A Bottleneck Model with Stochastic Queues', *Econometrica*, 63, pp. 327–70.

Daniel, J.I. (2001), 'Distributional Consequences of Airport Congestion Pricing', *Journal of Urban Economics*, 50, pp. 230–58.

Daniel, J.I. and Pahwa, M. (2000), 'Comparison of Three Empirical Models of Airport Congestion Pricing', *Journal of Urban Economics*, 47, pp. 1–38.

DeCota W. (2001), 'Matching Capacity and Demand at LaGuardia Airport', NEXTOR, Airline and National Strategies for Dealing with Airport and Airspace Congestion, p. 26.

Doganis, R. (1992), *The Airport Business*, London: Routledge.

EC (1997), *External Cost of Transport in ExternE*, European Commission, Non Nuclear Energy Programme, IER Germany.

EC (2001), *Concerted Action on Transport Pricing Research Integration – CAPRI*, Final Report, European Commission, Transport RTD of the 4th Framework Programme, ST-97-CA-2064, Brussels.

ECMT (1998), *Efficient Transport for Europe: Policies for Internalization of External Costs*, Paris: European Conference of Ministers of Transport, OECD-Organisation of Economic Co-operation and Development.

EUROCONTROL (2002), *An Assessment of Air Traffic Management in Europe During Calendar Year 2001: Performance Review Report*, Performance Review Commission, EUROCONTROL, Brussels.

EUROCONTROL/ECAC (2001), *ATFM Delays to Air Transport in Europe: Annual Report 2000*, CODA, EUROCONTROL, Brussels.

EUROCONTROL/ECAC (2002a), *ATFM Delays to Air Transport in Europe: Annual Report 2001*, EUROCONTROL, Brussels.

FAA (1998), 'Economic Values for Evaluation of Federal Aviation Administration Investment and Regulatory Decisions', Report FAA-APQ-98-8, Federal Aviation Administration US Department of Transportation, Washington, DC.

FAA (2003), *Airport Capacity Benchmarking Report 2001*, Washington, DC: Federal Aviation Administration.

FAA (2003a), *Aviation Policy and Plans (APO) – FAA OPSNET and ASPM*, Washington, DC: Federal Aviation Administration.

Fron, X. (2001), 'Dealing with Airport and airspace Congestion in Europe', *ATM Performance*, original paper presented at the EUROCONTROL meeting, March, p. 25.

Ghali, M.O. and Smith, M.J. (1995), 'A Model for the Dynamic System Optimum Traffic Assignment Problem', *Transportation Research B*, Vol. 29B, pp. 155–70.

Hall, R.W. (1991), *Queuing Methods for Services and Manufacturing*, London: Prentice Hall International Series, Prentice Hall International Ltd.

Janic, M. (2003), 'Large Scale Disruption of An Airline Network: A Model for Assessment of the Economic Consequences', 81st Transportation Research Board (TRB) Conference, January, Washington, DC.

Liang, D., Marnane, W., Bradford, S. (2000), 'Comparison of US and European Airports and Airspace to Support Concept Validation', 3rd USA/Europe Air Traffic Management R&D Seminar, Naples, 13/16 June, p. 15.

Mendoza, G. (2002), *New York State Airport Air Fare Analysis*, Washington, DC: Aviation Service Bureau, NYSDOT, p. 4.

Nash, C. and Sansom, T. (2001), 'Pricing European Transport System: Recent Developments and Evidence from Case Studies', *Journal of Transport Economics and Policy*, 35, Part 3, pp. 363–80.

Newell, G.F. (1979), 'Airport Capacity and Delays', *Transportation Science*, 13 (3), pp. 201–41.

Odoni, A., Bowman, J. et al. (1997), *Existing and Required Modelling Capabilities for Evaluating ATM Systems and Concepts*, NASA/AATT Final Report, International Centre for Air Transportation, Massachusetts Institute of Technology, Cambridge, MA, p. 206.

Odoni, A.R. and Fan, T.C.P. (2002), 'The Potential of Demand Management as a Short-Term Means of Relieving Airport Congestion', 4th USA/Europe Air Traffic Management R&D Seminar, Santa Fe, 3–7 December, p. 11.

PANYNJ (2003), *LaGuardia Airport: Traffic Statistics*, Report, The Port Authority of NY&PJ, New York, p. 2.

PANYNJ (2003a), *Schedule of Charges for Air Terminals*, Report, The Port Authority of NY&PJ, New York, p. 40.

Schiphol Group (2002), *Annual Report – 2001*, Amsterdam Schiphol Airport, Amsterdam.

Vickery, W. (1969), 'Congestion Theory and Transport Investment', *American Economic Review*, 59 (2), May, pp. 251–60.

Welch, J.D. and Lloyd, R.T. (2001), 'Estimating Airport System Delay Performance', 4th USA/Europe Air Traffic Management R&D Seminar, Santa Fe, 3–7 December, p. 11.

Chapter 7

Managing Congested Airports Under Uncertainty

Achim I. Czerny

Introduction

In order to control congestion many airports are slot constrained. An airline wishin to incorporate a slot constrained airport into its networks needs to have a respective permission (slot) to use that airport at a specified time. Because the number of slots is constrained, airline operations at the airport are limited and, consequently, demand and congestion can effectively be controlled and optimized. There is, however, another possibility to reduce congestion. An increase of take-off and landing fees can reduce airline demand until the optimal level of congestion is reached. Under certain conditions, either of these two approaches to manage congestion can generate the optimal results from a welfare perspective. The premises are, however, that airport slots are allocated efficiently among airlines and that the regulator has perfect information about benefits and costs of take-off and landing operations. Both of these premises are usually not fulfilled in reality.

First, at present, airport slots are generally allocated based on grandfather rights or, in other words, by history. This guarantees continuity of airline operations because airlines are allowed to constantly reuse slots which they have used in the past. On the other hand, allocation based on grandfather rights does not account for the willingness to pay of airlines and, therefore, hampers the efficient allocation of runway capacity (though secondary trading might result in efficient allocation of slots).

Second, regulators do not have perfect information. A regulator can neither be certain about the social benefits nor about the social costs of airport operations (Forsyth 1976). The air transport industry is characterized by variable demand which is susceptible to demand shocks and difficult to foresee. Thus, passenger benefits of airport operations are difficult to predict. The same holds true for the costs of airport operations. Although the airports' costs of operation and maintenance as well as the airlines congestion costs might be estimated fairly well, the measurement of passengers' and environmental costs of congestion is problematic. For these reasons a regulator has to deal with a considerable amount of uncertainty regarding the benefits and costs of airport operations when choosing the optimal measures to manage airport congestion. In the present chapter it is shown that the welfare performance of congestion pricing or slot constraints as instruments under uncertainty can be very different. Furthermore, in some situations congestion pricing might be preferred over slot constraints, while in others, slots might be preferred (Forsyth, 1976).

It should be noted that the effect of uncertainty on the choice of regulation instruments has not raised much attention from transport economists so far. In contrast, resource management under uncertainty was extensively analyzed in the field of environmental economics. Referring to pollution management (amongst others), Weitzman (1974) and Adar and Griffin (1976) showed that under uncertainty regarding benefits and costs of pollution the expected welfare depends on the regulator's choice between prices (e.g., pollution taxes) or quantities (e.g., emission standards) as instruments.

In this chapter the way in which uncertainty can affect the welfare performance of congestion pricing or slot constraints is described. Furthermore, the circumstances favouring one of the two instruments are explored. As a starting point a situation in which regulators have perfect information about market conditions is considered. This scenario is used in order to explain the basic ideas behind the concepts of congestion pricing and slot constraints. Then, a situation is investigated in which only congestion costs are uncertain but demand is well known by the regulator. It is demonstrated that uncertainty about congestion costs may not be relevant to the choice of regulation instruments. Next the focus is on demand uncertainty. In fact, under demand uncertainty the welfare performance of slot constraints and congestion pricing can differ according to the structure of demand and congestion costs.

So far, a simplifying assumption underlying the analysis is that demand and congestion costs uncertainties are not correlated. However, theoretical analysis suggests that both variables are negatively correlated. Observe that uncertainty about demand and congestion costs would be negatively correlated if, say, a higher than expected demand frequently comes together with lower than expected congestion costs. Furthermore, a negative correlation between demand and congestion costs has some important implications for the choice of regulation instruments. In fact, it tends to favour the use of congestion pricing.

Moreover, it is observed that the demand for different airports is interdependent because of airport services complementarity. Clearly, if the number of aircraft departing at one airport increases, then the number of aircraft approaching other airports also increases. Hence, it should be taken into account that the choice of regulation instruments at one airport might not only affect the level of airport operations at that specific airport but will also affect operation levels at other airports. In the following it is argued that the interdependence of airport operations, in the present chapter termed '*demand complementarity*', increases the relative attractiveness of using slots.

Altogether the present analysis shows that there is no straightforward answer to the question of whether congestion pricing or slot constraints should be favoured in order to maximize welfare. Congestion pricing can be the right choice if the regulator believes that a negative correlation between demand and congestion costs exists. On the other hand, slot constraints should be preferred if demand complementarity is of special importance.

Discussions about the right measures to manage congested airports generally emphasize congestion pricing and slot constraints as instruments. However, there are alternative instruments available specifically designed to address demand uncertainty. Some of these instruments are worth consideration because they possess the potential to further enhance welfare. Therefore, a sample of relevant instruments, other than congestion pricing and slot constraints, are described and discussed.

In the next section the concepts of congestion pricing and slot constraints in an environment of perfect information is explained. In this section it is assumed that the regulator has perfect information about both: airline demand and the costs of airport operations. The next section focuses on a situation with certain demand but uncertain congestion costs. The impact of demand uncertainty on the choice of regulation instruments is then analyzed. We then consider demand and congestion costs that are negatively correlated. Then, the relevance of demand complementarity for the choice of regulation instruments is analysed. Alternative instruments to allocate airport capacity under demand uncertainty are then presented. Finally, some conclusions from the preceding discussions are drawn.

Congestion Pricing and Slot Constraints under Perfect Information

In order to describe the basic setting of the present analysis and the concept of congestion pricing and slot constraints first a situation in which the regulator has perfect knowledge about demand and congestion costs is considered. For simplicity and to narrow the focus of the analysis, it is assumed throughout this chapter that the airport's variable costs are zero, i.e. airport usage does not affect the airports costs. Since a large share of airports costs is determined by fixed costs such as capital costs and wages, this assumption appears not too far from reality.

To illustrate the basic setting and the concept of regulation instruments Figure 7.1 is used.[1] Airport operations in terms of the number of flights are denoted by q. Figure 7.1 shows airline demand for airport capacity. The demand curve describes the relation between airport charges and airport usage. Furthermore, it determines the airlines' marginal willingness to pay for airport usage. Observe that the airlines' willingness to pay for a given number of airport operations is determined by the area below the demand curve. In the following, it is assumed that the airlines' willingness to pay is a valid measure for the social benefits generated from airport operations.[2] Then, the social benefits of airport operations given by q^* are shown by the dotted and the hatched area.

---

1   Notice that the model presented deviates from usual congestion models. Economists normally refer to total average and total marginal congestion costs curves, which the present chapter does not explicitly take into account. However, the model used in this chapter is equivalent to the traditional approach.

2   Notice that this only holds true if airlines are in perfect competition (Basso, 2005). In this case the demand curve is net of marginal airline costs.

Figure 7.1 also shows a curve named '*marginal external congestion costs*'. If the use of airport capacity generates congestion, then this increases social costs, termed here congestion costs. Congestion costs include the delay costs that passengers face. They also include the delay costs of airlines since delays increase fuel consumption and can incur larger crew costs. Moreover, higher fuel consumption can add to environmental damages. Note that an airlines' decision to approach a congested airport can lead to two effects. First, if that airline operates more than one flight at that airport then one additional flight can increase delay times for other flights operated by that airline. Second, an additional flight can increase delay times and congestion costs for flights operated by *other* airlines. These additional congestion costs incurred to other airlines are considered external costs.

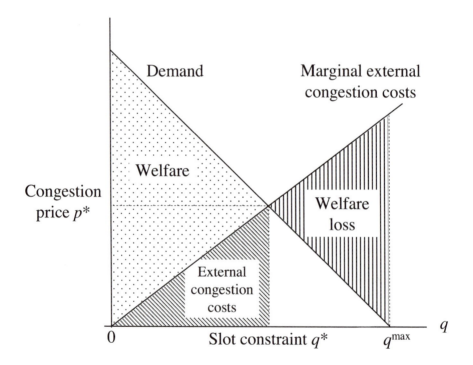

Figure 7.1   Congestion pricing versus slot constraints under perfect information

Against this background, the marginal external congestion costs curve in Figure 7.1 determines the external congestion costs imposed to other airlines and passengers if one flight is added to the airport. The marginal external congestion costs curve can be used to ascertain the total amount of external congestion costs. For a given amount of airport usage, the external congestion costs are given by the area below the marginal external congestion costs curve. In Figure 7.1 external congestion costs are given by the hatched area if airport usage is equal to q^*. Furthermore, the demand curve and the marginal external congestion costs curve can be used to determine the welfare generated from airport operations.

The benefits from airport usage are given by the area below the demand curve (for q^*: dotted and hatched area). On the other hand, the external congestion costs are given by the area below the marginal external congestion costs curve (for q^*: hatched area). Then, welfare can be determined by the benefits minus the external congestion costs (for q^*: dotted area).

Figure 7.1 can also be used to explore the effects of congestion pricing and slot constraints in a situation where the regulator has perfect information about airline demand and external congestion costs. However, first note what happens if neither of these regulation instruments are applied. In the extreme case where airport usage is completely free of charge the level of airport operations reaches q^{max}, which is the maximum demand, leading to a welfare loss determined by the hatched area (vertical lines). Although in practice airport charges are not equal to zero, this extreme example describes the problem of excessive airport operations, e.g. in the US (Kasper, Chapter 15 in this volume).

A first policy option to prevent welfare losses due to excessive airport operations is to increase airport charges until p^*. This is what is termed congestion pricing. In this case airport operations reach q^*. Notice, if the area below the demand curve depicts the benefits from airport operations, then the demand curve depicts the marginal benefits from airport operations. Now, observe that the marginal external congestion costs are exactly equal to the marginal benefits if airport operations reach q^*. Therefore, this is exactly the level of airport operations that maximizes welfare. For lower levels the increase in benefits reached by a marginal increase of operations would be higher than the increase in external congestion costs, which demonstrates that welfare could be enhanced by extending operations. On the other hand, for higher operation levels the increase in benefits reached by a marginal increase of operations would be lower than the increase in external congestion costs demonstrating that welfare could be enhanced by reducing operations. Thus, only for q^* operation levels maximize welfare, i.e. given that the level of operations is equal to q^* it is not possible to increase welfare by changing operation levels. Observe that maximum welfare is equal to the size of the dotted area in Figure 7.1.

A second policy option to manage congestion is to restrict airport access by introducing slot constraints. In this case airlines are required to have a slot to be allowed to take-off or land at a congested airport. Suppose that the number of slots is equal to q^*. Furthermore, suppose that slots are allocated to airlines with the highest willingness to pay; i.e. only airlines that generate the highest profits from using a congested airport get slots. To guarantee this, the slot allocation could be based on an auction (see the contributions of Menaz and Matthews or Button in this volume). Another possibility could be to distribute permissions by a lottery in combination with secondary trading. In both cases slot prices could reach p^*.[3] Notice that in this situation with perfect information, congestion pricing

----

3   If the demand for airport capacity is complementary, then auctions might be necessary to reach efficiency. On the other hand, secondary trading might not lead to an efficient allocation of airport capacity among airlines. The reason is that efficient pricing under complementarity might require non-linear tariffs (Parkes 2001). However, non-linear tariffs are difficult to implement under secondary trading.

and slot constraints lead to optimal welfare results. This shows that in a situation with perfect information the choice of regulation instruments is not relevant.

The Choice of Regulation Instruments Under Certain Demand and Uncertain Congestion Costs

In reality it is highly unlikely that regulators have perfect information about market conditions. Therefore, the present analysis is gradually extended in such a way that it captures different situations of uncertainty. This section concentrates on uncertainty about congestion costs. For this reason it is assumed that regulators have perfect information about demand but only expectations about marginal external congestion costs. Figure 7.2 illustrates this case. The regulator's expectations about external congestion costs are depicted by the *expected* marginal congestion costs curve.

In a situation of uncertainty regulators maximize expected welfare. Under congestion pricing this leads to an airport charge that is determined by $E[p^*]$. The expected level of airport operations is then given by $E[q^*]$. Furthermore, for $E[q^*]$ it holds that the marginal benefits of airport operations is exactly equal to the expected marginal external congestion costs. Hence, the congestion charge $E[p^*]$ does maximize expected welfare.

However, as depicted in Figure 7.2 the regulator made a mistake because the *real* marginal external congestion costs are lower than expected. For that reason the optimal congestion price p^* is also lower than the expected optimal congestion price $E[p^*]$. As a consequence, the expected level of airport operations $E[q^*]$ is lower than the optimal level of airport operations q^* and leads to a welfare loss determined by the hatched area.

Now, consider welfare results under slot constraints. The expected optimal level of airport operations is given by $E[q^*]$. Hence, if slot constraints are chosen such that they are equal to $E[q^*]$ then expected welfare is maximized. In this case airport operations are fully determined by slot constraints and, hence, will be equal to $E[q^*]$.

However, since expected marginal congestion costs are lower than expected, slot constraints are set too low, i.e. the level of airport operations maximizing welfare is given by q^* which is higher than $E[q^*]$. Observe that airport operations under congestion pricing are also equal to $E[q^*]$ because demand was correctly anticipated by the regulator. It follows that welfare is not affected by the choice of regulation instruments if demand is certain and external congestion costs are uncertain (Forsyth, 1976. Weitzman, 1974. Adar and Griffin, 1976).

This is essentially because airport operations cannot be affected by uncertainty, and thus the welfare loss is the same under congestion pricing and slot constraints.

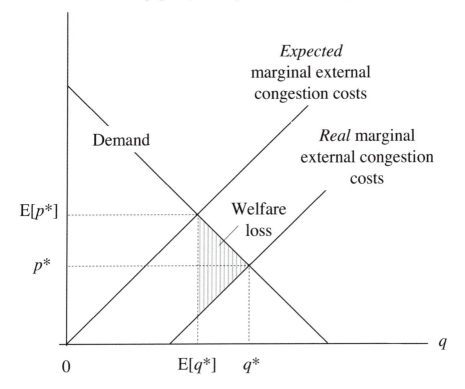

Figure 7.2 Regulation instruments under uncertain external congestion costs

The Choice of Regulation Instruments Under Demand Uncertainty

According to the analysis in the preceding section uncertainty about costs makes no difference with regard to the choice of regulation instruments. For that reason, the analysis turns to the case of uncertain demand. However, the effects of uncertain external congestion costs and demand uncertainty will be considered in the following section.

The assumption is made that the regulator has some expectations about demand but has perfect information about external congestion costs. Figure 7.3 illustrates this situation. It shows the *real* demand curve, the *expected* demand curve, and the (real) marginal external congestion costs curve. As before, the regulator maximizes expected welfare. Observe that the real demand curve lies above the expected demand curve. Hence, in this situation demand is higher than expected by the regulator.

Under congestion pricing the airport charge maximizing expected welfare is given by $E[p^*]$. The effect of congestion pricing on welfare is as follows. Since the regulator underestimates demand the airport charge is set too low. As a result airport operations are higher than expected. The higher level of airport operations compared to the expected level $E[q^*]$ implies additional benefits determined by:

$A + C$.

However, the increase in airport operations also implies an increase of external congestion costs determined by:

$B + C$.

The total welfare change is, then, determined by the sum of the additional benefits minus the additional external congestion cost which is

$$(A + C) - (B + C) = A - B < 0.$$

Hence, the total change in welfare loss due to the underestimation of demand is given by A minus B which is negative because A is smaller than B, i.e. $A - B < 0$.

Turning to slot constraints, it should be noted that under slot constraints airport operations are determined by $E[q^*]$ and will, therefore, be as high as expected. For that reason slot constraints can prevent welfare losses implied under congestion pricing due to excessive levels of airport operations.

Turning to Figure 7.3b, which is similar to Figure 7.3a, except that the slope of the marginal external congestion costs curve is smaller than before. Again, real demand is higher than expected leading to a level of airport operations that is higher than expected under congestion pricing. As before, the excessive use of the airport operations leads to a change in total welfare determined by A minus B. However, note that in this situation A is larger than B, i.e. $A - B > 0$. This means that the higher level of airport operations leads to a positive change of total welfare under congestion pricing. Hence, in this situation congestion pricing enhances welfare in comparison to slot constraints where airport operations are fixed at the ex ante optimal level.

Altogether, in the first situation (Figure 7.3a) slot constraints enhance total welfare, while in the second situation (Figure 7.3b) congestion pricing enhances total welfare. The difference between the two figures is that in Figure 7.3a the marginal external congestion costs curve is steeper than the demand curve (the slope of the marginal external congestion costs curve is greater than the slope of the demand curve in absolute values). In comparison, in Figure 7.3b the demand curve is steeper than the marginal external congestion costs curve (the slope of the marginal external congestion costs curve is lower than the slope of the demand curve in absolute values). Notice that if the marginal external congestion costs curve is flat, then changes in airport usage have a fairly constant effect on external congestion costs. In contrast, if the curve is steep, then changes in airport usage can have a high or low impact on external congestion costs depending on the level of airport operations.

In Figure 7.3a, the marginal external congestion costs curve is steep in comparison to the demand curve. The consequence is that the increasing level of airport operations resulting from low airport charges leads to a large increase in

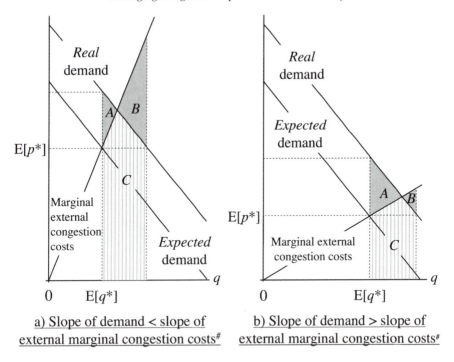

a) Slope of demand < slope of
external marginal congestion costs[#]

b) Slope of demand > slope of
external marginal congestion costs[#]

Note: The slopes are compared in terms of absolute values.

Figure 7.3   Regulation instruments under demand uncertainty

external congestion costs. Furthermore, since the demand curve is flat compared
to the marginal external congestion costs curve the effects changes of airport
operations have on benefits are moderate. Altogether, in this situation it is better
to manage airport capacity using slot constraints because positive changes in
airport operations strongly affect external congestion costs compared to its
moderate effect on benefits.

In Figure 7.3b opposite occurs. With congestion pricing the level of airport
operations is flexible. As a consequence, if demand is higher than expected then
the level of airport operations increases. If the marginal external congestion costs
curve is flat compared to the demand curve, then the effect airport operations
have on external congestion costs is moderate. In contrast, the effect changes of
airport operations have on benefits is more intense. Therefore, congestion pricing
enhances welfare compared to slot constraints.

However, real demand could also be lower than expected. In this situation
the relations described before are still valid. If the slope of the marginal external
congestion cost curve is steep compared to the slope of the demand curve, then slot
constraints are the better choice. On the other hand, if the slope of the marginal
external congestion costs curve is flat compared to the slope of the demand curve,
then congestion pricing is the superior choice. The line of reasoning is similar to
the arguments presented before in this section.

Altogether, this demonstrates that under demand uncertainty the choice of regulation instruments affects welfare (Forsyth, 1976; Weitzman, 1974; Adar and Griffin, 1976). Furthermore, the steepness or flatness of the marginal external congestion costs curve and the demand curve are important determinants of the welfare generated under slot constraints or congestion pricing. For a demand that is steep compared to the marginal external congestion costs curve congestion pricing enhances welfare compared to slot constraints. For a demand curve that is flat compared to the marginal external congestion costs curve slot constraints enhance welfare compared to congestion pricing. Given that the steepness of the curves is similar in terms of absolute values, the welfare effects of congestion pricing and slot constraints are similar, too.

Negatively Correlated Uncertainties Between Demand and External Congestion Costs

In the previous section it was assumed that demand is uncertain but external congestion costs are known by the regulator. Note that the presented results also hold for situations where demand and congestion costs are uncertain but not correlated. In contrast, in this section it is argued that uncertainties are in fact negatively correlated, i.e. high demand frequently implies low external congestion costs and vice versa. This changes the previous results.

To understand why demand and external congestion costs are likely to be negatively correlated recall that external congestion costs determine only a part of the total congestion costs. In general it holds that congestion grows if airport operations are extended. As a consequence, airlines operating additional flights experience higher congestion and congestion costs, which also affect all other airlines. Under the assumption that marginal external congestion costs are higher than expected, total congestion costs are likely to be higher than expected, too. Moreover, since congestion costs reduce airline profits, a situation with high congestion costs is likely to reduce airline demand for airport capacity. Therefore, it is argued in the following that demand will be lower than expected if marginal external congestion costs are higher than expected. The implications for the choice of regulation instruments are presented in the following.

Figure 7.4 is used to illustrate that congestion pricing is likely to enhance welfare compared to slot constraints if uncertainties of demand and external congestion costs are negatively correlated (which follows from Weitzman, 1974). Figure 7.4 shows the expected demand curve and the expected marginal external congestion costs curve. Observe that the slopes of these curves are equal in terms of absolute values. Hence, if any differences between the welfare performance of congestion pricing or slot constraints occur, then these differences cannot be explained by the slopes of demand and the marginal external congestion costs function.

To maximize expected welfare the regulator sets airport charges equal to $E[p^*]$ under congestion pricing or limits airport usage by $E[q^*]$ under slot constraints. However, the real marginal external congestion costs are lower than expected.

Furthermore, due to the negative correlation of marginal external congestion costs and demand this means that the regulator normally underestimates demand as is depicted in Figure 7.4. In this situation slot constraints lead to a welfare loss given by *A* plus *B* because airport usage is fixed below its optimum level. On the other hand, under congestion pricing the welfare loss is given by *B* only. This demonstrates that with negatively correlated uncertainties congestion pricing is likely to be the better choice. Note that this might not hold true if the demand curve is steeper than the external marginal costs curve. However, this example demonstrates that a negative correlation increases the comparative usefulness of congestion pricing compared to slot constraints. It is straightforward to show that the same relations hold if marginal external congestion costs are higher than expected.

The intuition being that under slot constraints the level of airport operations is fixed and airport charges are flexible, while under congestion pricing airport charges are fixed but the level of airport operations is flexible. With a negative correlation between uncertainties the flexibility that congestion pricing provides regarding airport operations enhances welfare. Observe that if the regulator overestimates the marginal external congestion cost curve then in the case of a negative correlation, demand is frequently underestimated. Both effects require extending airport operations compared to the ex ante optimal demand. Under slot constraints airport operations are fixed, but under congestion pricing airport operations are indeed extended which enhances welfare.

However, notice that congestion pricing is not always the right choice. The results derived in the previous section are still important. If the marginal external congestion costs curve is steeper than the demand curve than slot constraints can be the better choice.

The Effect of Complementarities Between the Demand for Airport Capacities

Thus far the fact that airport demand can be interdependent has been ignored. In principle, two types of interdependencies exist: substitutability due to airport competition or complementarity. Airports might compete for passengers or air cargo if they are closely located to each other. Another source for airport competition is the hub-and-spoke networks of airlines, because hub-airports can compete for transfer passengers. Under competition an increase of airport operations at one airport might decrease the demand for competing airports. However, due to the network character of the industry, airports often provide complementary services because flights connect different airports. Thus, under demand complementarity increasing levels of airport operations can increase the demand for other airports.

It can be shown that the comparative advantage of slot constraints compared to congestion pricing increases if demand complementarity is taken into account (Czerny, 2006). Under congestion pricing the charge for airport operations is fixed but not the level of airport operations, as under slot constraints. Therefore, under congestion pricing the amount of airport operations is uncertain. Moreover,

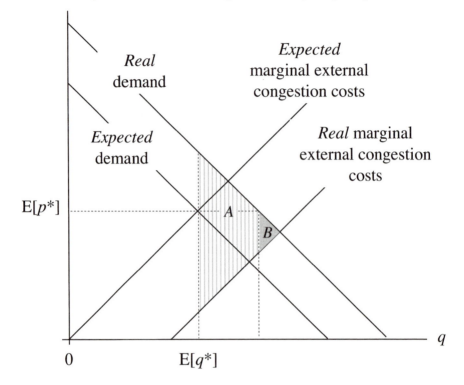

Figure 7.4   The choice of regulation instruments under negatively correlated uncertainties

under congestion pricing demand complementarity for airport facilities reinforces the effect of demand uncertainty on airport operations. Note that if demand at one airport is, say, higher than expected it also increases the demand at other airports. This in turn increases the demand at the former airport and so forth. However, if every airport uses slot constraints then airport operations are fixed and, as a consequence, demand shocks at one airport have no effect on other airports. Therefore, demand complementarity tends to favour the use of slot constraints compared to congestion pricing.

Alternative Regulation Instruments Specifically Designed to Address Demand Uncertainty

So far, the focus has been on slot constraints and congestion pricing as regulation instruments. However, there are many other regulation instruments available, which have been specifically designed to address demand uncertainty. All these instruments move away from a pre-set price or slot constraints. In contrast, regulation is designed such that the actual charges paid for the use of airport capacities is close to the marginal external congestion costs even if demand is different than expected. Thus, to the extent that regulation achieves a close (positive) correlation between airport charges and marginal external congestion costs, the deadweight loss resulting from demand uncertainty is reduced or even eliminated.

One regulation regime of this type was proposed by Roberts and Spence (1976) for pollution management. Applying their suggestions to airports, leads to a combination of slot constraints and congestion pricing (Brenck and Czerny 2002). It contains three elements: lower limits for airport charges, upper limits for airport charges and slot constraints.

The functioning of this regulation regime is as follows: first, slots are allocated by, say, an auction. If demand realizes as expected then airport charges will be equal to the expected level of airport charges. However, since demand is uncertain, airport charges will be either higher or lower than expected. If it is assumed that real demand is close to expected demand, the effect of combining regulation instruments is not different from the model of pure slot constraints. Airport charges are either higher or lower than expected but airport operations are determined by slot constraints.

Now assume that demand is much higher than expected. With pure slot constraints this would strongly increase airport charges. Furthermore, the welfare loss would be high because the level of slot constraints and, thus, airport operations was set much too low. However, with the above combination of regulation instruments the regulator guarantees upper limits for airport charges. Therefore, in order to avoid that airport charges exceed upper limits the regulator must offer additional slots. This increases the level of operations and, because slot constraints are set much too low, can enhance welfare.

On the other hand, if demand for airport capacity is much lower than expected slot constraints and, thus, the level of airport operations are set much too high. Recall that additional airport operations generate external congestion costs and if the additional benefits do not compensate these costs then welfare is reduced. In this situation welfare can be enhanced by reducing airport operations. Notice that this is exactly what the regulator needs to do in order to keep airport charges at lower limits. The regulator has to increase slot demand artificially by buying un-operated slots in order to keep airport charges at lower limits and to reduce congestion. Consequently, airport operation as well as congestion levels are reduced and welfare can be enhanced.

Furthermore, a closer look should be taken at the relation between airport charges and marginal external congestion costs under a combination of slot constraints with upper and lower limits for airport charges. Figure 7.5 illustrates the three regulation elements and marginal external congestion costs. The upper limit for airport charges is denoted by p_{max}, the lower limit by p_{min}, and slot constraints by $\bar{q}$. Airport charges are equal to p_{min} if airport operations are below $\bar{q}$. For airport operations higher than $\bar{q}$ airport charges are equal to p_{max}. Finally, if airport operations are equal to $\bar{q}$ then airport charges are in between p_{min} and p_{max}.

Note that airport charges depend on the level of airport operations and form a step function. Moreover, this step function better approximates the marginal external congestion cost curve in comparison to pure slot constraints (airport charges would be given by a vertical line) or pure congestion pricing (airport charges would be given by a horizontal line). For that reason, the combination of slot constraints and limits better manages demand uncertainty because airport charges are closely related to marginal external congestion costs which enhances the internalization of external congestion costs.

At this point a system is conceivable, which perfectly correlates airport charges to marginal external congestion costs in order to reach an optimal internalization of external congestion costs under demand uncertainty. In a static environment (only one time period) this could be realized by a simple modification of the auction design proposed by Grether et al. (1989). They suggested a 'sealed-bid, one price auction' in which each airline submits a sealed bid which indicates for each slot the maximum price they are willing to pay (see Menaz and Matthews, Chapter 3 in this volume). Bids are then arranged from the highest to the lowest.

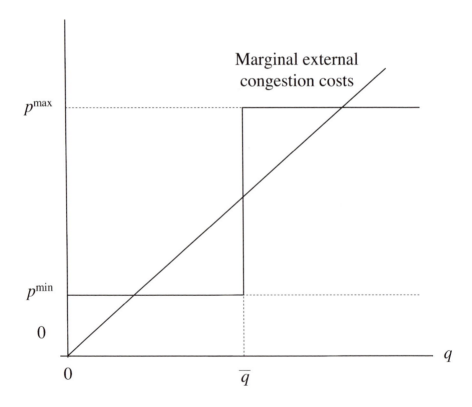

Figure 7.5   Airport charges under slot constraints in combination with upper and lower limits

Given that the highest 'x' bids are accepted the price paid for all slots is equal to the value of the lowest accepted bid. By choosing the number of accepted bids such that airport charges are equal to marginal external congestion costs a perfect correlation can be achieved, which implies a welfare optimum in the case of uncertain demand.

In contrast, in a dynamic environment where demand is continuously changing this approach can approximate optimality only if it is frequently applied, say, on a daily or weekly basis. However, frequent transactions increase costs for airlines and regulators. Against this background, Collinge and Oates (1982) presented a regulation regime that leads to a welfare optimum in a dynamic environment, reduces transaction costs and can be applied to airports (Brenck and Czerny 2002).[4] The distinctive feature of this system is that each slot is indexed by a specific number corresponding to a certain airport charge. Then airport charges are chosen such that they are equal to the marginal external congestion costs of operations given that all slots with a number less or equal to the number of the respective slot are used. This charge has to be paid if and only if slots are used, i.e. slots will only be used by airlines that are willing to pay the respective charge. Another elementary part of this system is secondary trading which guarantees that only the cheapest slots are used. Notice that the functioning of this system does not depend on the system of primary allocation as long as secondary trading generates an efficient allocation of slots.

Conclusions

At many airports congestion has become a significant policy problem. To control congestion two different policy options are generally proposed: slot constraints and congestion pricing. The former requires airlines to buy slots for the operation of flights at congested airports. The latter reduces the demand for take-off and landing operations of airlines by increasing airport charges. Under perfect information regarding demand and cost conditions both instruments can lead to optimal welfare results. Furthermore, uncertainty about external congestion costs does not necessarily affect choice of regulation instruments. The extreme example of uncertainty about the marginal external congestion costs curve and certainty about demand was used for demonstration purposes.

In contrast, under uncertainty about airline demand welfare can be affected by the choice of regulation instruments. In fact, the slopes of the demand curve and the marginal external congestion costs curve are important determinants of the welfare effects of congestion pricing and slot constraints. If the demand curve is steep compared to the marginal external congestion costs curve, then congestion pricing might be preferred. If the marginal external congestion costs curve is steep compared to the demand curve, then slot constraints might be preferred. It should be noted that this result is conditional on the allocation of capacity being equally efficient under slot constraints as under congestion pricing.

---

4   Collinge and Oates originally considered pollution management under uncertainty and not airports.

Furthermore, a negative correlation between demand and external congestion costs was assumed, providing another argument for the use of congestion pricing. On the other hand, the demand for different airports is interdependent because airport services are often complementary. This, however, provides an argument for the use of slot constraints. Hence, the presented results are ambivalent with regard to the choice of regulation instruments. Congestion pricing might be the right choice if the regulator assumes a negative correlation between demand and external congestion costs. In contrast, slot constraints should be preferred if demand complementarity is of special importance. Additionally, the slopes of the demand curve and the marginal external congestion costs curve have to be taken into account.

Alternative regulation instruments specifically designed to manage demand uncertainty were also discussed. These instruments have in common that they enhance internalization of external congestion costs and welfare. However, they will usually increase implementation costs. Hence, before implementing more sophisticated regulation instruments it should be verified whether implementation costs are likely to be covered by welfare gains.

References

Adar, Z. and Griffin J.M. (1976), 'Uncertainty and the Choice of Pollution Control Instruments', *Journal of Environmental Economics and Management*, 2 (3), pp. 178–88.

Basso, L. (2005), 'On Input Markets Surplus and its Relation to the Downstream Market Game', Working Paper, Sauder School of Business, University of British Columbia.

Brenck, A. and Czerny, A.I. (2002), 'Allokation von Slots bei unvollständiger Information', Diskussionspapier 2002/1 der Wirtschaftswissenschaftlichen Dokumentation der Technischen Universität Berlin.

Brueckner, J.K. (2002), 'Airport Congestion When Carriers Have Market Power', *American Economic Review*, 92 (5), pp. 1357–75.

Collinge, R.A. and Oates, W.E. (1982), 'Efficiency in Pollution Control in the Short and Long Runs: A System of Rental Emission Permits', *Canadian Journal of Economics*, 15(2), pp. 178–88.

Czerny, A.I. (2006), 'Congestion Pricing vs. Slot Constraints to Airport Networks', Diskussionspapier 2006/3 der Wirtschaftswissenschaftlichen Dokumentation der Technischen Universität Berlin.

Forsyth, P. (1976), 'The Theory of Pricing of Airport Facilities with Special Reference to London', thesis, University of Oxford.

Grether, D.M., Isaac, M.R. and Plott, C.R. (1989), *The Allocation of Scarce Resources: Experimental Economics and the Problem of Allocating Airport Slots*, Boulder, CO: Westview Press.

Parkes, D.C. (2001), 'Iterative Combinatorial Auctions: Achieving Economic and Computational Efficiency', PhD thesis, University of Pennsylvania.

Roberts, M.J. and Spence, M. (1976), 'Effluent Charges and Licences Under Uncertainty', *Journal of Public Economics*, 5, pp. 193–208.

Weitzman, M.L. (1974), 'Prices vs. quantities', *Review of Economic Studies*, 41 (4), pp. 477–91.

Chapter 8

Prices and Regulation in Slot Constrained Airports

Peter Forsyth and Hans-Martin Niemeier

Introduction

Much of the literature on airport slots centres on the slot allocation process and how efficient it is. With grandfathering of slots, and limited opportunities for secondary trading, there are real concerns as to whether slots are always allocated to the users with the highest willingness to pay. The granting, as well as the trading arrangements of slots, are both key determinants of how efficient the slot allocation outcome is. However they are not the only determinants. Of comparable importance are the prices which users pay to use the airport. This is an aspect which has been given relatively little attention in the slot literature.

In this chapter, we explore how slots, price structures and price regulation are linked to one another. It can be observed that all three aspects are linked closely to one another, that it turns out that the links are close, and that they all play an important role in determining how efficiently slots are allocated. We focus on three aspects of these linkages.

The first concerns the allocation of scarce capacity in a slot limited airport. The allocation which comes about will depend on the airport's price structure. For periods when slots limit output, efficiency requires that all users pay the same price. This rarely happens – in busy, slot limited airports, different users pay very different prices to use the airport at the same time. This leads to a misallocation of slots. This would be so even if slots were auctioned and traded in perfectly efficient markets.

The second aspect concerns the relationship between slot systems and price regulation (or more generally, constraints placed on the airport's prices). Slot systems provide a mechanism for making price regulation, which typically limits prices below market clearing levels, consistent with efficient allocation of available capacity. In this situation, slots perform the rationing task, not price levels.

In fact, we can distinguish between two types of airports. Airports which have excess capacity can have problems in raising sufficient revenues to cover costs. For these airports, slots are irrelevant, and differentiated prices provide the least distortionary means of covering costs. For those airports which are subject to excess demand and which are price regulated, market clearing prices are not feasible, and slots have to be relied upon to handle the rationing problem. For these airports, uniform, not differentiated, prices are efficient. Airports cannot be in both categories at the same time. Yet, in spite of this, we observe that virtually all slot limited airports have highly differentiated price structures. For airports which have only modest excess demand there may still be a cost recovery problem, yet they may use a slot limit system to better handle the use of capacity at the peak when there is uncertainty present. However, for busy slot limited airports, differentiated price structures are an indication of an inefficient solution to the capacity rationing and revenue raising problems.

The third aspect concerns how price regulation impacts on price structures. Different forms of regulation set up different incentives for the regulated firms. Rate of return regulation does not set up any incentives for firms to implement efficient price structures. Some forms of price caps can set up incentives to implement efficient price structures, though the incentives are sensitive to the exact form of the regulation. With airports, there is an added layer of complication due to capacity constraints and the presence of slots. We discuss some forms of price cap regulation and how they can affect price structures. Some price caps being used for airports can set up desirable incentive properties.

In the penultimate section we present some evidence on the structure of prices at slot limited airports. This indicates that these structures are highly differentiated, even for very busy airports. Price regulation and other institutional factors are not pushing these airports to implement more efficient price structures which could complement reforms to the slot granting and trading processes.

We commence by providing a brief background to slot controlled, regulated airports. Then, the characteristics of efficient price structures are outlined. The next section focuses on the role of slots in making regulated prices consistent with efficient allocation of the airport's capacity. We then look at how how regulation can set up incentives for the airport to choose efficient or inefficient price structures is explored. The penultimate section provides empirical evidence on price structures and regulation. We conclude by noting some implications which arise from the analysis.

Background

Slot coordination is part of a broader capacity allocation process. It is intricately linked to other parts of the process, including in particular, pricing. The implementation of a slot system has implications for both price structures and price levels. Many slot coordinated airports are now regulated, and this regulation in turn is a major factor in determining price levels and it may affect how price structures are set. Ideally, in designing an efficient approach to allocating scarce capacity at busy airports, all of these aspects should be considered together.

Our primary interest is in busy airports which are slot controlled. For some, such as London Heathrow and Frankfurt, slot limits are effective more or less all of the day. For others, such as Paris Charles de Gaulle and Amsterdam, slot limits are effective at the peak though there is spare capacity at the off peak. For many less busy airports, slot coordination systems are in place, not so much to ration demand as to even out the flow of traffic. Here we concentrate on the first two of these, where slots perform a rationing function. The fact that there is a slot control system in place to ration demand to capacity implies that other forms of capacity rationing, including congestion/queuing and prices are not being used, at least to their full potential.

Price levels and structures can affect the allocation of users to capacity. In a slot rationing context, price *levels* may not be important, but price *structures* will be, since they will affect which users obtain slots. Virtually all airports, including busy slot controlled ones, operate with a weight based, or passenger based, pricing structure. This structure is a tolerably efficient one when the problem is one of achieving cost recovery at minimum cost in terms of efficiency at an airport with excess capacity. However, when the problem is one of reducing demand to match scarce capacity, this ceases to be so. As will be argued in the next section, this price structure will reduce the efficiency of the capacity allocation process in the busy slot controlled airport.

Airports take a variety of institutional forms (for an overview, see Graham, 2003). Some are publicly owned, others are not-for-profit enterprises which are community owned, and others are privately owned. Most private airports are subject to some form of price regulation. Across most of these institutional forms, there is pressure on the airport to keep overall revenues close to costs. In the case of airports which face considerable excess demand, this amounts keeping prices below market clearing levels. Few, if any, busy airports are permitted to charge market clearing prices. Such prices at airports such as London Heathrow would lead to extremely high profits. To this end, some alternative means of limiting demand to capacity must be used so that delays do not become excessive – hence the role for slot limits. If busy airports are to be subjected to price regulation or to be required to keep prices to cost levels, a slot system is an essential complement to achieve efficiency in allocation of capacity.

In addition, the ownership and regulatory environment in which an airport operates will condition the incentives which it faces in setting price structures. With many ownership and regulatory forms, including public ownership and rate of return regulation, there may not be any clear incentives to adopt any particular price structure. However, some regulatory models create strong incentives to adopt specific price structures. Price caps are a case in point – the exact form of the cap creates incentives for the firm to implement particular price structures. Thus the choice of regulation will have important implications for the efficiency with which slots are allocated.

Price Structures and Slots

In the extensive debate over slot allocation arrangements – whether slots should be grandfathered, auctioned, and traded – there is the implication that the objective is to secure an efficient allocation of scarce slots. The allocation of slots which comes about is a result of two distinct processes:

- the slot allocation process, which determines how slots are allocated to users – by grandfathering, auctions, and secondary trading; and
- the structure of prices which are charged at the airport – which user pays what charges.

The debates in recent times have focused on the first of these points (for a discussion of this, see Bass, 2003; Boyfield, 2003; Humphreys, 2003; Starkie, 2003). However, both of these processes are, at least a priori, of comparable importance. It is possible to have a very efficient slot allocation process which leads to a highly inefficient allocation of slots – this would come about if the price structure is inefficient. In fact, given that the price structures of nearly all airports are inappropriate, this is the likely outcome. An efficient slot process is a necessary but not sufficient condition for efficient allocation of slots.

This is because busy slot controlled airports invariably have weight based or passenger based price structures. These can be moderately efficient ways of achieving cost recovery at airports which have spare capacity (for some discussion, see Morrison, 1982). However, for busy airports where allocation of scarce capacity is the problem, such structures lead to inefficient allocations.

The importance of price structures was recognized, and discussed extensively in the early analyses of airport capacity and congestion problems (Levine, 1969). This literature highlighted the fact that in the US, airports were being rationed by queues rather than prices (see for example, Carlin and Park, 1970). It recommended that prices rather than congestion be used as a rationing device. In addition, the authors noted that price structures which resulted in different users paying different prices for use of the airport at the same time would lead to inefficient allocation of capacity. It recommended that uniform prices be charged to all users during a particular time period.

The slot literature recognizes that slot limits, rather than price levels, will perform the task of rationing demand to capacity. It emphasizes the problem of allocation of slots. However, the issue of price structures is still very relevant, though it receives little attention. The prices charged to different users, as well as the slot allocation process, will determine who gains the slots. As long as different aircraft make the same use of the runway, uniform prices for the different users will promote efficient use of the facility.

In practice, prices are not usually uniform. Slots may be efficiently auctioned or traded, and all users may face the same slot price (where the slot price is understood to be exclusive of the airport charges). However, smaller aircraft will pay much less than larger aircraft for use of the airport at the same time. A 38 passenger aircraft might pay $1,106 per movement to use a busy, slot controlled airport, while a 396 passenger aircraft might pay $9,678 (see Table 8.1 below). Thus, if the value of the slot is $2,000, the all up cost of using the airport will be $3,106 for the small aircraft and $11,678 for the large aircraft. Such a price structure will promote an inefficient allocation of the airport's capacity, discouraging larger aircraft, with high willingness to pay, and encouraging smaller aircraft, with low willingness to pay.

Efficient Pricing at Slot Controlled Airports

Suppose that there is a slot constraint system in place at an airport, which is subject to excess demand for at least part of the day. Thus demand is rationed to capacity, K, by slots rather than queues or prices (see Figure 8.1). For the purposes of this chapter, it is assumed that slots are rationed efficiently, by auctions, trading or both – thus there is a going price for a slot (this assumption is relaxed in other chapters of this book). This could be the case when price levels are kept down, perhaps by regulation or perhaps as the result of an instruction from a government to a publicly owned airport. This is a very common situation in Europe and the Asia Pacific. There still remains a question of designing an efficient price structure. Two dimensions are prices at different times in the day, and prices for different users. Suppose also that the marginal costs of movements and passengers are zero. These assumptions can be relaxed without altering the nature of the results.

Airports with Spare Capacity Off Peak

Some airports face excess demand for a small part of the day. If there is no cost recovery problem, an efficient pricing structure would be a uniform price for all movements at the peak which just restricts demand to available capacity. Such a price structure may yield revenues to cover cost. If it does not, and cost recovery is a requirement, then an overlay of Ramsey prices will be needed.

Airports with Peaks and Price Regulation

It is possible that peak pricing with even modest levels of excess demand will lead to total revenues being above costs, and the airport may be required by the regulator to keep prices no higher than costs. If so, prices cannot be the sole capacity rationing device, and prices can be combined with a slot allocation system. An efficient constrained price structure would involve a zero charge for movements at the off peak, and a uniform charge for all movements at the peak, set at the maximum allowable level. A uniform airport charge would mean that, when combined with the slot price, all movements would face the same price, and runway capacity would be allocated efficiently.

This is shown in Figure 8.1. The airport operates with fixed capacity of K. The peak demand is shown as D_p and the off peak demand is shown as D_o. Capacity at the peak is insufficient to cater for all demand unless the price is very high. If there is a price cap of P_r per movement, prices cannot ration demand at the peak and slot rationing must be used. If there is a uniform peak and off peak price, output at the off peak will be X_o and at the peak will be K. The price of slot will be P_s. However if the off peak price is lowered to zero, and the peak price is raised to P^*_p, the highest consistent with meeting the price cap, output in the peak will be unchanged, though the price of slot will fall to P'. The output in the off peak will rise to X^*_o. The airport will gain more revenue and profit. Allocative efficiency will be maximized.

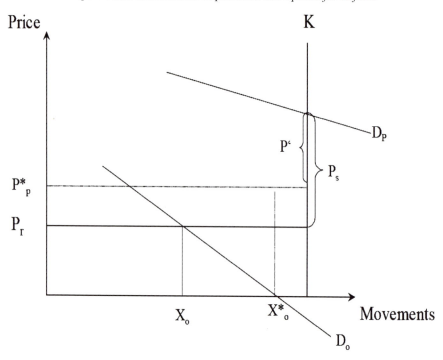

Figure 8.1   Moderately busy peak

In this situation, peak prices do not have any direct allocative function. However they have an important indirect function-optimal peak prices are such that they enable off peak prices to be zero while meeting the price regulation constraint.

Very Busy Airports

Some airports face excess demand for all the day, and prices would be insufficient to ration demand if they are regulated. Peak pricing becomes irrelevant, because all of the rationing function is served by the slot allocation system. If the marginal cost of handling passengers is not zero, the airport should impose a passenger charge equal to the marginal cost of handling passengers, and a movement charge which is as high can be within the price constraint. Again, the movement charge should be the same for all movements, or otherwise the slots will be allocated to users with a lower willingness to pay than others.

This case is illustrated in Figure 8.2. In this case it is evident that the price cap is set at a level which precludes prices from being used to ration demand even during the off peak (it may be feasible to use prices only to restrict off peak demand to capacity if peak prices are kept very low – though there is no point in doing so). Slots carry the entire allocative burden and prices are irrelevant. If the same prices are set during peak and off peak times, then the slot premium during peak times will exceed the one during off peak times. For a busy airport

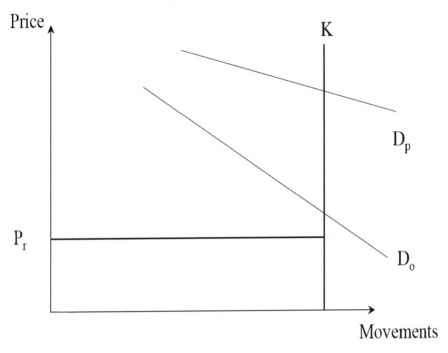

Figure 8.2   Very busy airport

such as Frankfurt there may be little advantage in differentiating peak and off peak prices, since prices are not needed for the rationing function.

Busy Airports with Differentiated Movements

Most airports face a range of aircraft sizes of movements using their facilities. Traditionally, airport charges have been weight based, though many airports nowadays are moving towards purely passenger based charging. Both of these structures result in smaller aircraft paying less for the use of the airport than a larger aircraft. When the airport is not busy and the problem of cost recovery exists, these structures make sense. However, when the airport faces excess demand, they cease to do so. Airports may face excess demand during part of the day or during all of the day in the case of very busy airports. The discussion here applies to both types of airport.

The important requirement for efficiency is that the charge for using the airport at a particular period be the same for all users, regardless of size or passenger load (when the costs of serving all users is the same). The relevant constraint is that of runway capacity, and this will be allocated most efficiently when all users face the same price. Assuming that slots are undifferentiated, the price for a slot will be the same for all movements, and to ensure a uniform total price, airport charge plus slot price, the airport charge will have to be the same.

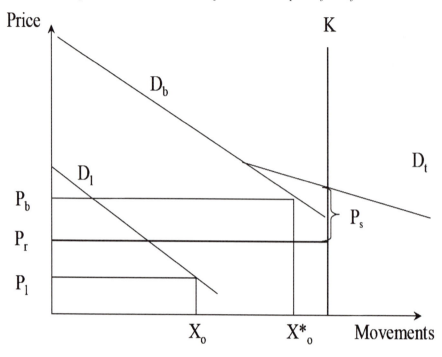

Figure 8.3   Pricing with differenciated users

This is shown in Figure 8.3. The demand curve of large movements is shown as D_b, and the demand of small movements as D_l. The demand curve for all movements is shown as the lateral summation of these, D_t. If a uniform price for movements, P_r is set, where this level is determined by the regulator's price cap, there will be excess demand to be rationed by slots. Here the slot price will be P_s, and the total price per movement will be given by $P_r + P_s$. It would be possible, within the price cap, to have differentiated prices. For example, a price P_l could be charged to small movements and a price P_b to large movements (subject to the average of these being no higher than P_r. There will still be excess demand and the price of a slot may change slightly to P'_s (not shown). The total price for a small movement to use the airport will now be $P_l + P'_s$, and this will be below the price for a large movement, $P_b + P'_s$. This will be an inefficient outcome, because the willingness to pay to use the airport of a marginal small movement will be below that of a marginal large movement. Efficiency will be maximized when all movements are charged the same price.

Efficiency Issues at Constrained Airports

Thus there are essentially two efficiency issues that it would be desirable for the price structure to address. These are:

1) The peak/ off peak pricing issue. In the capacity constrained airport with slot allocation, but with available capacity in the off peak, the important aspect is that the off peak price be set at marginal cost (in this case, zero). Slot allocation, prices at peak do not have any direct function in rationing capacity – if there is any excess demand, slots do the rationing. It is desirable that peak prices be set at a level, subject to the regulatory constraint, that enables off peak prices to be set at the efficient level – off peak prices do have an allocative role.

2) The allocation of slots amongst different users. It is desirable that the slots go to those who have the highest willingness to pay. This will only happen if passenger charges are set at marginal cost and there is a uniform movement charge, which when added to the price of a slot, results in a uniform all-up cost of using a slot.

Cost Recovery, Slots and Price Regulation

An important role for slots is implicit in the analysis of the previous section. Slot systems make it possible for price regulation, or limits on the profitability of airports, to be made consistent with achieving efficient rationing of capacity at busy airports.

Airports are often subject to a constraint such that revenues not exceed costs. This may be a requirement set by a government owner. It could be the objective of a not-for-profit community owned airport which seeks to keep charges as low as possible. Finally, it could be the result of regulation of a privately owned airport. This could be the case under explicit rate of return regulation, where prices are set so as to cover costs. Alternatively, price-caps might be set at a level such that the airport can expect to cover costs over the regulatory period (for discussions of airport regulation, see Forsyth et al., 2004).

These requirements pose a problem for pricing airports efficiently. Very often, charging the maximum price allowed will give rise to an excess demand problem. For example, as in Figure 8.2, no matter how prices are adjusted subject to the price cap, they will be too low to ration demand to capacity. An alternative rationing device must be used – queues are feasible, though they are inefficient. However a slot system can be implemented – if slots are auctioned or traded efficiently, the airlines will face a price for using the airport and a slot premium which together equal an efficient rationing price. In this context, price regulation does not promote efficiency – rather, it hinders it, by keeping prices below efficient levels.

In this situation, the slot system reconciles efficient allocation of capacity with the goal of keeping prices below market clearing levels. The overall level of airport charges has no allocative function, and regulators or owners can choose price levels to achieve other objectives. Thus, for London Heathrow airport, the regulator sets prices well below market clearing levels, but the slot system restricts demand to capacity. For this airport, slot prices are very high.

Apart from the situations where the airport is permitted to earn profits or operate at a loss, and the case where efficient pricing leads to exact cost recovery, airports will fall into one of two categories. The first is airports with a cost recovery problem – efficient prices will lead to revenues less than cost, and they will have to have recourse to second best pricing, such as Ramsey pricing, to enable them to cover costs. The second is airports which have an excess demand problem – maximum allowable prices will lead to demand in excess of capacity. To ration this capacity efficiently, they will have to rely on a slot system.

Thus, subject to the qualification noted below, if airports are allocating their capacity efficiently, we should expect to see them either using a slot system to allocate capacity, or a differentiated (Ramsey) pricing structure. They would not use both, because they either have a cost recovery problem or an excess demand problem – they cannot have both. In spite of this observation, slot controlled airports, which face excess demand at some part or all of the day, invariably have differentiated price structures. This indicates the existence of an efficiency problem – efficient slot allocation requires a move from differentiated to uniform prices.

It may not always be obvious which category an airport falls into. For very busy airports, the answer will be clear. However, for moderately busy airports, with excess demand at peaks but not at off peaks, it could be the case that they fall into the cost recovery problem category. If such an airport were to charge uniform prices for all users (though with different peak and off peak prices), it might find that its revenues fell short of costs, and demand short of capacity. Moving from differentiated prices to uniform prices lowers output at a given average price. To cover costs it may have to move away from uniform prices towards implementing a Ramsey pricing solution, albeit one with less pronounced differences in price than the initial price structure. In reality, there could be several airports in this situation – airports which have highly differentiated charges and a small degree of excess demand at the peak.

For other airports, moving to a uniform pricing structure will result in complete cost recovery at prices less than market clearing prices – these airports will have to rely on slots to ration demand to capacity.

Slot Constraints under Uncertainty

The discussion above presupposed that the purpose of implementing a slot limit system is to enable efficient rationing of capacity. Often this will be the case. However, it could also be the case that a slot limit system is implemented even though rationing prices would be feasible. This would be so if slots performed better than prices under uncertainty. There should be no presumption that prices work better, in terms of efficiency properties, than quantity limits when uncertainty is present (Weitzman, 1974). There can be cases when setting quantities in advance works better than setting prices in advance. This has been argued in the context of airports by Forsyth (1976) and Czerny (in this volume).

Thus an airport which has a cost recovery problem might still choose to implement a slot system. By setting prices below market clearing levels, it can ensure that its capacity is (almost) always fully utilized. It can set prices at a level at or just below the level which will clear the market in the situation of minimum likely demand. A uniform price will fail to raise sufficient revenues, but it can cover costs at little cost in terms of efficiency by moving to (approximately) Ramsey prices.

There are many, busy, airports which are not in this situation – for them, a slot limit system is necessary to allocate capacity efficiently in the face of price constraints. Efficiency requires that they charge uniform prices. However, for some less busy airports, slot limit systems can be consistent with differentiated pricing if the slot system is adopted to ensure high utilization of capacity under uncertainty.

Regulation and Incentives for Efficient Pricing

Perhaps the primary objective of the new forms of incentive regulation, such as price caps, was to give the regulated firm an incentive to minimize cost. Older forms of regulation, such as rate of return regulation, performed poorly in this respect because they were essentially cost plus forms of regulation – a firm which allowed its costs to rise was permitted to increase its prices. By setting prices in advance, independent of the firm's own costs, the firm was given an incentive to minimize its costs, since any saving would increase its profit (Beesley and Littlechild, 1989).

It was later noted that price caps could also could have desirable property of inducing the firm to adopt an efficient price structure. Certain types of price caps would induce the firm to implement Ramsey prices (Bradley and Price, 1988). Not all forms of price caps have the same effect, however. If the price cap is implemented with a price basket approach, i.e. one on which the firm is constrained to keep an index of its prices, with initially set weights, no higher than the prescribed level, it will move towards Ramsey pricing (Armstrong et al., 1994). However, even quite similar price caps need not have the same property. For example, if an average revenue price cap is set, whereby the average revenue (total revenue divided by an indicator of total output) is required to be kept no higher than a pre determined level, the resulting price structure will not be a Ramsey one. In certain contexts, suitably designed price caps will induce the firm to implement peak pricing (Brunekreeft, 2000).

With rate of return regulation, the price structure may be specified by the regulator. Even if it is not and the firm has the freedom to implement the price structure of its choice, it does not face any incentive to adopt an efficient structure, since all price structures yield it the same profit. The firm even has incentives to price peak traffic too low as this gives a rationale to invest in new capacity and produce capital intensively (Sherman, 1989). Much the same will be true for a government owned firm, which faces no particular incentives to implement a price structure which is efficient.

In some countries such as the UK, there is now a moderately long history of price cap regulation in a number of industries including airports, though it must be noted that this regulation is rarely pure price cap regulation (cost considerations affect allowable prices when caps are reset – see Toms, 2004). Guiletti and Waddams Price (2000) and Guilett and Otero (2002) have examined how price capped firms have adjusted their price structures, and find some evidence of adjustments, though the evidence is mixed. The regulated airport company, BAA, was one of the firms which adjusted its price structures, but it had already adopted peak pricing in 1978 under state ownership against the opposition of airlines (see Little and McLeod, 1972 and Toms, 1994). It should be recognized that there are limits on the strength of incentives, and that airports may be slow or reluctant to implement efficient price structures (see Starkie, 2005). Airports may be size rather than profit maximizers, and thus they may be reluctant to implement efficient pricing, especially when airlines are hostile to them. The majority of airlines which practise peak pricing oppose peak pricing by airports because it reduces their rent from scarce slots (IATA, 2000). Airports have been very reluctant in lobbying for peak and congestion pricing (ACI, 2003).

It is important to note that this literature has concentrated on the cases where capacity is adequate, and there is no problem in expanding the output of all of the firm's products. It would be very relevant to the case of an airport which faces no excess demand, and for which the second best, cost recovery price structure is a Ramsey structure. This case has been considered briefly by the UK Civil Aviation Authority (CAA, 2001). It may not be directly relevant where output is constrained for some or all of the day, as at busy airports. It is also not likely to be directly relevant to the case where prices do not perform the rationing function, at least for part of the day, and an additional mechanism, slot allocation, is used for this purpose.

This raises some central questions. Is it possible for price regulation to be implemented in a way which gives busy airports an incentive to adopt efficient price structures? Secondly, if it is, which precise forms of regulation will work best in this regard?

In this section we consider whether price regulation can be designed to give busy airports incentives to implement efficient price structures. The first case considered is that of peak pricing at airports which are busy for part of the day. The second case is where the airport is very busy all of the day. Both these cases involve homogeneous flights. In the next case we consider the situation where flights differ – there are large and small aircraft, and it is possible to price these differently. Finally, the case in which there are positive marginal costs of passenger handling is examined.

Peak Pricing in Moderately Busy Airports with Homogeneous Movements

An efficient price structure in this case is one which sets the price for off peak movements at marginal cost. This will be assumed to be zero, though this assumption can easily be changed (Hogan and Starkie, 2004, provide evidence of non zero movement related runway costs). The peak price does not matter for allocative purposes. A higher airport charge for use at the peak will mean that the value of a slot will be lower, however the all-up cost of using the airport at the peak will be the same. The price set for the off peak does have an allocative function, since higher prices discourage usage. Thus it is efficient to set this price at marginal cost (assumed to be zero here).

If the airport is subject to an average revenue price cap on *movements* it will have an incentive to maximize the total number of movements – it can achieve this by setting the off peak price at zero (see Figure 8.1). Indeed, the airport will have an incentive to set the price at even less than zero, since additional movements encouraged to use the airport will enable it to gain more revenue, which, up to a point, will be above the cost of attracting off peak movements. The regulator may need to specify that the airport is not permitted to charge prices less than zero. Much the same analysis will apply when the price cap is specified in terms of average revenues per passenger – more movements bring additional passengers, and there is an incentive to maximize the total passengers.

When there is a zero marginal cost of passengers, this regulatory rule will work well. However, when there is a positive marginal cost of passengers, the airport will face an incentive to set prices for off peak movements at less than the overall marginal cost, including the marginal cost of the additional passengers the aircraft movements bring. The loss incurred on the off peak movements can be loaded on to the peak movements, within the price cap. There is no incentive to do this at the peak since the number of movements and passengers is fixed.

Regulation with a price cap basket will produce similar results. Here the prices that enter the basket will be those of peak and off peak movements. By lowering the off peak price, additional revenue can be gained, through increasing the total number of movements (or passengers) and thereby increasing the allowable revenue.

Under rate of return regulation, there is no incentive for the airport to implement peak pricing. Given that revenue is fixed in relation to total costs, and costs are fixed, the airport gains nothing through encouraging better utilization of its facilities through off peak pricing (this also applies in other situations where price structures are an issue, for example with differentiated movements). Where there is a positive marginal cost of passengers, and the airport is allowed to recover the cost of these, it may have a slight incentive to increase movement numbers if the price it is allowed to charge is above the marginal cost of the movement.

Low off peak prices can have a moderately significant allocative effect. If the demand to use the airport is inelastic for all of the day, the allocative effect would not be large. There might be some switching from peak to off peak, but this is stimulated primarily by the non availability of capacity at the peak. Where low off peak pricing might make a difference is where there are some users with elastic demand. These could include regional airlines, and nowadays, low cost airlines. Currently, low cost airlines are moving to secondary airports, which are less convenient, because these are offering them lower prices than the metropolitan airports. If the metropolitan airports were to offer low charges at the off peak, these airlines might be willing to use them. Up to now there is not much empirical evidence, as most airports have not tried such a strategy, but the switching of low cost carriers between the London airports Gatwick, Luton, and Stansted and the current debate about pricing and investment at Stansted show the potential for low off peak charges (CAA, 2006).

Overall, price regulation through price caps will give airports an incentive to implement peak pricing, and to ensure that off peak charges are relatively low.

The problem which could well emerge is that the average revenue caps form of price caps will give the airport incentives to offer off peak prices which are too low, especially if there is a positive marginal cost associated with movements or the passengers they bring.

Very Busy Airports with Homogeneous Movements

With very busy airports, there may be something of a peak and an off peak, but peak pricing becomes irrelevant. This is because there will be excess demand even at the off peak, and it will be necessary to rely on slot allocation all of the time. There is no point in encouraging off peak use because there is insufficient capacity. The airport has no scope to increase revenue through its pricing. Price regulation will not induce a move to any particular structure of prices.

Differentiated Movements

Suppose that a busy airport is facing a range of sizes of movement, and that it is subject to price cap regulation. Suppose further that this takes the form of an average revenue cap, with the average revenue per flight being specified. Will this regulation give the airport an incentive to charge uniform prices during the excess demand periods?

The answer is no, because the structure of charges cannot alter the revenue received by the airport. This revenue is determined by the total number of movements, and this amount is fixed by capacity. The airport could encourage more large movements at the expense of small movements, but it will gain nothing from this, since its total revenue will be unchanged.

Alternatively, suppose that there is a cap on average revenue per passenger (a structure often adopted, for example at BAA's London airports). If this is the regulation it faces, the airport has an incentive to increase passenger throughput where it can, as this will increase total revenue. If it is faced with a choice between charge structures which are weight or passenger based, or a uniform charge structure, then it will opt for the latter. Such a charge discourages smaller aircraft with fewer passengers freeing up capacity for larger aircraft. In this case, an average revenue based cap has desirable efficiency properties.

It should be noted that the incentive this cap gives is to maximize passenger throughput, not welfare per se. With a restricted choice of pricing instruments, it will opt for a pricing structure that will have this effect. If the airport were able to adopt more sophisticated pricing structures, it could move away from efficient uniform pricing. Some small movements may have a high willingness to pay (perhaps flights on business traveller oriented routes), even though they may not have large passenger loads. Even though it is efficient for them to use the airport, at the expense of a larger movements with a lower willingness to pay (e.g. charter flights), the airport will prefer to serve the larger than the smaller movement since it can gain higher total revenue from it. An airport will thus have an incentive to develop a charging structure which discriminates against the smaller movement (eg by offering a schedule which gives a rebate according to the number of passengers). This response is feasible, though it is probably not much of a concern.

Regulation with Passenger Costs

It is quite likely that there will be positive marginal costs of serving passengers – indeed it is quite possible that passenger related costs may be of a similar order of magnitude to movement related costs. If this is so, the efficient price structure will be one of a uniform charge per movement plus a uniform charge per passenger. Again, the question arises as to what incentives price regulation will create.

If a price cap on average revenue per passenger is imposed, it is in the interest of the airport to encourage passenger throughput. As long as the average revenue per passenger permitted exceeds the marginal cost of a passenger, the airport gains extra profit from serving additional passengers. This is very likely to be the case since the allowable average revenue would normally be sufficient to enable recovery of all costs, not just passenger related costs.

The problem is that this will give the airport an excessive incentive to increase passengers. It will be in its interest to set a per passenger charge less than the marginal cost of the passenger, and it will recoup the loss of revenue by increasing the per movement charge. It will thus set up incentives which are stronger than optimal to switch to movements with large passenger loads. An airport might, for example, rely solely on per movement charges, even when the marginal cost of passengers is significant. Smaller movements with a high willingness to pay would be discouraged in favour of large movements with a low willingness to pay.

This discussion of regulation and price structures is, of necessity, a preliminary one. It highlights that (slightly) different regulatory rules create incentives for different price structures, and that some price structures encourage inefficient use of the airport. More attention to the precise design of regulatory rules is thus a priority.

Price Structures in Slot Limited Price Regulated Airports: Some Evidence

Airports have long operated with a very differentiated charging system, whereby large aircraft with large passenger loads are charged much more than smaller aircraft to use the airport at the same time. At most airports, charges are roughly proportional to weight and passenger loads. This is a structure which was appropriate when airports were not busy, but it is one which is quite inappropriate for the busy airports of today.

Table 8.1 provides data on airport charges for small, medium and large aircraft at European slot controlled airports. Three groups of airports are identified – airports which, like Frankfurt, are busy for nearly all the day; airports like Amsterdam, which have capacity constraints at the peak; and finally some airports like Helsinki which are not constrained even at the peak. In all but a few of these airports, there is a wide variation in charges between different airport types. On average, the very busy airports have a smaller range of prices than the

Table 8.1   Price structures ($US) and regulation at European Airports in 2004

Airports	B747 Charge 396 pax	A321 Charge 139 pax	CRJ50 Charge 38 pax	Ratio large/small	Ownership	Regulation	% of flights with 1–49 seats
Düsseldorf	6703	2268	833	8.1	P	Cost-based regulated. Since 2005 with revenue sharing agreement	11.9
Frankfurt	9678	3169	1106	8.8	P	Cost-based regulated with revenue sharing agreement	8.5
London Heathrow	8530	3652	1734	4.9	F	PC	0.5
London Gatwick	5907	2553	1235	4.8	F	PC	7.6
Madrid	6743	1757	492	13.7	G	Cost-based regulated	10.7
Paris Orly	8965	2400	666	13.5	G. P in 2006	Cost-based regulated. Since 2006 PC	11.9
Busy airports – average				9.0			
Amsterdam	11270	3043	820	13.7	G. P. is foreshadowed	Cost-based regulated	15.9
Brussels	8284	3067	855	9.7	G. F in 2005.	Cost-based regulated	22.6

Table 8.1 cont'd

Airports	B747 Charge 396 pax	A321 Charge 139 pax	CRJ50 Charge 38 pax	Ratio large/small	Ownership	Regulation	% of flights with 1–49 seats
Paris CDG	8965	2400	666	13.5	G. P in 2006	Cost-based regulated. Since 2006 PC	9.8
Copenhagen	11180	3328	924	12.1	F	PC	30.8
Dublin	5890	1442	410	14.4	G	PC	na
Rome Fiumicino	6596	1922	512	12.9	F	Charges are fixed until new regulatory is established	9.0
Lisbon	8388	2553	657	12.8	G	Cost-based regulated	15.0
Munich	7949	2551	864	9.2	G	Cost-based regulated	24.1
Milan Malpensa	6596	1922	512	12.9	P	Charges are fixed until newregulation is established	na
London Stansted	5826	2495	894	6.5	F	PC	na
Berlin Tegel	7529	2250	646	11.7	G	Cost-based regulated	na
Vienna	11896	3870	1211	9.8	P	PC with sliding scale	31.3

Table 8.1 cont'd

Airports	B747 Charge 396 pax	A321 Charge 139 pax	CRJ50 Charge 38 pax	Ratio large/small	Ownership	Regulation	% of flights with 1–49 seats
Airports with busy peaks – average				11.6			
Stockholm	8974	2566	615	14.6	G	Cost-based regulated	31.0
Rome Ciampino	6596	1922	512	12.9	F	Charges are fixed until new regulation is established	na
Helsinki	8262	2320	596	13.9	G	Cost-based regulated	na
Non-busy airports				13.8			

Key: G – Government owned, P – Partially privatised, F – Majority privately owned, PC – Price capped, cost-based regulated.

Sources: Air Transport Research Society (2005); Airports Council International et al. (2003b); Gillen and Niemeier (2006).

other two groups, mainly due to the effects of the London airports which have the smallest range of all.

An efficient pricing structure would not involve exactly uniform charges for all movements. This is because costs depend on passenger throughput in terminals as well as aircraft movements. Suppose that aircraft and passenger related costs were about equal – if so, half the charge for using the airport for an average sized aircraft would be passenger related costs, and half aircraft movement related costs. With such a charging structure, the ratio of the charge for the largest aircraft (Boeing 747) relative to the smallest (CRJ5) would be 3.0 for London Heathrow (as against an actual of 4.9) and 3.5 at Frankfurt (as against an actual of 8.8). What these calculations are saying in effect is that the variation in airport charges that are observed cannot be explained by passenger related costs. Airport charges are inefficiently structured for busy airports.

It is difficult to be definitive on the effects of regulation and ownership on pricing. Only the London airports have been fully privatised for an extended period. These airports are subject to price cap regulation, though with periodic cost based resets (Hendriks and Andrew, 2004; Toms, 2004). The London airports have the price structures which are closest to efficient ones. In addition, they are amongst the few which have peak/off peak differentials. However, both statements would have been true in the 1970s and 1980s under public ownership (Toms, 1994).

Most other airports are government owned or majority owned, and most are not subject to regulation which could be described as incentive regulation. For these airports, there are few if any pressures for them to adopt efficient price structures. In practice, with BAA as an exception, they have continued to implement weight or passenger based price structures long after the rationale for such structures has disappeared. Indeed, very few of the airports which experience peak problems have instituted peak pricing.(Doganis, 1992; Hague Consulting Group, 2000).

Price structures do appear to make some difference to the mix of aircraft using the airports. While the proportion of small aircraft at the busy airports is lower than at the less busy airports, the smallest proportions are recorded at the two London airports (Table 8.1). Indeed, the small aircraft are virtually priced out of London Heathrow.

Thus the empirical evidence points strongly to inefficiency in price setting by most busy airports. These airports are not setting prices which make efficient use of their limited capacity. Given the size of the differences in charges faced by different users – after adjusting for passenger related costs, some are paying five or more times what others are paying, the efficiency losses could be quite substantial. The efficiency costs of poor price structures well be greater than the efficiency costs of poor allocation and limited trading of slots.

This situation has remained so for many years, and there are no strong forces for change. Over time, moves towards privatization and incentive regulation, which are themselves slow, may induce some reform in airport pricing structures. However, even good regulatory arrangements may not be enough to guarantee that airports choose price structures which make efficient use of busy airport's scarce capacity.

Conclusions

There has been much interest in how well slot processes allocate scarce capacity at busy airports. There are good reasons for questioning whether, in the absence of auctions and with only limited secondary trading, the outcome is very efficient. By contrast, there has been much less attention given to price structures at busy airports, a factor which is just as important in allocating scarce capacity. Both slot arrangements and price structures determine the efficiency with which slots are allocated at airports.

Price structures at most airports have remained unchanged for several decades. While they may have been appropriate when airports had adequate capacity, they have become quite inappropriate for busy airports which have a problem in allocating scarce capacity to users. Airport price structures are far too differentiated between users, and discourage the use of large aircraft and encourage the use by small aircraft. They lead to significant inefficiencies in the allocation of available slots.

Price regulation is closely interlinked with slot processes and price structures. Two of the links are highlighted here. The first is the way in which price regulation at busy airports typically makes market clearing prices impossible.

Prices cannot serve their allocative function, and keeping prices down by regulation means that the task of keeping demand within capacity falls to the slot process. One of the major roles of slots at busy airports is they make it feasible to pursue regulatory objectives whilst, at the same time, making it feasible to allocate airport capacity in an efficient manner.

The second is the way in which price regulation may induce airports to alter their price structures. The incentives created by price regulation are quite sensitive to the form of that regulation. However some simple price regulatory structures can encourage efficiency in price setting. They can encourage the setting of peak/off peak differentials, and the setting of more uniform prices for different sizes of movements at the busy times. Both these would lead to more efficient use of the airport's capacity.

This said, however, few of the busy slot controlled airports operate in a regulatory and institutional environment which creates incentives for efficient price setting. Most airports are partly government owned and are not set to explicit regulation. There have been moves towards privatization and the implementation of incentive regulation, though these have not been extensive. Most busy airports are not under pressure to reform their pricing structures, even though their present ones are obsolete. Thus it is not surprising that price structures remain a serious source of efficiency in the allocation of slots at busy airports.

References

Airport Council International (2003a), *Airport Charges in Europe*, Geneva:ACI.

Airport Council International, Air Transport Action Group and International Air Transport Association (2003b) *Airport Capacity/Demand Profiles 2003*, Geneva: ACI, ATAG and IATA.

Air Transport Research Society (2005), *Airport Benchmarking Report-2005: Global Standards for Airport Excellence*, Vancouver: Centre for Transport Studies, University of British Columbia.

Armstrong, M., Cowan, S. and Vickers, J. (1994), *Regulatory Reform: Economic Analysis and British Experience*, Cambridge, MA: MIT Press.

Bass, T. (2003), 'The Role of Market Forces in the Allocation of Airport Slots', in Boyfield, K. (ed.), *A Role Market Airport Slots*, London: IEA, Readings 56, pp. 21–50.

Beesley, M.E. and Littlechild, S. (1989), 'The regulation of privatised monopolies in the United Kingdom', *Rand Journal of Economics*, 20, pp. 454–72.

Boyfield, K. (ed.), *A Role Market Airport Slots*, London: IEA, Readings 56.

Bradley, I. and Price, C. (1988), 'The Economic Regulation of Private Industries by Price Constraints', *Journal of Industrial Economics*, 37, pp. 99–106.

Brunekreeft, G. (2000), 'Price Capping and Peak-Load Pricing in Network Industries', Discussion Paper No. 73, University of Freiburg.

Carlin, A. and Park, R.E. (1970), 'Marginal Cost Pricing of Airport Runway Capacity', *American Economic Review*, 60, pp. 310–19.

Civil Aviation Authority (UK) (2001), *Peak Pricing and Economic Regulation*, Annex to CAA (2001), *Heathrow, Gatwick, Stansted and Manchester Airports Price Caps – 2003-2008: CAA Preliminary Proposals – Consultation Paper*, London: http://www.caa.co.uk.

Civil Aviation Authority (UK) (2006), *Airport Review –Policy Update*, London: http://www.caa.co.uk.

Forsyth, P. (1976), 'The Theory of Pricing of Airport Facilities with Special Reference to London', DPhil thesis, Oxford.

Forsyth, P., Gillen, D., Knorr, A., Mayer, W., Niemeier, H-M. and Starkie, D. (eds) (2004), *The Economic Regulation of Airports: Recent Developments in Australasia, North America and Europe*, Aldershot: Ashgate.

Gillen, D. and Niemeier, H.-M. (2006), 'Airport Economics, Policy and Management: The European Union, Rafael del Pino Foundation', *Comparative Political Economy and Infrastructure Performance: The Case of Airports*, Madrid, 18 and 19 September.

Giulietti, M. and Otero, J. (2002), 'The Timing of Tariff Structure Changes in Regulated Industries: Evidence from England and Wales', *Structural Change and Economic Dynamics*, 13, pp. 71–99.

Giulietti, M. and Waddams Price, C. (2000), *Incentive Regulation and Efficient Pricing: Empirical Evidence*, Research Paper Series No. 00/2 Centre for Management under Regulation Warwick Business School.

Graham, A. (2003), *Managing airports*, 2nd edn, Oxford: Butterworth-Heinemann.

Hague Consulting Group (2000), *Benchmarking Airport Charges 1999*, The Hague: Directorate General of Civil Aviation, Netherlands.

Hendriks, N. and Andrew, D. (2004), 'Airport Regulation in the UK', in Forsyth, P., Gillen, D., Knorr, A., Mayer, O., Niemeier, H.-M. and Starkie, D. (2003), *The Economic Regulation of Airports: Recent Developments in Australasia, North America and Europe*, Aldershot: Ashgate Publishing, pp. 101–1.

Hogan, O. and Starkie, D. (2003), 'Calculating the Short-Run Marginal Inrastructure Costs of Runway Use: An Application to Dublin Airport', in Forsyth, P., Gillen, D., Knorr, A., Mayer, O., Niemeier, H.-M. and Starkie, D. (2003), *The Economic Regulation of Airports: Recent Developments in Australasia, North America and Europe*, Aldershot: Ashgate Publishing, pp. 75–82.

Humphreys, B. (2003), 'Slot Allocation: A Radical Solution', in Boyfield, K. (ed.), *A Role Market Airport Slots*, London: IEA, Readings 56, pp. 94–106.

International Air Transport Association (IATA) (2000), 'Peak/Off-peak Charges', Ansconf Working Paper No. 82, ICAO, Montreal.

Levine, M.E. (1969), 'Landing Fees and the Airport Congestion Problem', *Journal of Law and Economics*, 12, pp. 79–109.

Little, I.M.D. and McLeod, K.M. (1972), 'The New Pricing Policy of the British Airports Authority', *Journal of Transport Economics and Policy*, 6, pp. 101–15.

Morrison, S.A. (1982), 'The Structure of Landing Fees at Uncongested Airports', *Journal of Transport Economics and Policy*, 16, pp. 151–9.

National Economic Research Associates (NERA) (2004) *Study to Assess the Effects of Different Slot Allocation Schemes A Final Report for the European Commission, D G Tren*, London, January

Sherman, R. (1989), *The Regulation of Monopoly*, Cambridge: Cambridge University Press.

Starkie, D. (2003), 'The Economics of Secondary Markets for Airport Slots', in Boyfield, K. (ed.), *A Role Market Airport Slots*, London: IEA, Readings 56, pp. 51–79.

Starkie, D. (2005), 'Making Airport Regulation Less Imperfect', *Journal of Air Transport Management*, 11, pp. 3–8.

Toms, M. (1994), 'Charging for Airports', *Journal of Air Transport Management*, 1, pp. 77–82.

Toms, M. (2003), 'UK-Regulation from the perspective of British Airport Authority', in Forsyth, P., Gillen, D., Knorr, A., Mayer, O., Niemeier, H.-M. and Starkie, D. (2003), *The Economic Regulation of Airports: Recent Developments in Australasia, North America and Europe*, Aldershot: Ashgate Publishing, pp. 117–24.

Weitzman, M. (1974), 'Prices vs Quantities', *Review of Economic Studies*, 41 (4), pp. 477–91.

PART C
Airline Strategies and Competition

Chapter 9

Do Airlines Use Slots Efficiently?

Jörg Bauer

Introduction

Largely driven by the ongoing globalization, world economic activity and foreign direct investments are increasing. However in order to conduct business globally a stable and dependable commercial air transportation network is paramount.

Therefore, the airline sector remains a growth business, regardless of the external shocks (9/11, Iraq, SARS), it has been subjected to in recent years. The two leading manufacturers of commercial aircrafts, Airbus and Boeing, both predict significant global growth rates at about 5 per cent per year, which is well above the long-term growth rate of global GDP (see Figure 9.1).

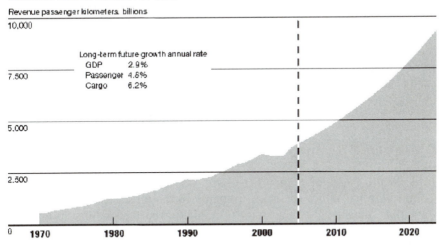

World Air Travel Continues to Grow

Revenue passenger kilometers, billions

Long-term future growth annual rate
GDP 2.9%
Passenger 4.8%
Cargo 6.2%

Figure 9.1   Predicted growth of the air-transport industry

Source   Boeing, 2005, p. 3.

As positive as these outlooks might be, the gap between the demand for worldwide air transportation and the availability of adequate airport facilities/

infrastructure and airspace systems to meet it is widening. The number of congested airports worldwide is growing. At present some 45 airports in the EU are slot coordinated[1] and about 75 per cent of all Lufthansa flights require the allocation of exactly timed take-off or landing rights (slots) at either end or both.

The imposition of piecemeal local government or airport regulations regarding the number of movements at capacity critical airports can however be inefficient in the context of an international air transport system. Schedules, by their global nature, always involve a large variety of airports, often in different countries or continents. Any solution likely to ease the problem in one location must be considered in a global context. Hence active participation of airlines and other stakeholders in the regulative process is mandatory.

The International Air Transport Association (IATA) developed the slot coordination process with a view to achieve a reliable and neutral planning process for the industry (airlines, airports and ATC) and to better manage the increasing number of operational disruptions, e.g., delayed flights which result in significant economic penalties for passengers, airlines and the economy. The slot coordination process has thus far been very successful in maintaining a high degree of coherence and stability in the international air-transport system. It provides airlines with the vital planning reliability needed to make significant investments in new aircrafts as well as the flexibility to adjust the schedules to changing customer demands. The interests of both airports and air-service providers regarding planning reliability and work-load balance are catered for by incentivizing airlines to use at least 80 per cent of each series of slots through the 'use-it-or-lose-it' rule.

IATA's comprehensive set of procedures was used by European law-makers as a basis to the EU regulation (EEC) No. 95/93 of 18.1.1993, which was further amended in 2004 by EU regulation (EC) No. 793/2004 of 21.4.2004. IATA's worldwide industry guidelines and the European regulations are highly complementary with respect to slot allocation and usage. Both ensure fairness, transparency and non-discrimination.

What is efficient Slot Usage?

In its White Paper of September 2001(COM, 2001, p. 36) the European Commission has indicated that in response to the forecasted growth path of the airline industry, it is important to rethink airport operations in order to efficiently use existing airport capacity. They state that slot allocation is one of the instruments that can help to handle capacity constraints (COM, 2001, pp. 37–8).

However '*efficient use of airport capacity*' has not been defined legally thus far. Largely depending on the perspective of the stakeholder in question a

---

1   These airports are termed Level 3 or SCR airports in comparison to Level 2 or SMA airports, where demand is not exceeding capacity and are therefore only schedule facilitated.

variety of interpretations exist. In general two dimensions of efficiency can be distinguished, technical and economic efficiency. The main objective of regulators is to maximize the technical efficiency of the administrative slot allocation system (see Chapter 4.).

Technical Efficiency

Although technical efficiency can be defined in simple terms, a large variety of metrics could be adopted to measure it. In general, the airport capacity is a composite of its runway(s), gate(s), terminal(s) and ramp capacity. As the limiting factor is most often the runway capacity, the allocated slot allows airlines to use the runway in a specific time-window for take-off or landing. Given, that the capacity of an airport is the maximum number of movements per hour,[2] a first key ratio would be the ratio of the number of actual movements per year divided by the number of maximum movements per year.[3]

Other dimensions of technical efficiency include:

- the total sum of maximum take-off weight (MTOW) in relation to the maximum possible sum of MTOW;[4]
- the number of passengers in relation to the maximum possible number of passengers (not taking into account if they are flying long-haul or short-haul);
- the ratio between allocated series of slots for a full season and total slot allocations (the difference being single allocations either for charter or ad hoc flights, e.g., general aviation).

However most of these metrics have the limitation that the maximum values can only be estimated as they heavily depend on the type of aircraft used (e.g., calculating the theoretical maximum number of MTOW by assuming solely 747 operations has very limited explanatory power.) In section 6 of this paper, the applied measure of technical efficiency is the number of used slots (ex post) in relation to the number of slots held by airlines after the slot return deadline. This measure is derived from the 'use-it-or-lose-it' calculation.

---

2   Commonly, airport capacity is defined by a matrix build of movement allowances per 10, 30 and 60 minutes for arrivals, departure and total movements. It can consist of up to nine different values.

3   Although this ratio appears simple to calculate, it has to be notes that the maximum number of movements allowed can vary widely depending on the inclusion of early morning and late night hours where no demand exists.

4   MTOW: maximum take-off weight of the aircraft, a measure on which landing charges are based.

Economic Efficiency

In economic theory, economic efficiency is defined as the sum of all allocative, productive and dynamic efficiencies. Maximization of welfare is defined as the maximization of the sum of consumer and producer surplus. Welfare creation equals the value of the goods and services to consumers as measured by their willingness to pay minus the production costs incurred by producers. Maximum efficiency and welfare is achieved if only the lowest cost producers are producing and the products are consumed by the highest paying consumers (Mankiw, 2004, p. 164ff). Hence with respect to slot usage, economic efficiency is difficult to assess. When regarding slots as a limited resource however, economic efficiency is arguably a more reasonable measure to evaluate their efficient use.

One method to assess the extent of welfare creation is to calculate the sum of all revenues collected from passengers for flights to and from the airport in question. By buying airline tickets, travelers indicate their willingness to pay for the utility derived from a particular service.[5] For example, a medium-haul flight with a high percentage of high-fare paying business passengers in a regional jet could use a particular slot more efficiently than a charter flight with a larger number of passengers on board, but less total revenue (see also NERA, 2004, p. 56 ff). These calculations become even more complex once network-effects and connecting passengers are taken into account. For example a small aircraft feeder flight with many passengers connecting on intercontinental services could be more efficient in terms of network revenues than a point-to-point short haul flight. Within the current legal framework, which only allows non-monetary slot exchanges, the creation of welfare is primarily achieved within each airline's network through the maximization of producer surplus (or profit).

Legal Framework of Slot Usage

In this chapter, the rules and regulations applicable to the usage of slots are reviewed. First, the industry guidelines as defined by IATA are presented. Then, the EU Regulation and the national rules of Germany and Spain are discussed.[6] It is shown that a comprehensive set of regulations exists, which enforces the use of scarce airport capacity in terms of technical efficiency.

---

5   To simplify matters it is assumed that production cost of all airlines are equal. In reality this assumption does not hold, as costs depend on a number of factors, such as the type of network (point-to-point or hub-and-spoke), average stage lengths, etc.

6   The local regulations of Germany and Spain are based on [EU 2004], Article 14 (5) and discussed here as examples. Portugal has also implemented sanctions and the United Kingdom (UK DfT, 2005) is in the consultation process to define such rules.

IATA Worldwide Scheduling Guidelines (WSG) (IATA, 2005)

The IATA procedures as stated in the WSG are intended as a best practice guide for worldwide application. Where states or regions have their own legislation in place, these regulations take precedence.

Guidelines to ensure the efficient use of airport capacity are given much emphasis within the WSG. In general, 'slots at an airport are not route, aircraft or flight number specific and may be changed by an airline from one route, or type of service, to another. It should be noted that any transfer, exchange or use other than that for which the slot was originally allocated, is subject to final confirmation by the relevant coordinator' (IATA, 2005, ch. 6.10).

Grandfather rights, which are the foundation of the current slot allocation mechanism, are an important principle allowing for the efficient use of airport capacity. These entitlements ensure that modified flight schedules are based on past, operationally feasible schedules and thereby guarantee airlines the overall continuity and coherence of their flight operations. In this highly interdependent and complex business environment, continuity has two further advantages:

- it allows passengers (especially the business travelers who constitute 30 per cent of all passengers) to get accustomed to specific airline schedules, thereby reducing passenger search and transaction costs, and
- it provides airports and air traffic control units with a high degree of certainty, which could not be attained if schedules were developed and slots allocated for each season anew.

Airport coordinators play a central role in the monitoring of actual slot usage, termed slot performance monitoring. As stated in the WSG the coordinators 'ensure that scarce resources are not wasted' (ibid., ch. 5.6).

Based on the slot performance monitoring, the use-it-or-lose-it (aka '80/20') rule is applied when calculating the grandfather rights for the next season. Using the slot holdings on the day after the slot return deadline as a basis, airlines are allowed to cancel up to 20 per cent of each series of slots without endangering their grandfather rights (ibid., chs 6.8.1.1 and 6.10.7). The 20 per cent margin was first defined by IATA on practicality grounds and later agreed upon in a political process by EU regulators, who balanced the interests of airports, airlines and other stakeholders.[7]

In order to manage schedule disruptions beyond airlines control (ibid., ch. 6.6.4), cancellations are permissible for the following reasons:

(a) Interruption of the air services of the airline concerned due to unforeseeable and irresistible causes outside the airline's control;
(b) Action intended to affect these services, which prevents the airline from carrying out operations as planned;

---

7   For a certain time, charter carriers were allowed to cancel 30 per cent due to their highly fluctuating business environment.

(c) An interruption of a series of charter air services due to cancellations by tour operators, in particular outside the usual peak period, provided that overall slot usage does not fall below 70 per cent. (Ibid., ch. 6.10.7)

The WSG are also very clear on the holding and returning of slots:

Airlines must not hold slots which they do not intend to operate, transfer or exchange, as this could prevent other airlines from obtaining slots. In this context 'operate' includes participation in a shared operation. ... Airlines that intentionally hold on to slots and return them after the slot return deadline (31st August for winter seasons and 31st January for summer seasons) may be given lower priority by the coordinator for the next equivalent season. (Ibid., ch. 6.10.3)

Besides holding allocated slots for too long, the WSG lists other incidents that would count as intentional misuse of allocated slots:

Airlines must not intentionally operate services at a time significantly different from the allocated slots. Airlines that do this on a regular basis will not be entitled to historical precedence for either the times they operated or for the times allocated. ... Airlines must not operate flights at a coordinated airport without the necessary slots. Any airline that does so will be requested by the coordinator to stop. If the airline concerned continues to operate without slots, the matter will be brought to the attention of the airport's coordination committee, or other suitable committee, which will decide on the action to be taken. (Ibid., ch. 6.10.6)

Further actions that constitute slot abuse include:

- requesting slots without the intention to use them; and
- requesting slots differently from the planned operations (e.g., larger aircraft, full season instead of part season and falsely using year-round or new entrant status) with the intention to gain a higher priority.

EU Regulation (EEC) No. 95/93 amended by Regulation (EC) No. 793/2004

EU Regulation (EEC) No. 95/93 was amended by Regulation (EC) No. 793/2004 in July 2004. Constituting phase 1 of an overall revision of the European slot allocation rules and regulations, the changes were mostly technical in nature. They particularly strengthened the position of the airport coordinators in monitoring the adequate usage of allocated slots.[8]

Similar to the WSG, the EU Regulation puts slot performance monitoring (EU, 2004, Arts 4 (6) and 5 (a)) and grandfather rights (ibid., Art. 9 (2)), which are also based on the use-it-or-lose-it rule (ibid., Art. 8 (2)), at its centre to ensure the best use of existing airport capacity.

---

8   The European coordinators association (EUACA) is developing common guidelines to harmonize the application of the EU regulation at community airports (see also EUACA, 2005).

Airlines are entitled to the same series of slots in the next equivalent scheduling season if they can demonstrate that the series has been operated as cleared by the coordinator, for at least 80 per cent of the time during the scheduling period for which it had been allocated. However, the EU exemptions are much tighter defined than in the WSG, since only:

a) unforeseeable and unavoidable circumstances outside the air carrier's control leading to:
b) – grounding of the aircraft type generally used for the air service in question
c) – closure of an airport or airspace
d) – serious disturbance at the airport concerned, and
e) interruption of air service due to action intended to affect these services which makes it practically and/or technically impossible for the air carrier to carry out operations as planned (…). (Ibid., Art. 10 (4))

justify the non-utilization of slots. However hazardous weather conditions (snow storm, thunderstorms, etc.),which may lead to flight cancellations, are not being considered beyond airline's control.

In addition, the amended slot regulation also includes means and measures to enforce the adequate and orderly use of slots (ibid., Art. 14).

For example, competent Air Traffic Management (ATM) authorities are entitled to reject flight plans (on the day of operation), if the air carrier intends to take-off or land without or with a different slot allocated. This is a very strong measure and has not existed before the amendment. It remains to be seen how ATM will apply its authority, since until now slots were solely a planning tool to ensure that airport capacity is not exceeded. The actual day of operation and the allocation of airway slots[9] has thus far never been taken into consideration.

Also, 'air carriers that repeatedly and intentionally operate air services at times significantly different from allocated slots … or use slots in a significantly different way … shall loose their status as referred in Article 8.2. The coordinator may decide to withdraw from that carrier the slots for the remainder of that season' (ibid., Art. 14 (4)).

The threat to potentially lose valuable grandfather rights, especially within a season when passengers have already booked flights, represents a very strong deterrent to airlines. Coordinators will have to apply this instrument carefully and not without a proper dialogue with the airline concerned. When judging airline behaviour, the three criterions of: (1) intention, (2) repetition and (3) significance need to always be applied. For example if the criteria of intention and repetition are not fulfilled as in the case of an unpunctual flight due to wind conditions or heavy traffic in the approach sector, it should not be counted as a slot misuse even if a significant gap exists between actual arrival time and the allocated slot time.

---

9   Airway slots are allocated by Eurocontrol on the day of operation on a first come first serve basis and must not be confused with airport slots considered in this chapter.

The EU regulation itself does not define the remedies or fines for misused slots, but leaves that to the discretion of member states. Member states 'shall ensure effective, proportionate and dissuasive sanctions to deal with the repeated and intentional operations of air services at times significantly different from slots allocated' (ibid., Art. 14 (5)). Interestingly, the issue of returning a presumably unused series of slots too late in the process is not covered by EU regulations. The respective dates are only mentioned as a basis for the calculation of the use-it-or-lose-it rule.

German Decree to Regulate Airport Slot Coordination (FHKV) (FHKV 2005)

The FHKV defines the implementation and details of the EU Slot Regulation in German law. In terms of regulating slot usage, FHKV states that slots have to be requested for take-off or landings for flights under instrument flight rules (IFR) conditions. IFR flights without an allocated slot are forbidden and slots that are held without the intention to use them have to be returned immediately. Violations of these rules are regarded as administrative offences punishable with fines of up to €50,000 (ibid., § 4 (1)).

Spanish Law 21/2003 (Safety/Security Law) supplementing Royal Decree 15/2001

Article 49 of Spanish law 21/2003 defines the offences in relation to the coordination and use of slots and corresponding fines are listed in article 55. The failure to return unused allocated slots by the deadlines established by EU Regulation (EC) No. 793/2004 may be fined with €6,000 to €90,000 for each series of slots. The operation without a slot may be sanctioned with a fine of €3,000 to €12,000 per flight. Airlines that operate intentionally and regularly at times different to those allocated may be fined with €3,000 to €30,000 per flight operated off-slot. Interestingly in the Spanish regulation the criterion 'significantly different time' from Article 14 of the EU slot regulation is missing. Additionally, airlines which undertake slot transfers not permitted by Regulation (EC) No. 793/2004 may be sanctioned with a fine ranging from €18,000 to €60,000 for each series of slots.

Slot Inefficiency – Misuse or Operational Necessity?

When reviewing current slot allocation procedures, industry observers often mention four practices, which allegedly waste scarce airport capacity thereby reducing the efficiency of the overall system. The practices most often mentioned are (a) overbidding, (b) late hand-back, (c) seasonality and (d) no-shows.[10] In

---

10   The latest consultation was initiated by ECAC (European Civil Aviation Conference), requesting information on 'Slot allocation procedures to promote efficient use of airport capacity' from their member states.

the following it is shown that these alleged malpractices are often not cases of fraudulent intent but rather directly result from the uncertain business environment in which airlines operate. Hence aiming for significant efficiency increases in the use of airport capacity while retaining a competitive free enterprise system in the aviation industry appears rather questionable. Striving for full-capacity usage of airports is as unrealistic as expecting 100 per cent full employment in a free market economy.

Overbidding

At present airport coordinators, being a government agency reporting to the Department of Transportation of their respective country, are mostly financed by the airlines residing in that country, in proportion to the share of annual aircraft movements.[11] Hence under the current administrative system, airlines are not paying directly for each new slot request, which, according to economic theory, could under conditions of uncertainty lead to the rational strategy of 'overbidding'.

In line with transaction theory, it is rational to request an additional slot as long as the (transaction) fee for requesting the slot and any additional costs (e.g., opportunity cost of reserving aircraft capacity or additional planning expenses) is lower than the expected monetary return attached to the request. This expected return is defined here as the probability to receive the slot multiplied by the utility value of using it. Currently for the most commercially interesting and thus constraint airports, not only the transaction fee but also the probability of allocation are close to zero.

Hence, the observed behaviour that slots are requested and, in the unlikely event of being allocated, are cancelled later at the slot return deadline should not be perceived as *gambling*. On the contrary, the main cause for the observation of late hand-backs (see Chapter 5.b) is the inherent uncertainty in the scheduling of the operations and allocation of slots in the airline industry as shown below.

Compared to other usage rights such as radio frequencies or offshore fishing rights (see NERA, 2004, p. 69ff.), slots are not independent from each other and are only one of many restrictions related to aircraft scheduling. Figure 9.2 depicts the required number of precisely timed slot series, depending on aircraft utilization and the number of capacity constraint airports within one aircraft rotation. The strongest links exist between departure slots and the corresponding arrival slots (the time in between is a mean value and called block time, which is largely predetermined by aircraft speed). Another interdependence exists between the arrival and the departure slot at one airport, particularly at spoke stations where the aircraft is '*turned around*'. This relationship is defined by the minimum ground time that is needed for deboarding, refueling and boarding.

---

11   In some countries, airports are financing the activities of airport coordinators.

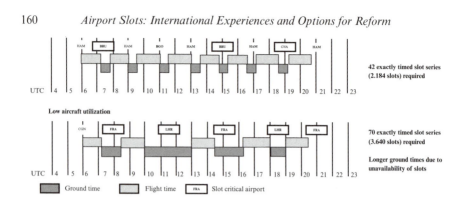

Figure 9.2   Interdependencies of slots due to aircraft rotations

Since aircraft are very costly assets (an A320 costs around €60m), it is mandatory that they are utilized to the fullest extend, even in a slot constraint environment. The strategies applied to mitigate potential slot allocation risks can be compared to hedging against these risks. The following exemplifies possible situations at four airports A, B, C, D:

a) slot is allocated at airport A but not available at B for the planned flight at a feasible time → slot at airport A is returned;
b) slot is allocated at A but only available at B at a different time → slot at A could be adjusted but due to the changes slots at C and D have to be returned;
c) slot is requested at very critical airport B and another one at C as a fall-back if slots at B cannot be allocated → when successful in B, Slot at C[12] is returned;
d) slots for a new rotation are requested at A, B, C, D → all slots are allocated except for C, therefore the rotation becomes uneconomical due to low utilization of aircraft. All slots at A, B, D are returned.

At the time these slot requests are submitted to the coordinators, the intent to use the slots is clearly present. But since submission occurs on average up to five months prior the start of the actual season, requests remain subject to change at any time until the season begins. As it is usually very difficult for an outsider to determine whether or not slot requests are genuine, it is understandable that suspicions of wrong-doing might arise.

Late Hand-back

The late return of slots is the second most mentioned root cause of inefficiencies in the current slot allocation system. It is argued that late returns might keep other airlines from using the returned slots, since the time-frame might be too

---

12   Contigency airports are most often not slot critical (or even schedule facilitated) compared to the primary target airport. Thus the return of slots usually does not prevent other airlines from operating.

short to start a new operation. Although the potential outcome seems obvious, the difficulty with this alleged malpractice is a clear definition of when the return of a slot has to be termed 'late'.

Simply applying the slot return deadline (SRD) as the prime criteria, as proposed by some stake-holders, is insufficient as it would violate the main purpose of the SRD. The following example will illustrate this argument: an airline holds a perfect departure slot at Frankfurt (FRA), however is still on the waiting list to get the corresponding arrival slot in Barcelona (BCN). Further assume that the airline was offered a slot an hour later, which it accepted in order to possibly swap this slot for the required slot with another airline. The airline must hold-on to both slots (in FRA and BCN) until after the slot return deadline in order to realize any potential improvements. If the airline succeeds (because some other airline returned its slot) it can operate the planned flight. If no or only an insufficient improvement is achieved, the airline has to return both slots a couple of weeks *after* the deadline even so it always intended to operate the flight. This rational behaviour is confused by some industry observers with *slot blocking*. The underlying problem, created by the interdependencies of slots, can currently not be solved in practice. A satisfactory solution would require a simultaneous and instant slot allocation mechanism for all airports and ailrines globally compared to the sequential process today.

Other reasons for late returns such as changes in demand, traffic rights, slots and fleet, (see also Figure 9.3) are often the result of operational and/or commercial uncertainties.

1) Changes in air travel demand: due to cyclical or short term changes in demand, capacity adjustments (in terms of aircrafts in use) may become necessary. This leads to the cancellation of the economically weakest flights. Exogenous effects triggering this process could be for example high fuel prices or natural disasters or terrorist attacks, which spread insecurity amongst the traveling public.

2) Traffic rights: negotiations over traffic rights are being undertaken by the Ministries of Transport of each country and can be very complex. The closing date and results is often subject to large uncertainty. If the negotiations cannot be closed in due time before the start of the season, the slots already allocated must be returned at short notice (e.g., the increase of the traffic rights FRA/ MUC – Moscow in the summer 2002 did not realize on time).

3) Slots: flights need corresponding departure and arrival slots for their operation. Frequently slot problems, for which scheduling solutions must be found, still exist after the closure of the slot conference. These solutions can consist of (1) slot exchanges with other airlines, (2) adjustments of the flight schedule and (3) later improvements by the coordinator due to schedule changes of other airlines. Depending upon the degree of risk, alternative scenarios are planned and slots requested. Whatever the situation after the SRD might be, some slots will be returned eventually.

4) Fleet: the economic performance of many operations depends on the availability of the right-sized type of aircraft (e.g., 50-seater or 150-seater). Implementation risks consequently arise either from the general availability

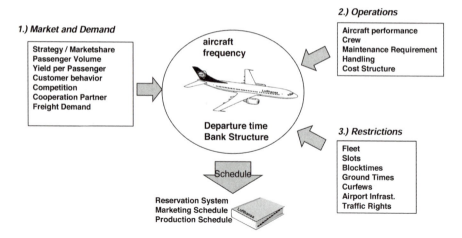

Figure 9.3   Internal and external factors influence network planning and scheduling

of the aircraft (e.g., due to late delivery from the manufacturer), or more specifically if the required aircraft is available on the desired day at the right time, which depends on the planned aircraft rotations. In order to maximise the use of expensive aircraft capacity weaker routes have at times to be canceled on short notice. The unused slots resulting from these cancellations have to be returned immediately since the intent to use them has vanished.

Seasonality

In general, seasonality can be differentiated by the length of the cyclical variation. Disregarding long-term economic cycles, the longest seasonal cycle is the division of the annual travel pattern into northern summer and winter seasons. Within a season, demand for air transportation varies widely by weekdays and on public holidays. Figure 9.4 shows that the majority of the summer season is largely characterised by bank holidays and vacation periods with depressed demand in between.

The 'use-it-or-lose-it' rule (aka '80/20' rule) was developed in order to manage these demand fluctuations, save on costs, lessen the negative environmental impact of operating empty planes and balance the needs of both airlines and airports alike.

The 'use-it-or-loose-it' rule provides airlines with the flexibility to cancel up to 20 per cent of a series of slots (six cancellations during summer and four cancellations during winter per series) without endangering their grandfather rights. Conversely airports are guaranteed the landing charges of at least 80 per cent of the slots operated per series, since airlines will not risk the loss of their grandfather rights. This almost perfectly balanced trade-off would however be inadequately distorted if the usage margin would be increased to 90 per

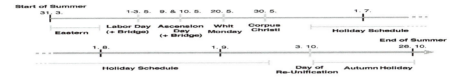

Figure 9.4   Public holidays and vacation periods affect most of the summer season and influence demand

cent as proposed by some stake holders in a bid to use runway capacity more efficiently.[13]

The impact of seasonality on a flight schedule can not be simply assesed by assuming that 20 per cent of all flights are cancelled. Even during public holidays airlines have skeleton schedules in place, which consist of at least one morning and one evening service. However due to weak load factors flights around mid-day are often cancelled first. Hub airports are less affected by seasonality, since intercontinental services and the general feeder function warrant a more stable demand pattern. In Chapter 6, the actual slot usage over the season will be shown with emphasis on the reduced demand on public holidays.

Besides seasonality, allowing cancellations through the 'use-it-or-lose-it' rule has yet another important purpose. It also serves as a buffer for ad hoc cancellations (e.g., engine break-down, rotational delays etc.), that cannot be avoided completely during airline operations. About 40 per cent of the allowance of cancellations is not used for seasonality, but serves as a protection against short term cancellations which would otherwise put grandfather rights at risk.

No-shows

Although there are difficulties in defining and measuring, when a slot is returned too late, the assessment of *no-shows* is easily achieved. A no-show is a slot that has been allocated but is not actually used by the airline. It can only be recognized ex post. For example, a slot was allocated for a departure on 23 April at 10 a.m. On 24 April, the airport records show that no departure flight was operated at 10 a.m. the previous day and that this slot had not been returned. The reason can be either negligence, intentional wrong-doing (both punishable offences under EU-slot regulations) or short-term cancellations on the day of operations by flight operations centres due to technical reasons or reasons beyond airline control. Occasionally simply not enough time is left to cancel a slot within the two hour period prior a planned departure. The focus is on overall operational stability and safeguarding of passengers already en-route from interruptions.

The magnitude of no-shows is overall marginal. Table 6.1 shows the number of allocated and operated slots at the end of the summer 2002 season at various European airports. The number of no-shows is most often around 0.5 percentage

---

13   It should be noted that landing charges are mostly calculated on a full cost base, hence airports recover their cost without regard to the actual usage of airport capacity.

points (the difference between columns V and VI). Hence reducing the no-show rate to zero will not solve the capacity shortage in Europe. Another malpractice warranting a mention are *go-shows*. Go-shows are flights operated without the required slots (e.g., Copenhagen in Table 9.1). Although less critised by airports as they earn revenue from it, go-shows have the potential to aggravate punctuality problems at an airport and thereby damaging the operational quality of other airlines.

Slot Usage in Practice

Slot Usage at Various European Airports

Table 9.1 shows the slot usage for a number of capacity-constraint European airports during the summer 2002 season.

At most airports the slot usage exceeded 90 per cent of the slots allocated after the slot return deadline. At some airports (Barcelona, Stockholm and Stansted) traffic even increased after the SRD. Only Milan Malpensa experienced a slot usage below 85 per cent, probably as a result of the restructuring of Alitalia at that time.

The high percentage of actually used slots as compared to slot holdings at the SRD is even more surprising, given that the SRD for the 2002 summer was on 31 January 2002, just five month after the tragic events in New York, which triggered the biggest crisis in the history of air-transportation. In the context of efficiency two possible explanations exist. Either the offsetting passenger demands at major hub airports (like FRA, LHR and CDG) stabilized the overall demand for air-traffic at those airports. As a result these airports were not as effected by cancellations as other minor airports (that are often not slot coordinated). Or alternatively hub airlines were simply safeguarding their grandfather rights, by using all allocated slots and thus not endangering their long-term hub operation by a short-term crises. Hence airport capacity during this period was used 'efficiently', while airlines had to bear the cost of operating temporarily uneconomical flights.[14]

Slot Usage per day

Summer Season: Figures 9.5 and Figure 9.6 show the slot usage per calendar day at a hub airport (Frankfurt) and a regional airport (Düsseldorf) during the summer season of 2004.

In the case of Frankfurt (see Figure 9.5) a very high slot usage of all airlines is observable. On average more than 95 per cent of slots allocated at the SRD. Within the season, the first half is lower in demand with four distinctive public

---

14   Later in the process, the EU slot regulation was changed to include a moratorium of the 80/20 rule in order to enable European carriers to cancel uneconomical flights and save cost.

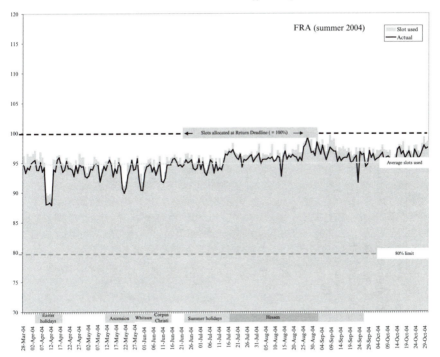

Figure 9.5 Slot usage at Frankfurt (FRA) during summer 2004

holidays (Easter, Ascension Day, Whit Monday and Corpus Christi). During this period demand dropped to 90 per cent. In August and September however slot usage peaked at about 100 per cent. Hence it is questionable if additional services for a full season could have been allocated by the coordinator.

At Düsseldorf (see Figure 9.6), the slot usage per day shows a more prominent variance. On average, slot usage is about 89 per cent, which is still substantially above the theoretical 80 per cent, showing that the 'use-it-or-lose-it' rule is not fully being applied by airlines. The drop in demand during the four public holidays in the first half of the season is more pronounced than at FRA. Merely 70 per cent of flights operated during Eastern and about 80 per cent during the other holidays.

During the stronger part of the season, slot usage widely varied between 89 per cent and 100 per cent. It is questionable if additional demand could have been accommodated. Since the seasonal demand pattern applies to most potential applicants for additional slots, interest in the available slots is arguably fairly limited.

The greater variation in Düsseldorf is a result of its function as a regional airport within the air-transport system. Due to the lower percentage of connecting passengers and the stronger reliance on business travelers, the passenger flows are less balanced. If there is a public holiday in Düsseldorf, for example, the trough in local demand can hardly be offset by passengers connecting from less affected

Table 9.1 Slot usage at capacity constraint airports in Europe (Summer 2002 in the aftermath of 9/11)

	I Total number of slots initially requested	II Total number of slots initially allocated (IATA Conference)	III Total number of slots allocated at the slot return date	IV Total number of slots allocated at the start of the season	V Total number of slots allocated at the end of the season	VI Total number of operations realized at the end of the season
Barcelona	1.31	1.14	1	1.03	1.02	1.02
Copenhagen	1.12	1.08	1	0.88	0.9	0.91
Düsseldorf	1.12	1.05	1	0.99	0.89	0.88
Frankfurt	1.06	0.99	1	0.96	0.95	0.95
London Gatwick	1.41	1.05	1	0.95	0.92	0.9
London Heathrow	1.22	1.01	1	0.99	0.99	0.98
London Stansted	1.24	1.19	1	1	1.03	1.03
Madrid	1.24	1.15	1	1.02	0.99	0.98
Milan Linate	1.5	1	1	0.98	0.87	0.85
Milan Malpensa	1.1	1.05	1	0.92	0.83	0.82
Munich	1.07	1.07	1	0.94	0.91	0.93
Paris	1.19	1.05	1	0.94	0.89	0.86
Amsterdam	1.09	1.02	1	0.97	0.92	0.91
Stockholm	1.32	1.22	1	1	1.01	1.01
Vienna	1.07	1.1	1	0.98	0.96	0.88

Source: own calculations, based on ACI, 2004.

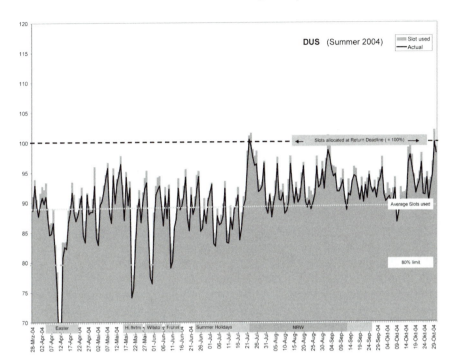

Figure 9.6   Slot usage at Düsseldorf (DUS) during summer 2004

regions in Europe or the world. The same holds true for low demand at the other end of the Düsseldorf flights (e.g., summer holidays in Italy or France).

Winter season   Figures 9.7 and 9.8 show the slot usage per calendar day at Frankfurt as a hub and Düsseldorf as a regional airport during the winter season 2003/2004.

In the case of Frankfurt (see Figure 9.7), a very high slot usage by all airlines is observable. On average more than 91 per cent of the slots allocated at the slot return deadline were also used. Besides the Christmas holidays and New Year, during which slot usage per day dropped to 60 per cent[15] due to the lack in travel demand, slot usage averaged about 95 per cent. If traffic volume had not been severely depressed due to the Iraq War and SARS during 2003, average slot usage figures would certainly have been higher.

At Düsseldorf slot usage per day shows similarly to in the summer season, a significantly larger variance. On average, the slot usage is about 86 per cent, which is still substantially above the theoretical 80 per cent if the 'use-it-or-lose-it' rule was fully applied.

---

15   It has to be noted that total usage by day is calculated differently than slot usage per series as required for the 'use-it-or-lose-it' rule.

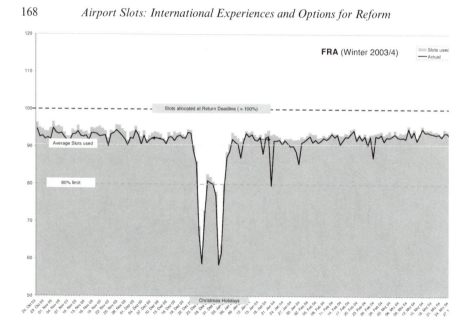

Figure 9.7   Slot usage at Frankfurt (FRA) during winter 2003/2004

Stronger seasonality in comparison to Frankfurt can also be seen in the period before the Christmas holiday. November and the first half of December are traditionally strong business months with a large share of business travel, hence slot usage was about 90 per cent with peaks of up to 100 per cent. However contrary to Frankfurt, traffic in Düsseldorf takes longer to climb back to levels prior to the Christmas holidays during January and even February.

During Christmas and New Year, travel demand and thus slot usage drops to as low as 55 per cent. During these times only the most basic core schedule is flown by airlines in order to provide travel options in line with reduced passenger numbers.

Slot Usage by Time of Day

The preceding section has shown that slot usage over the full season is averaging about 95 per cent at Frankfurt and about 90 per cent at Düsseldorf. This may suggest that there is a potential to accommodate roughly 5 per cent more flights thereby increasing the efficient use of constraint airport capacity. Figure 9.9 depicts the distribution of the returned slots by the time of day in Frankfurt in relation to the slot holdings after the SRD by the time of day, in this case all Fridays during summer season 2004.

The majority of slots returned after the slot return deadline were located at times of little economic interest (e.g., 5 a.m. and after 10 p.m.), since passengers dislike very early or late flights. During popular times, such as the mid-morning and early evening, slots are used to almost 100 per cent. During mid-day and in

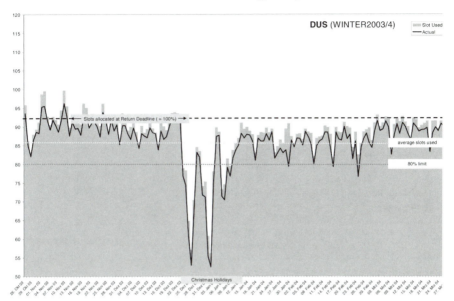

Figure 9.8   Slot usage at Düsseldorf (DUS) during winter 2003/2004

the afternoon, some slots were returned, hence decreasing usage to around 95 per cent.

Figure 9.6 shows a similar graph for the regional airport in Düsseldorf. In general, the same pattern can be observed as in Frankfurt. But similar to the analysis of the slot usage over the full season, Düsseldorf as a regional airport shows a greater variance in slot usage by time of day. The slots are mostly returned in the early morning and evening and to a lesser degree during the mid-day and afternoon hours.

Since Düsseldorf is not a banking hub, such as Frankfurt which attracts early morning intercontinental arrivals from North America and Asia, the economically useful times in the morning begin later than in Frankfurt. Due to the night restrictions in Düsseldorf, the evening hours show a higher slot usage than in Frankfurt. The greater variance can also be attributed to the higher turn-over in slots at Düsseldorf. The airline environment is less stable there, with more airlines ceasing operations and others starting services. This higher fluctuation leads to more complex schedule changes and a higher degree of unsolvable slot problems. Where these scheduling problems cannot be resolved by the airline, flights cannot operate and slots are returned later in the process.

Conclusions

In conclusion the presented paper has shown that airlines generally use their allocated slots efficiently. The primary driver of efficient slot usage is the requirement to provide passengers with the services they demand under conditions

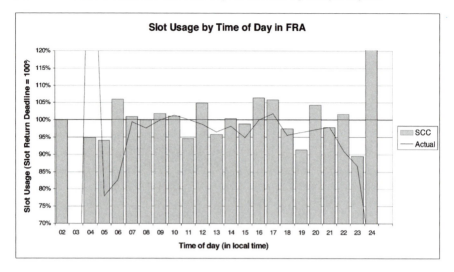

Figure 9.9   Slot usage by time of day in Frankfurt, Friday in summer 2004

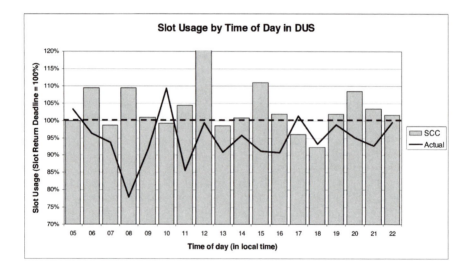

Figure 9.10 Slot usage by time of day in Düsseldorf, Friday in summer 2004

that allow airlines an acceptable profit from the significant investments made in equipment and routes. Furthermore an extensive regulatory framework exists, which aims at ensuring that airlines make the best use of the scarce infrastructure resources that many airports are today. However, slots at economically uninteresting times (either on a Christmas holiday or in the very early morning hours) are used to a lesser extent, as the demand elasticity for these slot are very low. Also rather than operating an unprofitable flight, airlines will often return any unfavorable slots if there is no probability of improvement. In this respect, the 'use-it-or-lose-it' provision serves both airlines and airports alike. It allows airlines enough flexibility to manage seasonal demand fluctuations and guarantees airports that their infrastructure will be used most of the time.

References

ACI (2004), *Study on the Use of Airport Capacity*, Airports Council International, Brussels: European Region.

Boeing (2005), *Current Market Outlook, The Boeing Company*, Seattle.: Boeing

COM (2001), *European Transport Policy for 2010: Time to Decide*, White Paper, COM (2001) 370 of 12.9.2001, Brussels: EU Commission.

EU (2004), *Regulation (EC) No. 793/2004 of the European Parliament and of the Council of 21 April 2004 amending Council Regulation (EEC) No 95/93 on Common Rules for the Allocation of Slots at Community Airports*, combined text, Brussels

EUACA (2005), *EUACA Community-Wide Guidelines on Slot Allocation and Schedule Coordination*, 2005, Brussels.

FHKV (2005), *Verordnung über die Durchführung der Flughafenkoordination*, latest change 6.6.2005, Berlin

IATA (2005), *World-Wide Scheduling Guidelines*, 10th edn, Montreal: IATA.

Mankiw, N.G. von (2004), *Grundzüge der Volkswirtschaftslehre*, 3, Stuttgart: Auflage.

NERA (2004), *Study to Assess the Effects of Different Slot Allocation Schemes – A Final Report for the European Commission*, January, London: NERA

UK DfT (2005), *Consultation on the Introduction of a Sanctioning Mechanism for Misuse of Slots in Line with EC Regulation 95/93 as Amended by 793/2004*, London: UK Department for Transportation.

Chapter 10

Slots and Competition Policy: Theory and International Practice

David Gillen and William G. Morrison

Introduction

In countries outside of the US, the coordination of air movements at many airports occurs through the allocation of take-off and landing 'slots'.[1] The majority of large airports in both the European Union and the United States are hub airports and the IATA slot allocation process has relied generally on grandfather rights.[2] The distribution of slots exhibits a similar pattern in these congested airports, with incumbent legacy carriers holding the lion's share of air movements. Typically, the second largest competitor at a legacy carrier's hub will have a (significantly) smaller share of the total number of slots available.

Competition issues related to airport access arise because of the implied barriers to entry given the allocation and scarcity of take-off or landing slots, as well as other needed infrastructure such as gates and terminal systems, often in combination with bilateral air services agreements between countries. However, such barriers can also exist in domestic markets between domestic carriers; US, Canada and Australia provide excellent examples. This form of barrier to entry can therefore be associated with proposed mergers or acquisitions in the airline industry given the potential consolidation of control over scarce slots at key airports. The aviation environment and different regulatory policies unique to various regions in the world have resulted in differing approaches and consequently different degrees of access to essential facilities across countries and across airports.

In this chapter, our purpose is to provide a relatively simple theoretical approach to framing the competition issues arising from airport slots from an

---

1   National authorities define slots in different ways ways, the EC definition of a slot as stated in the European Council Regulation (EEC) No. 95/93, is the most comprehensive. The European Council defines a slot to mean 'the permission given by a coordinator in accordance with this Regulation to use the full range of airport infrastructure necessary to operate an air service at a coordinated airport on a specific date and time for the purpose of landing or take-off as allocated by a coordinator in accordance with this Regulation'. See Chapter 4 in this volume.

2   This has also occurred at a small number of congested airports in the US (such as Chicago O'Hare), where slots are employed.

economic perspective along with an international survey of how these issues have been addressed by regulatory authorities. We begin with a simple linear demand approach to understanding the competition and associated welfare implications of airline market structure when airports are congested and capacity constrained. The next section provides a general comparison of policy approaches to airport access internationally, while what follows focuses on how regulatory authorities have defined air transport markets for the purposes of competition policy. We then provide a detailed international comparison of regulatory responses to mergers or acquisitions as they relate to slots and some concluding remarks are offered.

Competition and Economic Efficiency at Capacity-Constrained Airports

The links between potential detrimental effects of market power and slots primarily exist via the notion that slots create a barrier to entry under the administered slot system – a barrier that provides a substantial wealth transfer for those airlines lucky enough to have grandfathered control of the majority of slots. There are several implications that might then follow from the implementation of a market-based (traded) slot allocation system:

a) allowing slots to be traded could increase competition by either lowering the barriers to entry or by making strategic barriers to entry more expensive to create, in effect creating an opportunity cost to slot holders;
b) consumers would be the beneficiaries of a more competitive environment brought about by slot trading. That is, slot trading would improve consumer welfare.

To explore these ideas, consider the theoretical benchmark of a profit-maximizing airline with (exclusive) monopoly control over slots at a capacity constrained airport, as illustrated in Figure 10.1. The linear demand curve in figure 10.1 represents the relationship between q – the number of flight tickets demanded by travellers at any given price (p).[3] Now suppose that a single airline has monopoly control of this market and that it has constant operating costs of €x per passenger ticket. In this context, airport capacity is defined as the maximum number of passenger tickets available per period, which depends on both the number of slots available and average aircraft size (seat capacity per slot).

Suppose that there is a capacity constraint at the airport and that the maximum number of tickets available is less than the amount the incumbent airline would wish to sell to maximize its profits (q^*). Figure 10.1 represents this capacity constraint as $q_1 < q^*$ such that airline will charge price p_1 and will make operating profits (before any fixed costs) of $(p_1-x)q_1$, represented by area $[p_1BDx]$.

---

3   In order to abstract from the complexities of price discrimination, the diagram shows demand for a 'representative' or average flight ticket at any given average ticket price.

Further we can infer that the airline's maximum willingness-to-pay for airport slots is equal to its net economic profits: $\pi = (p_1 - x)q_1 - F$, where F = fixed costs. Finally, consumer welfare as measured by consumer surplus is the area $[Ap_1B]$.

If the airline is realizing cost savings due to density economies – operating at lower average costs than two or more airlines could (with smaller market shares) in this market, then the forced allocation of some slots to another airline will result in lost efficiency (higher average costs) on the production side. On the question of consumer welfare, we need some assumption concerning the mode of competition that would occur in the event of entry by one or more competitors. However, in the case shown in Figure 10.1, the airport's capacity constraint lies to the left of the monopoly output. Therefore, the lowest possible price even with an increase in the number of competitors is p_1 with no change in price or consumer surplus. Increasing the number of competitors does not improve either productive efficiency or consumer welfare.[4]

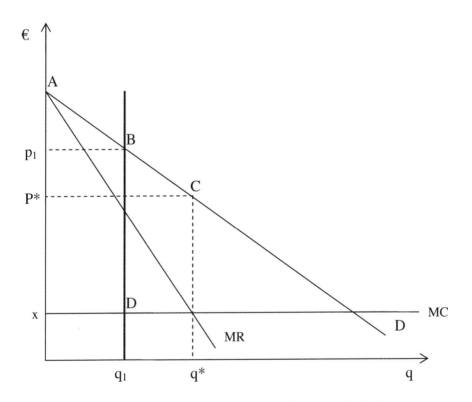

Figure 10.1 Monopoly control of flights at a capacity-constrained airport

---

4   Hazledine (2006) shows in a price discriminating oligopoly, average prices (fares) do not change with a change in the number of competitors but the amount of price discrimination does.

Figure 10. 2 illustrates the situation in which the airport's capacity constraint lies to the right of a monopoly airline's profit maximizing quantity and price. In this case, it is possible that a monopoly airline could charge p* and q* by using smaller aircraft while still using (and hence retaining) all the available slots at the airport. The question is how should such actions be interpreted? It is likely if not certain that smaller aircraft will be part of an optimal fleet mix for an incumbent hub-and-spoke carrier. However it is also possible that the incumbent could strategically 'hoard' slots under the administered system by continuing to use them with the sole purpose of preventing entry.[5] Thus, we must ask: 'What would the naturally occurring market structure be in the absence of any barriers to entry created by the administrative allocation of slots'? If the market structure is a natural monopoly (no strategic hoarding), then any initial reallocation of slots to potential competitors will eventually be transferred back to the monopoly incumbent if subsequent trading is allowed, with the price and output remaining at p* and q* respectively.

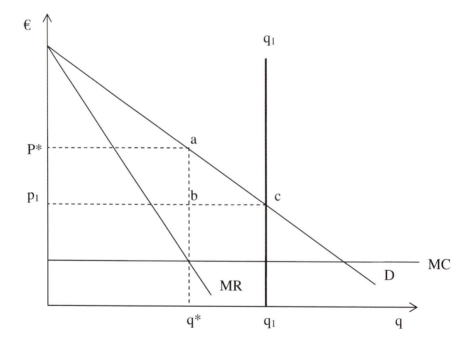

Figure 10.2 Duopoly price competition constrained by limited capacity at the airport

---

5   In reality airlines will also have slots 'babysat' by non-competing airlines (e.g., charter carriers).

As an alternative to the natural monopoly case, if entry by one or more competitors were to result in oligopoly competition, the greatest reduction in prices possible is p1 < p*, with a commensurate increase in consumer surplus.[6]

Therefore in the scenario illustrated in Figure 10.2, an efficiency case could be made for a market based system of slots with secondary trading, since the worst that can happen is that a natural monopolist would be forced to pay for valuable slots that the airline previously received free of charge. There would be a lump-sum reduction in the profits of the incumbent but no change in price or output. If there is a sustainable oligopoly market structure following a reallocation of slots, then the introduction of competition will reduce prices to air travellers and increase consumer surplus by as much as area [abc] in Figure 10.2.

There is some evidence to suggest that natural market structures are more concentrated that the structures created by administered slots. As shown in Table 10.1 below, the small set of capacity-constrained airports in the US at which slots are used and traded, concentration of control over access to the airport has increased since slot trading was introduced. As suggested by the theoretical discussion above, increased concentration may not be due to the fact slots are traded but rather to the value of slots in a hub and spoke network where capacity is constrained. The value of a hub and spoke network can increase with the size (and configuration) of the network due to both demand side and cost side influences. Networks are more valuable as the number of destinations increases, adding a destination increased possible connections by 2n, therefore larger networks find adding a destination of more value than smaller networks; larger carriers will therefore have a higher willingness to pay for slots in some cases. Note that consumer welfare may rise even with increased concentration.[7]

We can also see in Table 10.2 that the market share of at hub airports is higher at non-slot constrained airports and that slot constrained airports airlines market share is significantly less than their share of slots. Therefore, competition policy enforcement must exert care in distinguishing the basis of higher fares at slot airports; fares may be higher due to slot scarcity and not due to slot trading. Higher fares are not necessarily reflecting market power due to slot control but the value of scare slots to users (passengers).

On the question of whether airlines are likely to engage in slot hoarding, it should be noted that slots are actually part of the more general problem of airline dominance in an airport. In particular slots are only one form of several potential bottlenecks in passenger throughput at an airport, all of which could

---

6   By referring rather generally to 'oligopoly' competition, we are obfuscating some complex theoretical issues concerning the existence of equilibria in models such as the Bertrand (price competition) model. However the general point is that if some form of stable oligopoly competition is possible, then the result will be a price lower than P* with a commensurate increase in consumer surplus. The price p1 merely represents the greatest possible decline in price as a result of competition, given the airport's capacity constraint.

7   For a discussion of economies of spatial scope see Basso and Jara-Diaz (2005).

be used strategically by a dominant carrier. If competing airlines gain access to an airport through a reallocation of slots, the incumbent may still benefit from significant entry barriers through the location and number of ticketing counters it controls and its ownership and/or control of gates and bridges.

Table 10.1   **Percentage of slots held by carrier at US airports where slots are traded**

Airport	Holding entity	1986	1991	1996	1999
O'Hare	American and United	66	83	87	84
	Other established airlines	28	13	9	10
	Financial Institutions	0	3	2	3
	Postderegulation airlines	6	1	1	3
Kennedy	Shawmut Bank, American, and Delta[a]	43	60	75	84
	Other established airlines	49	18	13	14
	Financial Institutions	0	19	6	1
	Postderegulation airlines	9	3	7	1
LaGuardia	American, Delta, and US Airways	27	43	64	70
	Other established airlines	58	39	14	14
	Financial Institutions	0	7	20	10
	Postderegulation airlines	15	12	2	6
Reagan Washington	American, Delta, and US Airways	25	43	59	65
	Other established airlines	58	42	20	18
	Financial Institutions	0	7	19	14
	Postderegulation airlines	17	8	3	3

[a] In 1999, First Security National Bank replaced Shawmut Bank. First Security National Bank holds those slots pursuant to a trust as security for a loan to TWA, which uses some and leases others.
Source: GAO 1999.

Figure 10.2 can also be used to position our understanding of the relationship between slots and consolidation (mergers and acquisitions) in air travel markets.

Suppose we use p_1 and q_1 as our starting point and assume that currently there is an oligopoly market structure made up of three or more competitors. A merger between two airlines that were previously competitors at the airport will consolidate their joint control of slots and increase their market power relative to other airlines. Now the precise mode of competition becomes important in determining the competitive effects of the merger. In the extreme, if the market structure is Bertrand price competition between perfect substitutes, then consolidation will have no effect on the equilibrium price or output or on consumer surplus.

There could however be efficiency gains on the cost side. If however, as is likely, there is product differentiation then a merger is likely to raise the market price and reduce consumer surplus, leaving the calculation of net welfare (production and consumption) uncertain. Here it is worth mentioning that competition authorities do not necessarily view their role as one of maximizing economic efficiency, but rather as either maintaining or encouraging competition. The subtle difference here is an implied focus on competition rather than the *effects* of competition.

Table 10.2   Airlines' market share* at hub airports in 1999/2000

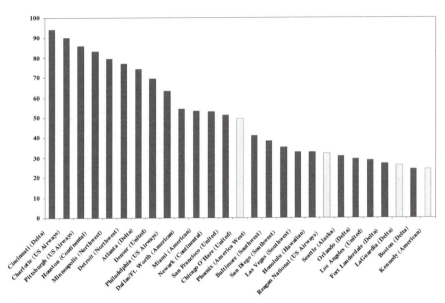

■ Airports coordinated on the basis of first-come-first-served.
□ Airports coordinated on the basis of slot trading.

Source: General Accounting Office.

Regulation and access to airport capacity

Access to an airport is impeded by either capacity constraints, when demand for slots exceeds supply, and/or by bilateral air service agreements between countries. In turn, the availability of landing and take-off slots depends in part on the regulatory policies unique to a region or state. Currently, the initial slot allocation processes are governed by administrative rules, the IATA slot allocation process, as opposed to market forces, wherein incumbent carriers are protected from competition under the provision of grandfathering rights. It is precisely because of a failure to have a market for slots that competition policy cannot be used in the important role of protecting competition in downstream markets.

European Commission

The availability of slots at airports, especially post-merger or alliance where the degree of competition declines ensures new carriers are not prevented from competing in the market. The current EC Regulation No. 793/2004 with respect to slot allocation addresses the issue of slot mobility in the context of alliances or acquisitions[8] and grants power to the slot coordinator to prohibit slot transfers or exchanges if these transactions are not in concordance with the requirements of the regulation. Most of the accepted mergers/alliances by the Commission must abide to very strict provisions. For example, if a new entrant cannot acquire the needed slots for its routes at the airport in question, the parties to the merger/alliance are required to return a certain amount of slots back to the slot pool.[9]

United States

In contrast to Europe, the slot policy in the US has predominantly sprung from federal regulations meant to address domestic concerns, with very little influence from international governmental bodies. In United States there were only four slot-constrained airports, namely JFK, LaGuardia (NY), Chicago O'Hare, and Washington National, otherwise known as the high density restrained airports (HDR). The primary initial allocation of slots at these airports, based on grandfather rights, severely discourages competition and the 80 per cent rule (also known as the use-it-or-lose-it rule) clearly favours incumbent carriers. However, post 1986, the Federal Aviation Administration (FAA) introduced a secondary market in slots allowing for monetary exchanges. Secondary trading has the potential to smooth out the distribution of slots amongst the different users of the air space (i.e., new, small, medium sized carriers and incumbent carriers).

Access to international routes is dictated by bilateral aviation agreements, one example being Bermuda II between US and UK. These agreements govern the number of carriers allowed to serve an international route, as well as the number of routes and sometimes frequency of flights permissible.

Some studies have pointed out there seems to be a link between the national flag carrier and a major national airport where the former would receive preferential treatment with respect to slots and gates at the expense of its competitors. In United States however his particular concern cannot arise since it does not have a single national flag carrier. In addition the Department of Justice clams the administrative rules for slot allocation in relation to international flights are in accordance with competitive guidelines.

----

8   http://europa.eu.int/eur-lex/pri/en/oj/dat/2004/l_138/l_13820040430en00500060.pdf.

9   This occurred in the KLM/Alitalia merger procedure.

United Kingdom

Three of London's five airports have been designated as highly congested airports or otherwise known as fully co-ordinated airports (Heathrow, Gatwick and Stansted). The excess demand for slots in combination with the slot allocation mechanism known as grandfathering rights has exacerbated difficulties for airlines in acquiring slots. The initial slot allocation mechanism is in concordance with IATA and EC rules where incumbent airlines are reallocated the same slots every season conditional on the carrier employing them 80 per cent of the time. New entrants have precedence over 50 per cent of the slots in the slot pool (the slot pool consists of slots that have not been reallocated to incumbent airlines or returned to the airport by carriers). Secondary trading with monetary exchanges is allowed and supervised by a slot coordinator. In a scenario such as this first mover advantages play an essential role in gaining a strong hold on competitive advantage. The significantly larger and more diversified portfolio of slots of incumbent carriers endows them with greater bargaining power in the slot trading market.

Although the access to slots at London's congested airports should not be underestimated evidence suggests there are some slots which become available each year either from capacity expansion or improved efficiency of capacity utilization. In 1998 and 1999 3 per cent of the slots at Heathrow were termed 'new slots' available for reallocation, while at Gatwick the figure increased to 7 and 6 per cent respectively. Furthermore airport substitutability has been substantial between Heathrow and Gatwick where as much as 67 city-pair routes with a high proportion of point-to-point business or leisure travellers are in competition with services from Gatwick.

Australia

In Australia there are no constraints placed on existing carriers with respect to services expansion on any domestic routes. Similarly, international routes are governed by liberal air service agreements with regards to multiple designation. In addition the largest markets in Australia either enjoy unlimited capacity, as in the case of New Zealand, or excess capacity (i.e., Singapore and the UK).

Access to airport capacity is governed by the provisions listed under Part IIIA of the Trade Practices Act 1974.[10] The provisions in this Act imply access at airports in Australia is negotiated between the airlines seeking admittance and airport service providers. The Commission is authorized to settle access disputes between the relevant parties and its determinations are enforceable in the Federal Court. The two major domestic airlines, Ansett and Australian carriers, 'own or operate domestic terminals under long term leases with the airport operator'.[11]

---

10   http://www.oecd.org/dataoecd/34/6/2489057.pdf http://www.comlaw.gov.au/comlaw/Legislation/ActCompilation1.nsf/0/03475BFBA45FA9B2CA257007001781 A3/$file/TradePrac1974Vol1_WD02.pdf.

11   Competition policy, p. 25.

The provisions developed in Part 13 of the Airports Act 1996 include detailed rules concerning airport service utilization.[12] There are four main types of schemes outlined in the Act, namely i) category exclusion scheme, ii) slot allocation scheme, iii) movement limitation scheme, and iv) schemes not covered by i), ii) and iii).

The category exclusion scheme means the airport disallows entry for a specified group of aircraft movements with the exception of emergencies or safety matters. The slot allocation scheme is supervised by a slot coordinator appointed by the minister. The movement limitation scheme is a cap placed on the total amount of slots enabled to service aircraft movements at an airport. The third category does not specify a particular technique, and neither does it prohibit the simultaneous application of two or more of the above schemes.

The only slot constrained airport in Australia is Sydney Kingsford Smith (KSA) airport. The government has capped airport entry at 80 movements per hour. Although a market for slots or leasing of slots has not yet been proposed slot swaps have been finalized. In order to prevent major carriers from exchanging regional flights in peak periods for interstate or international services, the Sydney Airport Slot Management Scheme allows slot swaps (for the slots stated above only) within the last thirty minutes of their original scheduled time.[13] Respecting the IATA Scheduling guidelines, once slots have been allocated to incumbent airlines based on historic rights, the remaining slots are returned to the pool. New entrants have precedence over 50 per cent of the slots in the slot pool.[14]

Germany[15]

In Germany there are six slot constrained airports: Frankfurt, Dusseldorf, Berlin, Stuttgart, Koln/Bonn and Munchen. The slot allocation methodology is in line with the recent EU 95/93 regulation and IATA scheduling procedures guidelines. As in most European countries, all slot transactions (i.e., slot swaps, slot acquisitions) are supervised by a slot coordinator or otherwise known in Frankfurt as the 'Flugplankoordinator der Bundesrepublik Deutschland'. The initial allocation of slots, which has not yet been disputed by the German Transport policy, is governed by grandfathering rights or use-it or lose-it rule. New entrants are given precedence to 50 per cent of the pool slots. The slot coordinator has the power to call back slots from air carriers if there are sufficient grounds to establish improper use of slots by incumbent airlines. Reasons for withdrawal of slots include hoarding of slots, utilization of slots at 'low loads' in order to pass the 80 per cent usage requirement.

As a response to the proposed merger between Lufthansa and Eurowings airlines, in 2001 the Bundeskartellamt cleared the merger on the condition the

---

12   http://www.comlaw.gov.au/ComLaw/Legislation/ActCompilation1.nsf/0/
BA147558EE841557CA256F7100502D82/$file/Airports1996.pdf.

13   http://www.ministers.dotars.gov.au/ja/releases/2001/january/a7_2001.htm.

14   http://www.comlaw.gov.au/comlaw/Legislation/ActCompilation1.nsf/0/
03475BFBA45FA9B2CA257007001781A3/$file/TradePrac1974Vol1_WD02.pdf.

15   http://www.bundeskartellamt.de/wEnglisch/CompetitionAct/CompAct.shtml.

two parties will relinquish three take-off and three landing slots to competitors if the latter cannot obtain these assets from other sources. The subsequent legal dispute, as decided by the Dusseldorf Higher Regional Court, resulted in favour of the Bundeskartellamt. However, the merged parties were not forced to comply with the exact time of the slots requested by competitors. In addition, under the merger rules a new competitor by the name of European Air Express (EAE) was protected from competition for six consecutive flight plan periods.[16] During this period, the merger parties were prohibited to increase flight frequency or seating capacity on the routes serviced by EAE.[17]

From 1 May 2004, in accordance with the EC Regulation 1/2003, the Bundeskartellamt is obliged to submit to Art. 81 or Art. 82 of the EC Treaty. Based on Art. 3 (2) of Regulation 1/2003 of the EC Treaty takes precedence over national regulations (see appendix).

Canada

The two airports in Canada subject to slot controls are Lester B. Pearson International Airport (LBPIA) and Vancouver International Airport. The former is slot coordinated for the entire operational day with coordination extended to included gate availability and terminal capacity while the latter is coordinated for gate and terminal facilities only.[18] At LBPIA, slot restrictions do not exceed 78 movements per hour or 20 movements in any one quarter hour. The slot assignment process is very similar to the EC slot regulation and is in concordance with the IATA scheduling Procedures Guide, wherein incumbent airlines benefit from grandfathering rights. The process is prioritized in this order: historic rights, slot pool allocation and a use-it or lose-it rule. The IATA guidelines are critized by the Canadian Commission Bureau as not adequately addressing 'the competition concerns that would emerge in a dominant carrier scenario'.[19] All slot transactions, such as slot swaps, are supervised by an impartial slot coordinator.

Under the Competition Act airlines are prohibited from:

pre-empting airport facilities or services that are required by another air carrier for the operation of its business, with the object of withholding the airport facilities or services from a market; to the extent not governed by regulations respecting take-off and landing slots made under any other Act, pre-empting take-off or landing slots that are required by another air carrier for the operation of its business, with the object of withholding the take-off or landing slots from a market; altering its schedules, networks, or infrastructure for the purpose of disciplining or eliminating a competitor or impeding or preventing a competitor's entry into, or expansion in, a market.[20]

---

16   http://www.oecd.org/dataoecd/34/6/2489057.pdf.

17   Lufthansa/Eurowings cleared subject to obligations, 24 September 2001 Bundeskartellamt.

18   Competition policy, p. 40.

19   http://www.competitionbureau.gc.ca/internet/index.cfm?itemID=1886 &lg=e#3-1.

20   http://strategis.ic.gc.ca/pics/ct/airline.pdf

The methodology used by the Competition Bureau to establish whether pre-emption of take-off and landing slots occurs consists of determining if the carrier's avoidable costs of offering the service are covered by its revenues.

Defining Markets for the Purposes of Competition Policy

To determine competition issues in the airline industry most countries have focused on 'city-pair' markets although the relevant market definition varies from one case to the next. There are situations where a city hosts a multiple airport system which raises the question of airport substitutability and the consequent possible improvement in competition. Across countries, demand side substitutability has lead to distinctions amongst customers/passengers, such that competition issues are analyzed within different markets.

European Commission

In the past, the EC has used the point-of-origin and point-of-destination (O&D) approach in defining the relevant market.[21] From the customer's point of view, different combinations of point-of-origin and point-of-destination would represent separate markets. The process of establishing the degree of competition in the specified market consists of an analysis of all possible substitutes to the particular serviced route, which might include other modes of transport such as road, train or sea, as well as indirect flights or overlapping flights between the two airports.

Time-sensitive vs. non time-sensitive passengers   The EC has been known to distinguish between 'time-sensitive' and 'non time-sensitive' passengers. One altering factor between the two groups is that 'time-sensitive' customers would be willing to pay a premium to guarantee their seat on a flight at a particular time. It is important to recognize the underlying characteristic of 'time-sensitive' consumers could overlap with both business and leisure passengers. While one might intuitively attribute 'time-sensitiveness' to business clientele, leisure passengers could exhibit the same preferences.

Direct vs. indirect flights   The passenger type is exceptionally important in determining whether direct and indirect flights offered by independent competitors are substitutable (ECA, 2004). Indirect flights may be considered inferior to 'time-sensitive' passengers owing to the duration of the flight and connecting time. In the case of United Airlines/US Airways proposed merger, the Commission concluded that indirect flights for long-haul routes could be considered substitutes to non-stop flights if they are marketed on the computer reservation system (CRS) as connecting flights for the same city-pair and if the difference in travel time

---

21   http://www.europa.eu.int/comm/competition/antitrust/cases/decisions/38477/en.pdf.

between the two flights is small.[22] For the short-haul routes, in the example of Lufthansa/AuA, the Commission acknowledged that the number of non time-sensitive passengers willing to travel by indirect flights (due to a price gap between the two services) would be trivial, and dismissed indirect flights as substitutes on this basis. In contrast, within the Air France and KLM[23] merger procedure, the EC concluded non-stop and one-stop services should be considered substitutes for the two city-pairs (Bordeaux-Amsterdam and Marseille-Amsterdam). The unique differentiating factor in this merger was the relatively higher frequency of indirect flights offered by Air France which made the service more attractive to time-sensitive passengers (i.e., possibility of a one-day return trip).

Airport substitutability Another rather complex issue of market analysis is airport substitutability.[24] Cities with multiple airports systems such as London, home to five airports (Heathrow, Gatwick, Stansted, Luton and London City), raise issues as to the degree of airport substitutability and depend highly on passengers' value of the final destination in that particular city. In the case of British Airways and SN Brussels Airlines alliance, the Commission concluded 'services from/to London Stansted do not exercise sufficient constraint on the other services from/ to London. Similarly, services from/to Charleroi and Antwerp might not exercise sufficient constraint on services from/to Brussels National'.[25] In the British Midland/Lufthansa/SAS merger, the EC regarded alternate airports as poor substitutes to business travelers. In contrast to these findings, in the case of BA/ Iberia/GB Airways, the Commission found secondary airports met the criteria of substitutes for service from UK to Spain, albeit restricted to non time-sensitive passengers only. The same distinction was made for service within the Spanish territory, namely Sevilla/Jerez and Valencia/Alicante, where alternate airports could adequately constrain service for non time-sensitive passengers but not for time-sensitive consumers.

The EC approach to charter flights and low-cost carriers According to the EC, charter flights although closely related to regular flights, are not considered adequate substitutes to scheduled services, at least for time-sensitive passengers. Illustrative of the rationale behind this conclusion are the cases of KLM/Alitalia, BA/Iberia/GB Airways, BA/City Flyer and Air Canada/Canadian Airlines.

With respect to low-cost flights, then similarities between these and traditional flights depends on a number of factors, including not only the airfare but also the frequency of service. In the case of BA/Ibera/GB Airways, the EC found some degree of substitutability between low-cost and traditional services on

---

22   Ibid.

23   http://europa.eu.int/comm/competition/mergers/cases/decisions/m3280_en.pdf.

24   Airport substitutability is significant in measuring the degree of competition in that market.

25   http://www.europa.eu.int/comm/competition/antitrust/cases/decisions/38477/ en.pdf.

routes between London and Spain, but only in the context of non time-sensitive passengers.

United States

In the US, the Department of Justice (DOJ) concluded the relevant geographic market in the context of airline competition is at least as small as a city-pair. Drawing on merger rulings, secondary airports within multi-airport cities should not necessarily be considered substitutes to primary airports. In the case of the proposed American Airlines/ British Airlines (AA/BA) merger, analysis conducted by DOJ concluded that Gatwick airport is not an adequate alternative to Heathrow airport owing to the clear preference exhibited by time-sensitive passengers for the latter destination.

In relation to passenger classes, DOJ has distinguished between time-sensitive and non time-sensitive customers, as well as business and leisure travelers all in the context of either direct or indirect flights. As business passengers are known to place a high value on their time, a distinction has been made between direct and indirect service markets for time-sensitive travelers only. Using AA/BAA as an example, a comparison between non-stop and indirect flights to London has shown a large portion of the business passengers (or those who have a high willingness to pay) use direct services which has lead DOJ to conclude non-stop and stop products form separate markets.

Other factors relevant to competition issues include alternative modes of transport, such as high speed train service. Although in United States, high speed train service does not exist, evidence suggests pricing of short haul routes may be constrained by inter-modal competition (i.e., rail service or driving).

United Kingdom

The relevant market definition with respect to mergers and alliances is developed on a case by case basis in UK. In some situations inter-modal competition is available such as the high-speed rail link between London to Paris and Brussels. The Commission has distinguished between time-sensitive and non-time sensitive travelers, mostly although not exclusively, respectively known as business and leisure passengers. Business passengers are estimated to experience a relatively greater opportunity cost with respect to time. Distinctions have also been drawn between direct and indirect flights within the context of short haul routes and denser routes between UK and US, where the market for time-sensitive passengers has been limited to direct services only. Whereas on long routes, such as London-Australia, and on routes where direct flights are relatively infrequent, the Commission concluded connecting services/products are adequate substitutes for all travelers. These differentiating characteristics with respect to markets are extremely important in approximating the effects of a merger or alliance on competition.

New Zealand and Australia

In New Zealand, the process carried out by the Competition Commission in determining the relevant market definition includes the following five components:[26]

1) a product dimension analysis of the goods or services supplied and purchased;
2) the customer characteristics in the market;
3) the geographic area of the supplied goods and services;
4) production or distribution chain analysis;
e) temporal dimension of the market.

In contrast to New Zealand and other countries, the Australian Commission has only considered international markets wherein the relevant market definition has been based on broad geographic areas rather than on the common city-pair or airport-pair approach. The reason behind this selection is related to the large distance between Australia and other destinations which strengthens the substitutability between direct and indirect flights. To illustrate the feasibility of alternative routing a passenger travel directly from Sydney to Bangkok or he/she could take a route via Kuala Lumpur or Singapore with only an insignificant increase in travel time. Furthermore, the pricing of indirect routes constrains significantly the prices of direct flights. But if a situation arises where the Commission must address domestic service routes in Australia a city-pair definition of the market would be appropriate. In addition, the Commission has not distinguished between different types of customer classes. In the case of the BA/Qantas Joint Services Agreement (JSA) the Commission found the characteristics of different customer types to fall in different segments of the same market rather than constituting different markets. To exemplify this point, the difference in price of a business class ticket on some carriers and an economy class ticket on others from Australia to London is not significant.

Slots and Consolidation in the Airline Industry

In the past, although historic rights have not been disputed, competition analysis of mergers and acquisitions proposals have lead national authorities to impose regulatory restrictions on the slot holdings of the relevant parties on a case by case basis.[27]

---

26   http://www.comcom.govt.nz//PublicRegisters/ContentFiles/Documents/5110. pdf.

27   In some situations however, incumbent airlines possess a relatively larger amount of slots acquired through grandfathering rights if compared to the slot holdings of two merged carriers, yet slot restrictions on competition grounds are imposed only in the latter case.

European Commission

Similar to the provisions outlined by the UK Competition Commission, the EC has sought to encourage competition by requiring the merged parties to divest slots or/and placing a cap on flight frequency. The EC has also recognized in some cases a reduction of slot holdings by itself does not necessarily lead to new entry and Sabena/Swissair serves as an example. Complementary measures to slot reduction imposed by the EC include access to code-share and frequent flyer programmes as well as interlining. Moreover, if the merger or alliance results in a substantial overlap on routes with few competitors and/or potential entrants the best response may be to disallow the merger.

The effects of alliances and mergers between airlines could be quite complex in the aviation industry. Consequences of market structure in this industry have to be analyzed at the network level as well as the 'route' (O&D) level. For example, although a merger would reduce competition for a given route it may increase competition between major network alliances such as Skyteam and Star.

United States

Some of the structural constraints placed on the parties of the merger/alliance include divestiture of routes or slots to encourage entry by new carriers. When American airlines acquired US-London Airways the authority obligated American airlines to divest certain US-London routes. The Department of Justice encouraged new entry by supporting the divestiture of a substantial amount of Heathrow slots in the case of AA/BA merger. However, in some cases new entry is highly unlikely. An apposite example is Dallas-Ft Worth-Heathrow where the route is served by two alliance partners with hubs at both ends. In special cases such as these, the DOJ has advocated the removal of antitrust immunity for the route operations under a given alliance.

United Kingdom

The UK Competition Commission, upon analysis of merger and alliance proposals, has advocated either a return of the proportion of slots or a flight frequency cap. For example, the CC's response to the proposed alliance between BA and AA included a reduction in the proportion of slots, especially on the routes where competition declined the most, such as BA's hub at London and AA's major hubs at Dallas, Miami and Chicago. Furthermore, in situations with excess demand for slots the CC recommended to allocate slots to carriers based on the potential of providing effective competition to the BA/AA alliance. The CC also recognized preventive measures for the abuse of market power are essential on hub-to-hub routes and proposed a cap on BA/AA's flight frequency for a six month period if new entry occurs.

In response to the BA/City Flyer Express merger and the worrisome increase in hub strength at Gatwick, the CC advocated a slot cap for a period of five years, in order to ensure new carriers have adequate access to peak period slots.

Australia

In 1995, the Commission approved a Joint Services Agreement between Qantas and British Airways for a period of five years conditional on certain airfare restrictions. The airlines were not allowed to increase airfares above the increase in consumer price index. Interestingly enough developments in aircraft efficiency, technology as well as markets have rendered these restrictions unnecessary. A look at airfares five years after the agreement shows a 40 per cent decline relative to authorized level imposed by the Commission.

Germany

In Germany, the Bundeskartellamt has applied similar remedies in an attempt to address competition concerns. For the merger between Eurowings and Lufthansa, the Bundeskartellamt imposed a return of slots by the merged entity if new entrants cannot obtain access at the relevant airports through other processes.

Summary and Conclusion

In the majority of cases, administrative rules preclude the use of competition laws in protecting downstream competition. There are active markets in the US and UK, and the DG of Competition for the EU is eager to have secondary trading beginning in the EU, however slot co-ordinators are much less enthusiastic about any such prospect.

Although there may be opportunity for carriers to raise their rivals costs using slot strategies, existing provisions in the competition laws of the US, Canada, Australia, UK and EU dealing with abuse of dominant position, mergers and acquisitions appear adequate to protect against the anti-competitive behaviour and the potential effects of slot concentration.

The use of regulations such as slot caps – limiting the number of slots a given carrier or alliance might control at a given airport as a substitute for competition policy is inferior and assumes that slot concentration is per se bad and that any agent (carrier) with a large proportion of slots will abuse their market power. On the contrary, as Starkie (2004) shows, slot concentration may well be welfare enhancing.

There is also the prospect of third parties participating in slot markets. As Table 10.1 shows banks have been owners of slots particularly after bankruptcy proceedings. More recently we have seen Jet Blue acquire 4 slots at Chicago O'Hare despite not serving the airport and this airport not fitting the airline's current business model. In a similar vein Ryanair's bid for Aer Lingus if successful will leave it with slots at Heathrow. These events raise interesting competition questions concerning whether LCCs could purchase and use slots strategically to affect the competitive position of a rival airline. As a hypothetical example, it could be argued that Easyjet would benefit strategically by purchasing slots at LHR and leasing them to Lufthansa or some other competitor of BA, thereby weakening BA's general competitive position. In this example, 'refusal to supply' might be applicable as a competition policy issue if Easyjet refused to lease back the purchased slots to BA. However there is a significant variance across competition laws governing refusal to supply.

A final conjecture concerns the implications of slot allocations in a world of increasingly 'open skies'. The continued trend towards freer trade in international air travel will promote consolidation of airlines and therefore an increased concentration of slots. Since slot concentration by itself may not be a bad thing, this trend strengthens the argument for market-based slot allocation mechanisms in order that competition policy rules that can be consistently applied across airports and eventually across jurisdictions.

References

Basso, L.J. and Jara-Díaz, S.R. (2005), 'Calculation of Economies of Spatial Scope from Transport Cost Functions with Aggregate Output', *Journal of Transport Economics and Policy*, 39, pp. 25–52.

Civil Aviation Authority (2001), *The Implementation of Secondary Slot Trading*, November: http://www.caa.co.uk/docs/5/ergdocs/slotsnov01.pdf.

Commerce Commission (2003), *Air New Zealand/Qantas Final Determination*, 23 October: http://www.comcom.govt.nz//PublicRegisters/ContentFiles/Documents/5110.pdf.

Commerce Commission (2002), *Part IV Inquiry into Airfield Activities at Auckland, Wellington, and Christchurch International Airport*, 1 August: http://www.comcom. govt.nz//RegulatoryControl/Airports/ContentFiles/Documents/Executive_Summary0. PDF.

Competition Bureau (1999), Letter from Konrad von Finckenstein to the Honourable David Collonette, 22 October: http://www.competitionbureau.gc.ca/internet/index. cfm?itemID=1885&lg=e#1-1.

Czerny, A.I. and Tegner, H. (2002), 'Secondary Markets for Runway Capacity', essay prepared for the second seminar of the IMPRINT-EUROPE Thematic Network: *Implementing Reform on Transport Pricing: Identifying Mode-Specific Issues*, Brussels, 14/15 May: http://www.imprint-eu.org/public/Papers/IMPRINT_Czerny&Tegner. pdf.

Dresner, M., Windle, R. and Yao, Y. (2002), 'Airport Barriers to Entry in the US', *Journal of Transport Economics and Policy*, 36, Part 2, September, pp. 389–405.

European Competition Authorities (2004), *Mergers and Alliances in Civil Aviation*: http:// europa.eu.int/comm/competition/publications/eca/report.pdf.

FAA/OST Task Force (1999), *Airport Business Practices and their Impact on Airline Competition*, October: http://ostpxweb.dot.gov/aviation/domav/airports.pdf.

Federal Aviation Administration (2004), *Airport Competition Plans*, September: http:// www.faa.gov/arp/financial/aip/guidance/PGL0408AtchB.pdf.

GAO (1997), *International Aviation: Competition Issues in the US-UK Market*, June: http:// www.gao.gov/archive/1997/rc97103t.pdf.

Hazledine, T. (2006), 'Price Discrimination in Cournot-Nash oligopoly', *Economics Letters* 93, pp. 413–20.

House of Commons of Canada, Bill C-27, Second Session, Thirty-seventh Parliament, 51–52 Elizabeth II, 2002–2003: http://www.parl.gc.ca/LEGISINFO/index.asp?

Lang=E&Chamber=C&StartList=2&EndList=200&Session=11&Type=0&Scope=1&query=3366&List=toc-1.

New Zealand, The Commerce Act 1986: http://r0.unctad.org/en/subsites/cpolicy/Laws/New-zealand.pdf.

Office of Legislative Drafting and Publishing, *Trade Practices Act 1974*, Act No. 51 of 1974 as amended: http://www.comlaw.gov.au/comlaw/Legislation/ActCompilation1. nsf/0/03475BFBA45FA9B2CA257007001781A3/$file/TradePrac1974Vol1_WD02. pdf.

Office of Legislative Drafting, Airport Act 1996, Act No. 42 of 1996 as amended: 1 March (2004): http://www.comlaw.gov.au/comlaw/Legislation/ActCompilation1.nsf/0/BA147 558EE841557CA256F7100502D82/$file/Airports1996.pdf.

Organization of Economic Co-operation and Development (2000) *Airline Mergers and Alliances*, 1 February: http://www.oecd.org/dataoecd/1/15/2379233.pdf.

Organization of Economic Co-operation and Development (1998), *Competition Policy and International Airport Services*, 14 May: http://www.oecd.org/dataoecd/34/52/1920318. pdf.

Starkie, D. (2004), 'In Defence of Slot Concentration at Network Hubs: A Note' (mimeo).

Bel, G. and Fageda, X. (2006), 'Airport Management and Airline Competition in OECD Countries', in Fichert, F., Haucap, J. and Rommel, K. (eds), *Competition Policy in Network Industry*, Berlin: LIT-Verlag, pp. 81–9.

Chapter 11

The Dilemma of Slot Concentration at Network Hubs

David Starkie

Introduction

A number of the worlds' major airports have a high proportion of their capacity
utilized by a single airline, or alliance of airlines. At major US airports, for
example, it is common for the leading carrier to account for three-quarters or
more of the flights; at European airports the proportions are generally smaller
but often exceed 50 per cent (see Table 11.1). This has lead to concerns that such
high levels of capacity utilization by few airlines ('slot concentration') will impact
adversely on competition and prices and, in turn, has inclined policy makers
and regulatory bodies towards pro-active competition measures. In the US, for
example, the General Accounting Office (GAO, 1996) judged that the buy-sell
market in slots established at certain US airports in 1985, had failed because
it was argued to have done little to reduce slot concentration and introduce
competition into the market. It went on to recommend that slots should be redistributed. In Europe, the current slot allocation regulation (Regulation 95/93,
Article 10) requires that at slot coordinated airports[1] preference be given to new
entrant carriers in the allocation of unused, returned or new slots (that together
constitute the slot pool)[2]; there is also a reluctance on the part of the European
Commission to accept unrestricted slot trading because of the fear that it will
reinforce dominant positions; and approval of proposed alliances between carriers
has often been accompanied by conditions requiring slot divestiture.

In this chapter, I suggest that these positions require re-examination; it develops
a number of issues raised in Starkie (2003a). The effects of slot concentration are
complex and there are a number of economic efficiency considerations that suggest
that a (high) degree of concentration of slots at an airport might be beneficial.
On the other hand, there are the understandable concerns that concentration can
reduce competition leading to a reduction in welfare. This tension between the
advantages and disadvantages of slot concentration at airports has long been

---

1   A listing of coordinated airports will be found at the appendix, 'Characteristics
of Slot Coordinated Airports 2002–2004'.

2   The European Commission has commented that 'new entrants [should be
given] the possibility to grow [at coordinated airports] so as to reach a critical mass to be
considered as viable competitors against the "big" carriers'.

Table 11.1   Concentration at selected European and US network hubs

		Airport		Proportion of flights by three leading carriers						One firm HHI
Europe	Munich	MUC	Lufthansa	64.0	dba	7.8	Air France	1.8		4,096
	Paris	ORY	Air France	61.3	Iberia	7.5	easyJet	5.6		3,758
	Frankfurt	FRA	Lufthansa	60.1	BA	3.2	Condor	2.3		3,612
	Milan	MXP	Alitalia	59.1	Lufthansa	7.1	Air France	3.6		3,493
	Paris	CDG	Air France	57.9	Lufthansa	5.2	BA	4.0		3,352
	Madrid	MAD	Iberia	56.7	Spanair	13.7	Air Europa	6.8		3,215
	Amsterdam	AMS	KLM	50.7	Transavia	5.3	easyJet	3.5		2,570
	London	LHR	BA	42.3	BMI	11.5	Lufthansa	4.5		1,789
US	Houston	IAH	Continental	85.7	AA	2.4	Delta	2.1		7,344
	Charlotte	CLT	US Airways	84.0	Delta	3.5	AA	3.2		7,056
	Dallas	DFW	AA	83.2	Delta	2.3	United	2.0		6,922
	Detroit	DTW	NW	80.3	Delta	3.3	AA	2.5		6,448
	Atlanta	ALT	Delta	74.8	Air Tran	13.4	AA	2.3		5,595
	Newark	EWR	Continental	66.2	AA	7.2	Delta	4.1		4,382
	Denver	DEN	United	54.9	Frontier	18.0	Great Lakes	8.6		3,014
	Chicago	ORD	United	47.5	AA	38.7	Delta	1.9		2,256

Source: adapted from *Airline Business*, June 2005.

recognized and commented upon in previous chapters.[3] Where this chapter is different is that it incorporates recent developments in the theory of congestion pricing that, it is suggested, have an important bearing on the balance of the argument. It also argues that higher fares at slot constrained airports do not constitute a *prime facie* case that a dominant position is being exploited and that there are a number of possible reasons apart from market power why we might expect higher air ticket prices at major hubs.

The next two sections outline the economic factors favouring slot concentration. This is followed by a consideration of the possible effects of concentration on consumer prices. A concluding section considers the implications for policy.

Networks and Natural Monopoly

Characteristics of Hub Networks

Most major and, as a consequence, often congested airports are hubs for an interconnecting network of air services and it is well known that (two-way) service networks connected through hubs produce powerful connectivity externalities (Oum and Tretheway, 1995). In the case of two-way networks, as the number of nodes attached to the network increases so do the number of possible connections and thus its value to the consumer: the number of possible connections increases by the square of the number of connections (minus one). There is, therefore, the potential for a significant scaling effect so that the consumer benefits from an exponential growth in the number of possible destinations s/he can fly to.[4]

In addition to increasing connectivity between nodes, the other important effect of a hub network is to increases the traffic density across the connecting flows and this also benefits the passenger by allowing for increased service frequencies. On the supply-side, an increase in traffic density is also associated with empirically established economies (Caves et al., 1984). Increasing the flow of traffic on a particular route for example, allows fixed station costs associated with the node to be spread over more passengers thereby reducing unit costs.

On the other hand, there are offsetting factors. For those passengers who, in the absence of a hub network, would have had the opportunity of a direct service to some of their preferred destinations (and no services to others), there is a price to be paid; longer journey times (but probably less schedule delay)

---

3   Borenstein (1989, p. 362), for example, comments: 'Though the link between airport dominance and high fares seems clear, a welfare analysis of increased airport concentration must also include the benefits that may accrue from hub operations ... These possible benefits ... should be weighed against the higher prices that seem likely to result'.

4   This is qualified when the nodes differ in their traffic generation potential. As Krugman (1999) has pointed out (referenced in Miller, 2003), connections are usually made to the larger cities (nodes) first so that in practice, as more nodes are connected to the hub, the scaling effect decreases.

and the inconvenience of having to change flights. Overall, however, there is a presumption that both passengers and airlines benefit from a business model based on hub networks. The issue here, however, is not the size of the overall net benefit; it is whether the characteristics of the hub network impart a degree of natural monopoly so that single firm delivery is more efficient. If this is the case, *ceteris paribus*, greater slot concentration at the hub is also more efficient.

Hub Networks as Natural Monopolies

The connectivity externality associated with hub networks is not in itself an argument in favour of service delivery by the single firm. To gain the connectivity externality the passenger does not require each connecting service to be operated by a single airline or alliance; the connectivity of the network could in theory be established with each hub connecting service operated by a different airline with slots allocated accordingly.[5] Equally, there is no obvious gain if the station costs of all the network *nodes* feeding through a hub are the responsibility of the single airline because there are no evident economies from combining these station costs. Consequently, the traffic density effect at route level would be captured equally well if each separate route in the hub network was operated by a different carrier.

Nevertheless, there is a demand and supply side gain to be achieved by single firm delivery across the whole of the network servicing a hub. In the case of the former, the benefit is from an improved quality of service. This improvement in service quality is the result of two factors. First, passengers have a preference for making *on*line connections rather than interlining between carriers, so that as the proportion of possible online connections grows, service quality increases (Bailey and Liu, 1995). Second, service quality also improves if connecting times between services are minimized. Arguably, such times are more likely to be minimized for the maximum number of passengers, the larger the proportion of services provided by the singular airline. This is because of information asymmetries: to minimize total connection times in a non-concentrated network requires each airline to be fully informed of the flow dynamics on other parts of the network that it does not operate and also to act cooperatively on the basis of such information.[6]

On the supply side, until recently empirical studies had suggested that there were no evident production economies from increasing the size of networks so

---

5   This is facilitated by the IATA interlining system.

6   An interesting example of this problem and attempts to address it, has arisen in relation to the forthcoming opening of Terminal 5 at London Heathrow airport. The movement of British Airways into T5 will release space in two of the existing four terminals leading to an opportunity for a general reallocation of airlines to terminals. To assist in this task, the airport operator, BAA, commissioned work to assess what allocation would minimize connect times, taking into account flows of interlining passengers and the preferences of alliance partners to be co-located (Economics-Plus, 1995). The analysis was based on the interlining data then available but inevitably failed to capture the dynamics of interlining.

that there were no efficiency gains if a singular airline providing services across all the routes in a particular hub-based network.[7] However, the more recent empirical literature on airline economics has indicated that, in addition to economies of traffic density, there are economies of network scale possibly associated with the better utilization of aircraft fleets (Ng and Seabright, 2001; Brueckner and Zhang, 2001).[8] In addition, Jara-Diaz et al. (2001) have suggested that previous approaches to measuring the effects of scale were confounded because an increase in network size adds more products (points) to the network; measurement, therefore, should take into account economies of spatial scope. Initial analysis on these lines using Canadian airline data suggests that there might be pronounced spatial scope economies, especially for the expansion of smaller networks (Basso and Jara-Diaz, 2005). With finite slot capacity at a hub this cost advantage can be obtained only by slot concentration.

Thus, to summarize, there are considered to be net gains to passengers from networking services through an airport hub. Although these connecting services could be provided by as many airlines as there are routes serving the hub, the quality of the network is improved if it is possible for passengers to make online connections and, if these connections are organized within the firm, they are more likely to be optimized. This, in turn, suggests that service quality increases the greater the proportion of nodes linked by one carrier (or, but to a lesser degree of quality, an alliance of carriers); in turn, with finite airport capacity, this implies slot concentration can increase consumer welfare. Larger networks operated by the single carrier from a hub are also likely to introduce supply-side economies of scale or scope (possibly associated with improved equipment utilization and with better utilization of fixed establishment costs associated with the hub itself), again suggesting a degree of natural monopoly and thus efficiency gains from slot concentration.

Congestion Externalities

At airports that act as network hubs, peak period congestion is a common feature so that an additional service imposes at the margin costs of delay on existing flight operations. The traditional prescription in these circumstances is to achieve an efficient level of use (and allocation of slots) by introducing (short-run) marginal cost pricing to internalize the delay externality. Recent economics literature on the cost of airport congestion has, however, led to a major revision in thinking about the airport marginal cost issue and this new approach also has interesting connotations for how we might view slot concentration.

The new literature argues that the traditional approach to congestion externalities at airports has the effect of overstating the marginal costs of delay (Brueckner, 2002 and 2005; Mayer and Sinai, 2003). This is because the traditional

---

7   See, for example, Gillen et al. (1990) and Caves et al. (1987),

8   There might also be scale effects associated with establishment costs at the hub.

approach does not take into account the fact that the delay costs that an airline imposes (by choice) on its own operations when it adds an additional flight to the peak at a particular airport are, in effect, internalized and therefore do not constitute an externality. Through adjustments to peak/off-peak fares, the process of internalization takes into account the impact of rescheduling on passenger travel times (longer in the peak because of the added delay) and on the operating costs of all other flights that the airline operates. It is only with respect to the costs added to the flight operations of other airlines that the externality arises and therefore the marginal cost charge should reflect only this latter.

There are a number of consequences that follow from this insight.

- The traditional measure of marginal delay costs only applies to the airport case where, in the extreme, each service is operated by a different airline (the 'atomistic' case).
- Where, in the opposing polar case, all services are provided by a monopoly airline, there are no delay externalities, even when the airport appears to be congested; the efficient congestion charge is zero.
- If, as is typically the case, the airport has a mix of carriers operating different peak period quantities, different carriers should pay different levels of congestion charge; generally carriers with a larger number of flights will pay less.
- The different levels of congestion charge will encourage slot concentration.
- More slot concentrated outcomes will have the effect of reducing total congestion externalities; more slot concentrated airports will have on average lower delay externalities (Starkie, 2003a).

This revised view of congestion externalities was developed in the context of airports where there are no administered slot controls and airlines are free to add peak capacity by scheduling extra flights; the context is essentially that of a queuing model (see Daniel, 1995). However, at a few US airports, together with most of the significant airports in Europe, the number of slots that can be used on an hourly basis is subject to a restriction, so that airlines have, in effect, to pre-book access; the object is to reduce the congestion externality by administrative means (see Forsyth and Niemeier in this volume). This raises the issue of whether the revised conceptual approach to congestion externalities requires modification, or ceases to apply, at such slot controlled airports. This would clearly be the case, for example, if the administrative rules had the effect of eliminating congestion.

Generally, this is not the case; the controlled number of flight operations per hour is usually fixed at a level which still results in some delay, albeit usually limited.[9] In addition, under the incumbency rules that apply under the EU slot allocation regulation at slot controlled airports, an airline holding a portfolio of slots is free to re-allocate its services between slots, to replace services and, by changing the equipment mix, to add capacity in the peak (and reduce it in

---

9   At London Heathrow, for example, the agreed average delay is five minutes but, in practice, the variance around this mean is considerable.

the off-peak) by altering the quantity of seats offered. There remains therefore considerable flexibility for the airline to exercise choice in the context of peak/ off-peak operations, taking into account the impact of re-allocating seat capacity between time periods on travel times and fleet operating costs, and thus, at least in part, to internalize the peak period delay externality. Consequently, at slot controlled airports also, we would expect different carriers to pay different levels of charge reflecting different marginal cost of delay imposed and we would expect the differential tariffs to result in greater slot concentrated, in turn, leading to lower congestion externalities in total.

Concentration and Prices

The objective of regulators interventions in the allocation of slots at major airports is to enable competitive entry and, in the case of the EU, this object is reflected in the preferences that must be given to entrants when allocating new slots. But in exercising this preference there is an opportunity cost involved; the benefit foregone by not allowing the incumbent network carriers access to the prescribed slots (unless potential new entrants decline their option). These benefits, as suggested above, include reduced congestion externalities, the higher quality of service to passengers associated with online transfers and the potential for lower costs as a result of density, scale and scope economies at the level of the firm. The repost would be, however, that the potentially lower costs of the larger and more densely used network do not translate into a *reduction* in prices paid by airline passengers; in spite of potential economies of density, scale and scope, a number of studies have shown that fare yields on average are higher rather than lower, at slot constrained airports where there are large incumbent carriers, with the implication that large carriers are exploiting their dominance.[10]

There are, however, a number of points to consider before it can be concluded that the higher fare yields at hubs do indicate an exploitation of market power. First, providing for transfer passengers including baggage adds complexity and thus costs to airline operations and with more transfer passengers there is a greater incentive to use expensive air-bridges rather than non-contact stands in order to speed up transfers. Second, hubs are usually large airports and larger airports are generally more costly to operate from because of the spatial separation of activities. For example, taxiing distances can be greater. Third, the organization structure of a hub can add to airline costs because of the ground staff requirements associated with bunching arrival and departure flights to minimize connecting times. Fourth, the temporal concentration of on-line transfer opportunities also adds to demands for extra (peak) capacity which might, and should, be reflected

---

10   The GAO (1999, p. 10), for example, pointed out that there were 13 airports serving large communities where passengers in 1998 paid, on average, over 8 per cent more than the national average fare. Seven of these airports were hubs for major airlines.

in higher airport charges or in terminal lease fees if, as seems likely with major airports, there are diseconomies of scale in capacity.[11]

A number of factors, therefore, suggest that there are some operational costs that might be higher for airlines operating network hubs from major airports. Whether the factors pushing up operating costs offset the economies of density/scale/scope associated with a connected network is a moot point. If fares are higher from concentrated hubs, a basic issue is whether and to what extent this is a cost reflective outcome so that passengers are, in effect, paying higher prices for a differentiated and, ultimately, more costly quality of service, one that is associated with higher service frequencies and on-line connections.

There is a further important consideration. Many hub airports are increasingly operating at or near capacity, either operational capacity or declared capacity, particularly during peak periods. In these circumstances, prices (or quality deterioration) will have to be used to limit demand to capacity available. If the airport is pricing efficiently such rationing prices will be reflected in higher (peak) period charges, hence in higher costs to the airlines and, in turn, in higher fares charged to passengers for travel at peak periods. But for various reasons congested airports often charge airlines inefficiently low prices.[12] Nevertheless, it does not make sense for the airlines to pass on sub-optimally low airport charges in the form of lower fares. If they were to do so, service quality would deteriorate (a growing number of frustrated customers would be unable to obtain a booking at posted prices). The sensible airline will maintain its fares at market clearing levels. This will result in high fare yields;[13] but these high yields will reflect scarcity and not monopoly rents.

There are, therefore, a number of good reasons why we might expect higher fare yields at airports dominated by hub carriers. The existence of higher yields *per se* does not indicate that dominant carriers are exploiting their market power. On the contrary, for airlines to extract true monopoly rents there has to be evidence that output (as reflected either in the inefficient use of slots or in low aircraft load factors) is being deliberately restrained below capacity and the analytical work done so far on this point does not support such a conclusion (Kleit and Kobayashi, 1996).[14]

Conversely, if you cannot increase output because of capacity limits, the proclivities of regulators to (re)allocate slots to potential competitors should make

---

11   Starkie and Thompson (1985) argued from first principles that there were diseconomies of scale at major airports. Pels`s (2000) empirical work supports this view with respect to terminal development.

12   See Starkie (2003b) for a discussion why airports seem reluctant to use efficient charging schemes.

13   For example, the GAO (1999, p. 20) found that: 'Airfares at six gate constrained and four slot constrained airports were consistently higher than airfares at non-constrained airports that serve similar sized communities …' but it tended to draw the conclusion that this must be due to exploitation of market power.

14   Use-it-or-lose-it clauses in EU regulations and regulations governing the use of slots at high density airports in the US are also intended to guard against this practice.

little difference to the overall level of fares at the hub; some fares may go down as capacity is shifted to allow competition on particular routes but, generally, this will be at the expense of other routes from which capacity is withdrawn and which as a consequence experience lower frequencies or the loss of a service. The effect of a reallocation of capacity in these circumstances is not to introduce competition across the network but, in large measure, to reassign the scarcity rents.[15]

Conclusions

This chapter has suggested that high levels of capacity utilization by a single airline at major hub airports might constitute an efficient outcome and that recent developments in the theory of congestion pricing applied to airports has made this outcome more likely. The efficiency gains from concentration appear to have been largely ignored by policy makers who, instead, have focussed on the negative aspects of slot concentration, particularly on the reduced opportunities for potential service competition at the *route* level. In doing so, they have also neglected the competitive pressures in many city/hinterland-to-city/hinterland markets facilitated by the use of secondary airports with spare capacity. These generally less crowded airports have been used by airlines with a business model different from that used by the typical network carrier. It is based on simplified low cost operations and, in Europe especially, point-to-point-services.[16] Low cost airlines using this business model have had a major impact on competition in the US domestic[17] and intra-European markets[18] and, in the process, they have made rules such that in the current EU slot Regulation giving preferential allocation of pooled slots to new entrants, unnecessary for the purpose of achieving competitive outcomes.

Thus, in judging an appropriate level of slot concentration there is an important balance to be struck between the benefits of concentration and the disbenefits associated with an increase in market power (increasingly tempered in short-haul markets by entry from low cost carriers using secondary airports). The challenge for the regulators is to balance the benefits and disbenefits and to do

---

15   At major congested airports there is rent-seeking behaviour. At the time of writing, the *Financial Times* (US edition, 3 November 2005) reports on the prospect of a deal between the EU and the US which could remove restrictions on further US carriers accessing London Heathrow. It comments: 'A deal would open London Heathrow to more competition [*sic*].'

16   Although the low cost airlines have developed operational hubs, usually these do not provide for network traffic. Ryanair and easyJet for example, the two largest European low-cost carriers, provide for no connecting traffic at their respective hubs. Passengers can transfer between flights but have to purchase separate sector tickets. Ryanair has concentrated exclusively on what were, at the time of entry, secondary airports (although Dublin was an exception) and has made no use of the major network hubs, but easyJet has made some use of them (see, for example, Paris Orly in Table 11.1).

17   See, for example, Morrison (2001).

18   See, for example, Davies et al. (2004).

this on an airport specific basis without recourse to *per se* rules. For the purpose of dealing with such matters normal anti-trust legislation relating to mergers and acquisitions and the abuse of dominance will probably suffice. As matters stand it is by no means evident that the level of slot concentration to be found, for example, at major European network hubs (between 40 and 60 per cent of slots in the hands of the leading carrier), is inefficiently high once account has been taken of the natural monopoly characteristics of network hubs, the reduction in congestion externalities of increased concentration and the growing competitive threat from low-cost carriers using secondary airports. The prevailing, generally moderate, degree of slot concentration might, in fact, prove to be too low if efficiency and, thus, welfare is to be maximized.

References

Bailey, E, and Liu, D. (1995), 'Airline Consolidation in Consumer Welfare', *Eastern Economic Journal*, Fall, pp. 463–76.

Basso, L.J. and Jara-Diaz, S.R. (2005), 'Calculation of Economies of Spatial Scope from Transport Cost Functions with Aggregate Output with an Application to the Airline Industry', *Journal of Transport Economics and Policy*, 39, Part 1, pp. 25–52.

Borenstein, S. (1989), 'Hubs and High Fares: Dominance and Market Power in the US Airline Industry', *RAND Journal of Economics*, 20 (1), Autumn.

Brueckner, J.K. (2002), 'Airport Congestion When Carriers Have Market Power', *American Economic Review*, 92 (5), pp. 1357–75.

Brueckner, J.K. (2005), 'Internalization of Airport Congestion: A Network Analysis', *International Journal of Industrial Organization*, 23, pp. 599–614.

Brueckner, J. and Zhang, Y. (2001), 'Scheduling Decisions on an Airline Network', *Journal of Transport Economics and Policy*, 35 (2), pp. 195–222.

Caves, D.W., Christensen, L.R. and Tretheway, M.W. (1984), 'Economies of Density Versus Economies of Scale: Why Trunk and Local Service Airline Costs Differ', *RAND Journal of Economics*, 15 (4), Winter, pp. 471–89.

Caves, D.W., Christensen, L.R., Tretheway, M.W. and Windle, R. (1987), 'An Assessment of the Efficiency Effects of US Airline Deregulation Via an International Comparison', in Bailey, E.E. (ed.), *Public Regulation: New Perspectives on Institutions and Policies*, Cambridge, MA: MIT Press, pp. 285–320.

Daniel, J.I. (1995), 'Congestion Pricing and Capacity of Large Hub Airports: A Bottleneck Model with Stochastic Queues', *Econometrica*, 63, pp. 327–70.

Davies, S., Coles, H., Olczak, M., Pike, C. and Wilson, C. (2004), 'The Benefits from Competition: Some Illustrative Cases', *Economic Paper* No. 9, UK DTI.

Economics-Plus Ltd (1995), 'Heathrow Terminal 5: Allocation of Capacity', report to BAA plc.

Gillen, D., Oum, T.H. and Tretheway, M. (1990), 'Airlines Cost Structure and Policy Implications', *Journal of Transport Economics and Policy*, 24, pp. 9–34.

Jara-Diaz, S.R., Cortes, C. and Ponce, F. (2001), 'Number of Points Served and Economies of Spatial Scope in Transport Cost Functions', *Journal of Transport Economics and Policy*, 35, pp. 327–41.

Kleit, A. and Kobayashi, B. (1996), 'Market Failure or Market Efficiency? Evidence on Airport Slot Usage', in McMullen, B. (ed.), *Research in Transportation Economics*, Greenwich, CT: JAI Press.

Mayer, C. and Sinai, T. (2003), 'Networks Effects, Congestion Externalities, and Air Traffic Delays: Or Why All Delays Are Not Evil', *American Economic Review*, 93, pp. 1194–215.

Miller, R.C.B. (2003), *railway.com* London: Institute of Economic Affairs.

Morrison, S. (2001), 'Actual, Adjacent and Potential Competition', *Journal of Transport Economics and Policy*, 35 (2), pp. 239–56.

Ng, C. and Seabright, P. (2001), 'Competition, Privatisation and Productive Efficiency: Evidence from the Airline Industry', *Economic Journal*, 111 (473), pp. 591–619.

Oum, T., Zhang, A. and Zhang, Y. (1993), 'Inter-Firm Rivalry and Firm-Specific Price Elasticities in Deregulated Airline Markets', *Journal of Transport Economics and Policy*, 17 (2), pp. 171–92.

Pels, E. (2000), *Airport Economics and Policy: Efficiency, Competition and Interaction with Airlines*, University of Amsterdam/Tinbergen Institute.

Starkie, D. (2003a), 'The Economics of Secondary Markets for Airport Slots', in Boyfield, K. (ed.), *A Market in Airport Slots*, London: Institute of Economic Affairs.

Starkie, D. (2003b), 'Peak Pricing and Airports', in Helm, D. and Holt, D. (eds), *Air Transport and Infrastructure: The Challenges Ahead*, Oxford: OXERA Publications.

Starkie, D. and Thompson, D. (1985), *Privatising London's Airports*, London: Institute for Fiscal Studies.

Tretheway, M.W. and Oum, T.H. (1992), *Airline Economics: Foundations for Strategy and Policy*, Centre for Transportation Studies, University of British Columbia.

US GAO (1996), *Airline Deregulation: Barriers to Entry in the Airline Industry*, GAO/RCED-97-4.

US GAO (1999), *Airline Deregulation: Changes in Air Fares, Service Quality and Barriers to Entry*, GAO/RCED-99-92.

Chapter 12

How the Market Values Airport Slots: Evidence from Stock Prices

David Gillen and Despina Tudor

Introduction

The regulatory parameters enacted by various national authorities governing access to airports by air carriers, amidst the background of privatization and rising congestion, have changed dramatically over time. The success and failure of these regulatory policies in providing adequate access at airports has in many cases lead to a market in slots, both in the United States and in Europe, although in Europe the legal nature of these developments is controversial and in the US, Congress has removed all rules governing airport access.

The scarcity of airport slots, otherwise known as the landing and take-off rights at an airport during a specified period of time, is reflected by the air carriers' and others willingness to pay for these resources.[1] Based on recent airline slot transactions, some studies have estimated the value of peak-period slots at London Heathrow airport to fall in between £6–£10 million (House of Commons Northern Ireland Affairs Committee, 2004–2005). In addition, the value of these assets is directly affected by the growth of air travel and indirectly shaped by the regulatory policies aimed at ameliorating congestion and improving access to airport facilities, gates and runways.

In the US, at the four airports which had high density rules (HDR),[2] the regulatory bodies have experimented with slot controls, secondary trading, some form of peak pricing and slot lotteries. With the exception of secondary trading,[3] these measures have not succeeded in providing adequate solutions to the congestion and access problems faced by airports and airlines respectively over the last four decades. Although the academic response to congestion has predominantly supported market based measures, such as auctions or peak pricing, currently the monetary secondary trading approach employed in the US is the only known scheme to rely to some degree on market forces.

---

1   As we illustrate below, banks are major holders of slots.
2   LaGuardia (NY), Chicago O'Hare, John F. Kennedy and Washington National Airports.
3   The performance of secondary trading mechanism is also subject to critique from academics.

This evolution of regulatory policies coupled with a growing intensity in airline competition as air carriers fight for access on the most profitable routes have set a scene where the most powerful incumbent air carriers, such as British Airways, grab valuable slots from ailing airlines in an attempt to strengthen their presence at their hub airports. For example, London Heathrow is estimated to be one of the most slot constrained airports in the world with a very attractive expected return on investment – the CAA has estimated a profit of £2 million for a typical short-haul service. The legendary purchase of £18 million per slot at London Heathrow by both United and American airlines from the struggling Pan Am in 1991 could easily be surpassed in the near future.

The acquisition of slots through monetary exchanges is however restricted in Europe where the movement towards market-based mechanisms for the allocation of scarce airport resources has demonstrated a greater dependency on administrative instruments relative to the US experience. This reliance on administrative rules in mainland EU could be partly attributed to the economic integration of the European Union (EU) and also to the fear of airline competition being compromised should entry at an airport be left unregulated. As a result, although both the Federal Aviation Administration (FAA) and the European Commission (EC) exercise the grandfather rule[4] for the initial allocation of slots or otherwise known as the primary market, the distinctive characteristic separating these two jurisdictions as well as the UK occurs in the secondary market. In the US air carriers are allowed to trade slots since 1986, that is slot swaps, as well as engage in monetary slot transactions. This is also true in the UK which has had an active slot trading market since 2001.[5] However, in Europe, the slot secondary market is restricted to slot swaps only. Nevertheless, a recent staff working document developed by the European Commission in 2004 exemplifies serious deliberations of market forces outcomes by the regulatory authorities (Commission of the European Communities, 2004).

The prohibition of monetary exchanges of slots under the current system, overlooking an environment of increasing congestion and limited access to airports, has in effect placed a severe constraint on the ability of air carriers to acquire the needed slots. Naturally this restriction has lead to the creation of a grey market in the EU. Thus major airlines, such as British Airways (BA) and Qantas Airways, have undertaken contractual agreements with other air carriers where the purchase of landing and take-off slots is disguised as a slot swap. Of greater importance, however is the magnitude of the prices for the most valuable slots (known as peak slots) at some of the most congested airports such as London Heathrow, LaGuardia New York, and Reagan National, which are reported to have reached remarkable levels. These events are relatively uncommon and news coverage in Europe as elsewhere is limited and spotty owing to the induced camouflaged nature of the affair by the current legislation. The purpose of this

---

4   If a slot is used 80 per cent of the time in the previous season by an air carrier the party is entitled to the same slot in the following season.

5   Slot swaps occur in a number of countries where airports are slot constrained and are slot coordinated.

event study is to investigate the degree of impact of these occurrences on the stock prices of airlines.

The chapter is organized in the following way: the following secetion includes a literature review,[6] Section 2 provides a background on European slot legislation, The next two sections respectively address in turn the grey market and the United States' experience. Then we look at different slot valuation methods, followed by the empirical analysis, while the final section comprises concluding remarks.

Event Studies: A Literature Review

Although there is a large body of literature on event studies in the fields of economics and finance, there are a few such studies investigating the correlation of various business events and stock market reactions in the airline industry. In addition, we are not aware of any event study analyzing the effect of monetary slot transactions between air carriers on stock prices. Some of the important event studies within the scope of the airline industry include Borenstein and Zimmerman (1988), Whinston and Collins (1992), Eckel et al. (1997), Zhang and Aldridge (1997), Singal (1996), Gillen and Lall (2003) and Carter and Simkins (2004).

All of the studies above are based either on the single-index standard event-study model or minor derivations of it. The basic event-study model is simply the regression of the stock price of a firm against the market return variable as well as a number of dummy variables used to measure the abnormal effects on the dependent variable during a specified event window. The econometrics can be relatively sophisticated clustering and serial correlation.

Contingent on the nature of the effect of the event in question, event studies could be broadly categorized as being based on either univariate or multivariate regression models. There are several potential techniques to estimate abnormal returns using the standard model of which the most widely used are: a single-index model, the market model, and the capital asset price model. However, Henderson (1990) points out the various techniques yield similar results. A multivariate regression model is typically used when a particular event has a simultaneous effect on all industry firms and the stock return residuals are not independently and identically distributed. This methodology has its roots in the work conducted by Zellner (1962) on seemingly unrelated regressions (SUR). Illustrative of a simultaneous equations multivariate regressions approach are the more contemporary studies of Binder (1985) and Schipper and Thompson (1985) which are based on Zellner's (1962) earlier work. Examples of relatively recent studies employing this technique include Zhang and Aldridge (1997), Singal (1996), Eckel et al. (1997), and Carter and Simkins (2004).[7] For an application of the standard event-study model using a single index see Gillen and Lall (2003).

---

6   Although not directly related to this topic but to other event studies in the airline industry, focusing on methodology.

7   These studies are outlined below.

Of particular importance is establishing the right time frame for the event window. Typically, studies select a 10-day period prior to the event which is mirrored post the occurrence of the event, such that the event window consists of 20 days in total with the day of the event sitting at zero. The estimated period of 'normal' returns generally runs between 120 and 210 days prior to the start of the event window. The work by Zhang and Aldridge (1997) measured the effect of mergers/alliances on stock prices in the Canadian airlines industry, with a study period between years 1992–1993. The study comprised a sensitivity analysis wherein the chosen time frame of a one-day event window was tested against a two-day event window and found the results were 'robust' using the latter frame. In addition, the authors examined the impact of using different frequencies – weekly returns as opposed to daily returns. If the exact day of the event is not known with certainty using a daily frequency may place the 'market's response to the event outside the observation period'. The sensitivity analysis also addressed the problem of 'clustering',[8] a characteristic typically found in financial and especially stock returns data. In order to account for the 'clustering' feature the model was estimated using GARCH (1,1) structure. However, the results of the GARCH (1,1) estimates were similar to the OLS results.

Other scholars have based their work on a two-index model. The rationale for the inclusion of a second index, normally an industry return, in the regression equation is substantiated by the linear relationship between the second additional index and the market return per firm/airline, such that the former variable may capture other relevant movements in the industry. The work by Eckel et al. (1997), which examines the effects of privatization on the performance of British Airways, with a focus on airfares and stock prices, is illustrative of such a technique. For further reference also see Singal (1996).

European Slot Legislation Background:

The slot allocation mechanism in the primary market, both in Europe and United States, is based on the grandfather rights ordinance. As described earlier this technique penalizes incumbent airlines if the slots allocated are not used 80 per cent of the time in each season. The penalization consists of slots being retracted to the slot pool and redistributed to other airlines based on slot pool rules and preferences; in many cases the priority is to new entrants. This so called 'punishment' is very important, in that although ownership rights are not clarified by the current slot legislation some rights do exist and these rights ironically also act as the platform for the observed grey market.

Although the EC Council Regulation (EEC) No. 95/93 that came into force in 1993 prohibits monetary transactions in slots, there are known contradictory court rulings in the UK that contribute to the observed discourse. One famous example is the *Guernsey Case* where the court legitimized the monetary exchange

---

8   Clustering occurs when changes of a given magnitude are followed by changes of similar magnitude but an unpredictable sign.

of slots between airlines as long as the slots are exchanged between air carriers rather than transferred in one direction from one to another (Civil Aviation Authority, 2001). In Europe, such developments, that guarantee more or less protection against future contestations of similar events in the grey market, have encouraged the swapping of off-peak for commercially valuable slots. The monetary transaction that compelled the Guernsey Transport Board to take BA to court reportedly involved a payment of £16 million for four daily pairs of slots at Heathrow airport (Economic Regulation Group, 2001).

The 2004 amendment to the EC Council Regulation (EEC) No. 95/93 on the allocation of slots at community airports defines a slot as 'the permission given by a coordinator in accordance with this Regulation to use the full range of airport infrastructure necessary to operate an air service at a coordinated airport on a specific date and time for the purpose of landing or take-off as allocated by a coordinator in accordance with this Regulation'.

Note, in contrast to the North American definition which states a slot is 'a reservation for an instrument flight rule take-off or landing by an air carrier of an aircraft in air transportation' (United States Code, title 49, subtitle VII (49USC41714)), the EC Council has included all airport infrastructure necessary to complete a take-off or landing under the term slot, which might involve airport bridges as well as staff in addition to the runway space. The European definition is in line with IATA's interpretation which states

> a slot is defined as the scheduled time of arrival or departure available for allocation by, or as allocated by, a coordinator for an aircraft movement on a specific date at a coordinated airport. An allocated slot will take account of all the coordination parameters at the airport, e.g. runways, aprons, terminals, etc. (IATA Worldwide Scheduling Guidelines, section 5.3)

The slot allocation process under the EU legislation is defined succinctly in the EC No. 793/2004 Regulation. Some of the more important clauses are found within Articles 4, 8, and 10. The member states are required to designate a neutral airport coordinator qualified to supervise slot allocation process, transactions, as well as administer penalties to negligent or ill-intentioned parties in accordance with the EC Regulation (Article 4). In addition the coordinator must take into consideration other rules and guidelines, both international and domestic, conditional on their compatibility with the community law. Any slot exchanges or transfers are considered illegal if completed prior to the approval of the slot coordinator (Article 8). Further, in order to ensure competition is maintained, the service provided by new entrants is protected under Articles 8 and 10. For example, section 3c of Article 8a dictates 'slots allocated to a new entrant as defined in Article 2(b) may not be exchanged as provided for in paragraph 1(c) of this Article for a period of two equivalent scheduling periods, except in order to improve the slot timings for these services in relation to the timings initially requested' (ibid.).

Likewise, Article 10 of the EC document ensures new entrants are able to gain access to an airport by allowing the first half of the slot pool slots to service these requests and only subsequently are the remaining slots distributed to incumbent airlines. The failure to fulfill these requirements on the part of air carriers is subject to penalties, as enforced by Article 14:

> air carriers that repeatedly and intentionally operate air services at a time significantly different from the allocated slots as part of a series of slots or uses in a significantly different way from that indicated at the time of allocation and thereby cause prejudice to airport or air traffic operations shall lose their status as referred to in Article 8(2).

In practice however, the daily number of slots at Heathrow airport available to new entrants each year has been estimated to be about 270 per week down from over 500 per week in 2000 (GAO, 2004). Although the document makes reference to the efficient use of airport capacity and includes precautionary clauses toward this end, it fails to consider the most important instrument in achieving it – placing a value on the scarce resource.

There are three recurrent themes amongst scholars and regulators with respect to the grandfather rule criterion. The first one attests the grandfather rights ordinance stifles competition and nurtures inefficiency in the airline industry. Although incumbent airlines, upon failure to fulfil the requirements of the 80/20 rule, are subsequently obligated to relinquish the relevant landing and take-off rights to the slot coordinator in order to be placed in the slot pool,[9] some argue air carriers have an incentive to strategically utilize these slots even if it would be unprofitable to do so.[10] In addition, even though evidence with respect to this practice, otherwise known as 'slot hoarding', is minimal, other scholars argue the slot pool landing and take-off rights are not only insufficient in amount but are also characterized as possessing a low commercial value. Naturally, this translates into a crippling effect on the ability of new entrants to countervail the market power of incumbent airlines.

The second view maintains the regularly observed slot abuses, against which better safeguards are recommended, are not the result of anti-competitive behaviour but rather examples of 'the business environment in which airlines are required to operate' (Bauer, 2005, ch. 9). Some of these inefficiencies which are generally cited as airline misconduct include 'late hand-backs' and 'no-shows'. Firstly, owing to the complementarity quality and the uncertainty associated with receiving the corresponding slots, it could be argued airlines are compelled to hold on to these assets, sometimes past the return deadline, while they wait for confirmation for landing rights at other airports. Secondly, no-shows could occur due to negligence and intentional wrong-doing, but both are punishable by the slot legislation. However, these events amount to an insignificant 0.5 per cent of total capacity and a solution to eliminate this behaviour would not greatly improve capacity utilization (ibid.).

---

9   Slot pool refers to the group slots, either newly created or not utilized by airlines, over which new entrants are given preference.

10   In the absence of opportunity cost of usage airlines are able to extract scarcity rents.

Conversely, it is also argued the grandfather rule reduces scheduling costs and ensures continuity of service. It is true the uncomplicated constitution of the historic precedence rule does increase certainty over future slot holdings, which invariably leads to improved business planning and investment decisions, yet a closer inspection unveils a system that fosters distorted incentives and inertia. It is a fundamental flaw to support a system on the basis of conventional wisdom alone[11] – a perfect example of the current approach which fails to pinpoint the central problem: the inefficient allocation of resources.

The Grey Market

An analysis of any market should be based on three principles: equity, efficiency, and competition. As all three elements are interrelated if one is compromised the other two are invariably subject to change. The grey market for slots in mainland Europe does not satisfy any of the three requirements of an ideal market and post the Guernsey Case decision, little can be done to eliminate it. In the words of the judge that ruled slots at Heathrow could be freely exchanged for money, he stated that he reached his decision 'on what I believe to be the clear meaning of the relevant words in the [EU] Regulation 95/93' (Boyfield, 2003). However, most scholars would agree it is not a question of eliminating the grey market as it is of improving it, possibly through an amendment to the EU Regulation 95/93. Both the CAA and the UK government are of the opinion the bilateral trades and exchanges of slots between air carriers improve the day-to-day operations at airports, and one supporting example is that over 30 per cent of slots at Heathrow are modified every year for operational purposes (CAA, 2004). Recent evidence from the UK sows that secondary trading at Heathrow has resulted in a 90 per cent increase in average seats per aircraft, a 12 times increase in average sector length and an increase by a factor of 22 of available seat km per slot.

Presently there are several unfavourable characteristics of the grey market that could be corrected through legalization. The slots are traded mostly among the members of each alliance and as alliances expand so does the effortlessness to maintain this cycle. Moreover, it is safe to assume there are high transaction costs associated with the grey market. The lack of a public notice for each intended sale or purchase means many potential buyers and sellers are excluded from the trading table. This selection among interested parties to a transaction has a negative effect on the market price and it may be the case these prices do not reflect a slot's true value. Furthermore, the uncertainty that accompanies the events in the grey market, especially amongst foreign air carriers, has lead to acquisitions of the entire business of airlines. In response to the fear expressed by policymakers over legalization of the grey market, the CAA argues the extant competition law is sufficient to correct or prevent anti-competitive behaviour, such as intra-alliance transactions and slot hoarding, which would have a far

---

11   If it has worked in the past it will continue to work in the future

greater chance of being correctly applied if the monetary slot transactions were not shrouded in secrecy.

United States' Experience

In contrast to the stringent attitude toward slot allocation found in European regulations, the enactment in the United States of the monetary secondary market in slots exemplifies an understanding of the workings of opportunity cost. Even so, some are of the opinion if the European Commission would have its way it will follow in the United States' footsteps toward much wider use of commercial mechanisms. However, there are divergent views between the EC and the European Union Association of slot co-coordinators who favour the status quo.

As noted, the FAA is gradually steering access to airports towards the free market end of the spectrum, but not without complications. Even though the four HDR airports in US still heavily rely on forms of slot controls (since Congress removed the HDR) the FAA is currently experimenting with introducing slot lotteries (LaGuardia NY) and monetary trading of slots. The latter mechanism (which is the most relevant to our discussion), has the potential to allocate slots to air carriers who value them the most upon implementation of various safeguards.

Some of these concerns as well as safeguards were expressed by The United States Department of Justice (DOJ) (Department of Justice, 2005). The DOJ has recently submitted a congestion and delay reduction proposal to the FAA wherein the monetary secondary trading of slots is criticized for being transparent and uncertain. Transparency is usually seen as a positive attribute, however, in the case of a slot lease, whereby a leasor possesses the contractual right in some cases to 'call back a slot', new carriers are reluctant to enter the market through this means. In addition the issue of property rights of slots, that has been commented on for some time in the literature, creates uncertainty and risk for the parties engaged in slot transactions. Some of the lessons learned from the US experience are to establish clear and enforceable legislation on property rights and introduce various corrective measures to adjust incentives in the secondary market, such as blocking the identity of the parties subject to these transactions.

Slot Valuation Methods

Within the present institutional slot arrangement the factors that establish the value of a landing/take-off right are distorted by the lack of legal clarity of property rights. The risk variable associated with the uncertainty of obtaining slots alters the perceived opportunity cost of these assets. Naturally, the value of a slot at an airport is dependent, *inter alia*, on the substitutability of airports, transport demand, time of day, future capacity, airline regulation, and access to complementary infrastructure. However, when slots can be withdrawn from air carriers the value generated by these elements is diminished as is the efficiency of the market. Furthermore, in many studies the term efficiency is often construed to mean a maximization of producer surplus as opposed to total surplus (economic efficiency).

Upon the arrival of the Buy-Sell Slot Rule in the United States, introduced by the OMB in 1986, approximately four measures meant to address the 'value of a slot' were developed in the marketplace (see Spitz, 2005, ch. 13). Establishing the true value of slots is a difficult task owing in part to the unique characteristics of these assets but also to the network nature of the aviation industry (hub-and-spoke vs. point-to-point). These attributes have stood in the way of reaching an accord in connection with the most appropriate valuation structure.

In abstract terms the value of a slot for an airline is equal to the expected return of the asset. In turn, the expected return of the slot depends on the amount the air carrier is able to extract from the passenger surplus. Therefore, the air carrier is willing to pay a fee for the slot pair as long as the fee plus the opportunity cost of the investment are not higher than the expected profit for a given route. But what is often not mentioned in the literature is the 'option value' (Department of Justice, 2005). For incumbent air carriers, 'the value of a slot includes not only the current value of operating a slot at the slot-constrained airport, but also the option to change their operational pattern in the future by adding flights' (ibid.). This last characteristic, in the context of rising congestion, has a particularly negative effect on competition.

Bearing these complications in mind and the fact that a single conventional definition with respect to slot valuation methods does not exist, we will attempt to decompose the advantages and disadvantages of a few accepted assessments, namely:

* market price valuation;
* earning potential valuation;
* valuation based on long-term lease rates;
* operational network opportunity values;

The first methodology, market price valuation, relies on an average price per slot which is determined, as already indicated, by a number of factors: location, time of day (i.e. peak slots), number of slots, future capacity, airline regulation and transport demand. The weight given to these factors may depend on comparable sales and valuations at other slot constrained airports as well as on the accounting treatment of other complementary assets.

As can be seen in Tables 12.1 and 12.2, the prices of slots vary only slightly across the most congested airports. Also it is important to recall that only under the EU legislation is a slot defined to include the necessary infrastructure to complete a take-off or landing. In the United States, in a 1990 study carried out by the DOT found the price of slots doubled at the four high density airports if slots were bundled with gates (Secretary's Task Force on Competition in the US Domestic Airline Industry, 1990). Although these observed properties ensure consistency, the blanket approach of calculating an average price per slot by time period or attempting to homogenize prices across similar airports may fail to reflect the

Table 12.1 List of events related to the monetary slot transactions of Airlines

List of events related to the monetary slot transactions of airlines	
Press Date	**Description of events**
21-Jul-99	The event took place on July 20, 1999. British Airways was given permission by the government to acquire CityFlyer airline for £75 million. The acquisition implied an expansion of ownership rights over CityFlyers slots, implying an increase from 26 to 38 percent of slots at Gatwick airport.
25-May-99	Virgin Atlantic airlines bought 7 slots at Gatwick airport from AB airlines for 2 million stg. The date of the actual event is the same as the press date.
10-Oct-00	Continental airlines made an offer to United airlines of $215 million for 119 jet slots and 103 commuter slots at London Heathrow airport.
10-Jan-01	On the exact day of the press release AMR Corporation and its whollyowned subsidiary, American Airlines announced three key transactions: the acquisition of TWA assets (inclusive of 173 slots) for $500 million; it had also acquired 36 slots, 14 gates, 66 owned aircraft and an additional 20 leased aircraft from US Airways for a total of $1.2 billion; thirdly it had agreed to acquire a 49% stake in DC Air for $82 million.
Early in April 2001	Virgin Atlantic Airways bought 2 slots at London Gatwick airport for €2.4 million from Virgin Express Holdings.
25-Feb-02	British Airways bought four slots at London Heathrow airport for £3.15 million from Balkan airlines.
18-Aug-02	British Airways bought 7 slots at London Heathrow for £30 million from SN Brussels airlines. The actual event took place on July 12, 2002.
23-Sep-03	British Airways made a £35 million loan to Swiss Airlines in exchange for access to 8 slots at London Heathrow airport.
Nov-03	British Airways bought 4 slots for £12 million at London Heathrow from United Airlines . The actual event took place on October 10, 2003.
25-Mar-04	Alitalia airlines purchased Gandalf company and 224 slots, some at Paris CDG airport, for €7.1 million.
20-Jan-04	Qantas Airways bought 2 pairs of slots for £20 million at London Heathrow airport from Flybe regional airline. The actual event took place in January 2004.
20-Jan-04	Virgin Atlantic bought 4 pairs of slots at London Heathrow airport from Flybe regional airline. The actual event took place in January 2004.
26-Oct-04	AirTran Airways announced on Oct. 26 it will assume lease on 14 gates and acquired 8 slots at Ronald Reagan Washington National and 19 time-controlled take-off and landing slots at LaGuardia airport for $87.5 million from ATA airlines.
17-Dec-04	Southwest won the auction with a bid worth $89.9 million, securing 6 gates at Chicago's secondary airport and slots at both LaGuardia (NY) and Ronald Reagan National airports. The actual event took place on December 16, 2004.
16-Mar-05	Republic Airways Holdings bought 113 Ronald Reagan Washington National airport slots and 24 LaGuardia (NY) slots for $110 million, with a plan to lease them back to US Airways. The announced took place on March 14, 2005 and

Table 12.1 cont'd

List of events related to the monetary slot transactions of airlines	
Press Date	Description of events
18-Aug-90	United Airlines purchased 12 slots at Ronald Reagan Washington Airport from Trans World Airlines for $19.3 million.
15-Oct-90	American Airlines bought 14 slots at LaGuardia (NY) and Ronald Reagan Washington National airports for an undisclosed price estimated to fall in between $7-$14 million from Midway airlines.
19-Feb-91	Delta Airlines purchased 9 slots at Ronald Reagan Washington National airport for $5.4 million, and 7 slots at LaGuardia (NY) $3.5 million from Eastern airlines. The actual event took place on February 15, 1991. Total acquisition was valued at $41.4 million, inclusive of 18 Eastern gates at Harstfield International Airport.
06-Feb-91	United airlines bid $54 million for 21 Eastern takeoff and landing slots and 3 gates at O'Hare International Airport. The actual event took place February 15, 1994.
15-Nov-91	Continental airlines bought 64 slots at LaGuardia (NY) airport and six A-300 for a total of $85 million from Eastern airlines. The actual event took place on February 15, 1991.
15-Nov-91	US Airways bought 62 jet slots and 46 commuter slots at LaGuardia (NY) airport, as well as 6 slots at Ronald Reagan Washington National airport for a total of $61 million. The actual event took place November 14, 1991.
11-Mar-91	United Airlines paid Pan Am £18 million per slot at London Heathrow airport. The deal was cleared off on the same day as the press date, with a total cash injection for Pan Am of $290 million.
13-Aug-91	The winning bid took place August 12, 1991. Delta Airlines took over some of Pan Am's assets: these include most of Pan Am's transatlantic services, plus its European operations based at Germany's busy Frankfurt airport. The deal is expected to add $1 billion per year in revenues.
06-Mar-91	American Airlines paid Trans World Airlines £18 million per slot at London Heathrow airport. TWA agreed in December 1990 to sell its London Heathrow landing rights to American airlines but the deal was approved on March 5, 1991
10-Jul-92	American airlines bought 40 slots and three gates from Trans World Airlines for $150 million at Chicago O'Hare airport, on the same date as the press release.
07-Dec-97	British Airways purchased 8 daily slots at London Heathrow airport from Air UK, now KLM Royal Dutch airlines, for $16.3 million. The actual event took place December 1, 1997.

Table 12.2 Abnormal returns

Event date	Airline	D1	D2	D3	D4	D5	D6	CAR
1 18 Aug 1990	United Airlines	0	0	7.19				7.19
		(–1.166)	**(1.831)***	(0.779)				
2 15 Oct. 1990	American Airlines	0	0	0				0
		(2.562)**	**(–2.425)****	(–1.142)				
3 19 Feb. 1991	Delta Airlines	–1.72	–6.1	7.6	0	1.96		1.74
		(–0.028)	(–0.101)	(0.126)	(0.257)	(0.033)		
4 6 Feb. 1991	United Airlines	0	5.66	–1.46	3.13	–1.87	–4.14	1.32
		(–1.499)	(0.564)	(–0.146)	(0.312)	(–0.019)	(–0.004)	
5 11 Mar. 1991	United Airlines	3.77	4.73	–8.54				–0.04
		(0.384)	(0.482)	(–0.871)				
6 13 Aug. 1991	Delta Airlines	–8.71	0	0				–8.71
		(–0.015)	(0.565)	(–0.268)				
7 6 Mar. 1991	American Airlines	0	0	–2.29	–8.2			–10.49
		(–1.563)	(0.851)	(–0.057)	(–0.206)			
8 10 July 1992	American Airlines	0.02	0.02	–0.01				0.03
		(1.072)	(1.064)	(–0.554)				
9a 7 Dec. 1997	British Airways	8.95	0	0	0	0	6.32	15.27
		(0.065)	(0.147)	(0.074)	(0.077)	(0.198)	(0.046)	
9b 7 Dec. 1997	KLM Royal Dutch Airlines	4.32	3.47	–5.36	2.81	–8.83	4.31	0.72
		(0.468)	(0.375)	(–0.582)	(0.305)	(–0.957)	(0.467)	
10 21 July 1999	British Airways	0	6.82	0				6.82
		(0.0761)	(0.032)	(0.139)				
11 25 May 1999	AB Airlines	3.01	0	3.01				6.02
		(0.005)	**(–3.638)*****	(0.005)				
12 10 Oct. 2000	Continental Airlines	–0.01	–0.01	0.01				–0.01
		(–0.224)	(–0.431)	(0.296)				
14 25 Feb. 2002	British Airways	–3.59	0	0				–3.59
		(–0.016)	(0.076)	(0.109)				
15 18 Aug. 2002	British Airways	7.95	–7.27	0	9.3	0		9.98
		(0.061)	(–0.559)	(–1.092)	(0.721)	(–0.986)		
16 23 Sep. 2003	British Airways	0	0	0				0
		(1.340)	(–1.313)	(1.372)				
17 3 Nov. 2003	British Airways	–4.65	–4.67	–7.47	6.21	–3.1	0	–13.68
		(–0.267)	(–0.268)	(–0.043)	(0.356)	(–0.018)	(0.735)	
18 25 Mar. 2004	Alitalia Airlines	–0.01	0.01	0				0.01
		(–1.495)	(2.161)	(0.539)				
19 20 Jan. 2004	Qantas Airways	0	0	0				0
		(–1.327)	(1.067)	(–1.197)				
20 26 Oct. 2004	AirTran Airways	–2.75	–6.16	0				–8.91
		(–0.215)	(–0.482)	**(2.109)****				
21 17 Dec. 2004	Southwest Airlines	5.24	5.48	5.03	5.47			21.22
		(0.135)	(0.141)	(0.129)	(0.141)			
22 16 Mar. 2005	Republic Airways Holdings	0	0	2.26	0	0		2.26
		(0.354)	(–0.347)	(0.051)	(0.904)	**(–1.769)***		

Note: CAR is the cumulative abnormal return.

full cost of runway use as affected by congestion and noise externalities. Also, this measure does not accommodate the gain in value of bundled slots relative to individual slots nor does it consider the incumbents' option value.

The *earnings potential valuation* determines the price of a slot using the operating margin or the incremental earning power afforded by slot access as its basis. In this case, the slot usage is grouped by stage length while the calculation of the average RPK per landing/departure is carried out for each hour of the day. In addition, the calculation of the average yield per pax is then applied to RPK to give average revenue per arriving and departing flight by hour. This is done for all carriers and weighted by the percentage of slots. Evidently, the slot value will vary with the number of slots and the airport to which they provide access. Although this would be a more solid approach in estimating future earnings per slot it fails to internalize noise and congestion externalities. Moreover, similarly to the market price valuation measure, the value gained by bundling slots and the strategic consideration of network effects are ignored.

The third methodology, which is also known as the economic value of a slot, uses a projected future income stream approach. In other words, the value of a slot to the incumbent airline is equivalent to the discounted present value of the net profit stream from the fare premium it is able to charge (Dempsey, 2001). One disadvantage in this scenario that also pertains to all of the above valuation studies, is the slot holder has a strong incentive to overstate the slot value owing to a number of factors, such as strategic and competitive issues, financial weakness of large buyers and/or large sellers, liquidation issues, increasing effect of LCC entry over time, and wedge between observed lease rates and marginal profit opportunities, all of which lead to a jump in air fares (see Spitz, 2005, ch. 13). This measure also neglects to account for complementarity effects of slots and the air carriers' option value.

Notice how all of the above measures have determined profitability on a given route by restricting the analysis to revenues and costs. The 'operational network opportunity value' defines profitability to include the effects on multiple origin-destination markets for any given segment flight as well as the value added of a slot to the network with and without flight (ibid.). The option value and the complementarity effects mentioned in the opening paragraph are both captured under this arrangement. As illustrated, this methodology is superior in many respects to the ones described above, yet in a system with high congestion costs the net marginal contribution to an air carrier is not equal to the socially efficient price – owing in part to the difficulty of attributing delay costs to oneself versus to others. One other drawback is a potential economically irrational refusal on the part of air carriers to sell slots even when buyers are willing to pay more than the incumbent's commercial value, due in part to the option value.

This last point has lead to numerous parties expressing a concern over the effects of secondary trading on competition on the grounds that the sale and the lease of a slot empower incumbent air carriers with undeserved advantages over new entrants. As a barrier to entry, incumbent airlines are motivated to outbid smaller carriers or refuse to enter in slot transactions with potential competitors. However, both in the United States and Europe the merger and competition laws (such as the UK Fair Trading Act and Merger Regimes Act) act as avenues to prevent or correct anticompetitive effects in a slot or related market.

Efficiency and Elasticity Considerations

Most of the proposed schemes, both in practice and theory, are either concerned with technical efficiency or only claim to fulfill allocative efficiency. There are many studies that reinforce the introduction of operating licenses arguing it will maximize welfare but we know 'whether the scarcity reflects a true resource constraint or is imposed artificially, the allocation of a fixed number of operating licenses by a market mechanism will not, in general, be efficient' (Borenstein, 1988). In other words regulators aim to improve technical efficiency at airports which may be satisfactory if the only concern is maximizing producer surplus.

To recapitulate, a slot is valuable only in relation to another slot.[12] From this statement it is reasonable to suppose the value of a slot may differ substantially across markets and airlines. If a landing slot may be used to service various routes, then we can generally assume its value depends on: i) 1988 profitability functions per route, as well as on: ii) the degree of passenger elasticity per route. For example, contingent on the category of consumers an air carrier targets on a particular route, namely leisure versus business passengers, elasticity could vary significantly across these groups. Moreover, a firm's profitability depends on a number of elements, such as higher yield passenger demand, marginal and fixed costs, aircraft size, service quality, economies of density etc. Taking all these factors into account, full service and low cost air carriers experience distinct profit margins while network carriers (owing in part to economies of density) value slots at their hub airport more highly than other carriers. Under these conditions a scenario could easily be envisaged where air carrier A will outbid air carrier B even though the total surplus carrier B could generate is higher than the total surplus generated by carrier A (this is amply documented in Borenstein 1988).

In response to the arbitrary nature of administrative measures scholars have proposed alternative arrangements, *inter alia*, combinatorial or sealed bid auctions. These are covered in great detail in NERA (2004) and DotEcon (2001). However, in NERA (2004) the definition of efficiency on which these schemes are founded is very narrow and limited to maximization of producer surplus: 'the system is efficient, if the slot has been allocated to the airline that values them the most', although it is clear that the application of this definition will not guarantee that the total surplus will be maximized (Borenstein, 1988). Other studies conducted by the CAA and DotEcon have repeatedly relied on the same delineation of efficiency.

---

12   A landing/take-off right at an airport is valueless without a corresponding slot at a different location.

Empirical Analysis

There are three possible implications about the efficiency of the market from the stock market reactions in relation to the monetary slot transactions between air carriers, which we interpret using the following hypotheses:

1) if the abnormal returns across the event window are significant it can be concluded markets may *not* be efficient because the changes in the stock price imply the air carrier either pays too little (+) or too much (–);
2) if the abnormal returns across the event window are not significant it could be assumed markets are efficient since the amount paid for a slot just equals the increase in profit resulting from the slot;
3) markets may be efficient or inefficient with non-significance but the monetary slot transactions are not used in a way to improve the air carrier's profits.

We note that the efficiency does not refer to social efficiency but rather the ability for the market to capitalize information and transactions to reflect the expected future profits and hence stock value. A further consideration is to extend the concept of efficiency to consider the returns of both transaction partners to evaluate the efficiency (in terms of profit maximization) of a slot transaction. If both firms have positive abnormal returns, this would be an indication that the transaction was efficient. On the other hand, if one transaction partner has positive and the other negative abnormal returns, then it is difficult to come to a definite conclusion. Finally, if both transaction partners have negative abnormal returns, then this might indicate that the transaction was indeed inefficient.

Potential gains from monetary slot transactions are captured by the dummy variables across the event window. If the price of the slot transaction does not reflect the asset's true value, it is represented by a positive sign (+) if the price is below, or by a negative sign (–) if the price is above the efficient level.

To measure the effects of monetary slot transactions on the stock prices of air carriers, we estimate abnormal returns using a standard market-model regression. This technique has been employed widely in event studies, for example, MacKinlay (1997) and Bonin (2005). The events studied are spread over the period between 1990 and 2005. The sample consists of daily stock prices, generally with an estimation period of a year prior to the event, collected from Datastream database. For the standard base model the daily stock price is regressed on the total market index and dummy variables. The binary variables are included to capture the abnormal returns over the period of the event window. The event window varies across air carriers between three to six days, meant to accommodate the actual event day as well as the press date. The regression equation takes the following form:

$$R_{it} = \alpha_i + \beta_i R_{mt} + \sum_{k=1}^{Te} \gamma_k D_{kit} + \varepsilon_{it} \tag{1}$$

where R_{it} represents the change in daily stock price of firm i on day t, α_i is the intercept coefficient of firm i, R_{mt} captures the change in the total return index with β_i capturing the parametric relationship between the aggregate stock index and the daily change in stock price of the firm , for day t, D_{kit} is a binary variable taking the form of one only for the event days inside the event window k and zero otherwise, and γ_k represents the average abnormal returns for the period of the event window only. The monetary slot transactions are estimated to have a positive effect on the stock price for the buying firm, as it exemplifies an expansion or an improvement in operations. Conversely, the seller is expected to experience a downturn in the market valuation of assets depending on whether the sale price was deemed to be at or below market values.

Data and Analysis

The standard return model incorporates 21 events in total affecting 12 air carriers. Most of the monetary slot transactions, on the simple basis of bilateral trades, occur as either part of a take-over or in conjunction with the sale of other assets such as gates and aircraft. A study conducted by the CAA on the present UK (grey) market in slots states the circumvention of restrictions on trading has prevalently relied on the acquisition of the entire business of airlines (CAA, 2001). Typically the bilateral deals are characterized by a trade of assets for cash, although sometimes the payment may consist of 'forgiveness' of debt or loans to ailing air carriers in exchange for access to slots. Thus leasing[13] of slots does arise and is considered more attractive by the leasor than a sale as the holder retains control over the asset. However, for competitive reasons leasing to new entrants is a rare event, especially in larger point-to-point markets (see Table 12.1 below). Alternative identified means to trade slots in the grey market include barter trading or 'double coincidence of wants' which means the value of a slot is acknowledged through avenues other than money. One such example is code sharing arrangements between air carriers. However, this study is based on transactions with compensations in cash or forgiveness of debt.

Most of the airlines under observation are classified as traditional or full-service carriers with only five LCCs. In the Appendix, under Case Studies, we have included a more detailed description of the events and airlines listed in Table 12.1. Also for reference to regression results see Table 12.2. The dependent and independent variables were modified once unit roots tests were run to avoid spurious regressions. To ensure stationarity for the relevant variables we have taken the growth rate of both the stock price as well as the total return index. It is also worth noting, we have only considered the events that passed the significance test at the five and one percent levels and were forced to drop some of these events from the analysis as not all air carriers are publicly traded.

The null hypothesis of $\alpha_0 \neq 0$ was rejected for 18 events out of a total of twenty one. This suggests the market in slots is indeed efficient with transacting prices

---

13   Short-term agreement with early termination clauses.

fairly reflecting the value of the slot to the transacting firms. The transaction between American Airlines and Midway Airlines (listed under event 2 in Table 12.3) passed the significance test at the five percent level. Across the three event window the calculated abnormal returns were significant only for the day prior to the actual event date as well as on the event day 0 which suggests stock market prices adjust instantaneously. The abnormal returns for transaction between Virgin Atlantic and AB Airlines (listed under event 11), were significant at the 1 per cent level at time zero only. The press release of this event coincided with the actual event day, again illustrating instantaneous adjustment of prices to relevant information. However, the abnormal returns of the third significant monetary slot transaction between AirTran and ATA Airlines (listed under event 19), occurred past the actual/press day with a lag of one day.

The negative correlation between the abnormal returns and the stock prices for events 2 and 11 indicate the landing/take-off rights were overvalued, while for event 19 the results suggest the price paid by AirTran to ATA Airlines was too low. Ironically, two out of the three significant events occurred in the US secondary market. Although two examples are not sufficient to form a rule, the discrepancy between the sale price and the efficient level could be result of the uncertainty associated with the lack of property rights.

Overall, if we are to apply the sample to the population, the results show approximately 86 per cent of the time monetary slot transactions will not affect an airline's profitability. The results also support the third hypothesis but we do not have enough information to state with certainty whether hypothesis two is true or false.

Concluding Remarks

This study uses an event study methodology to assess the effect of monetary slot transactions on investors' reactions. The analysis provides new insights on the heavily disputed issue of slot market efficiency. In general we find slot transactions involving monetary compensation do not have a significant impact on stock prices with approximately three out of twenty-one events contradicting the market efficiency hypothesis. However, the results of this study do not necessarily provide proof that markets are efficient but they certainly suggest buyers and sellers have a very good idea on the value of slots.

The question of whether markets are efficient is a much more difficult task and cannot be answered by one empirical assessment. However, that said the results do indicate that slots trade in such a way that carriers achieve an efficient result in that stock prices are not affected; the price paid must reflect future discounted value. The result of this study could prove useful to European policymakers in their decision over whether to legalize the current grey market for slots. If we found that stock prices were affected by an increase in the number of slots held, one might become concerned of monopolization but this is not what we found. We also see that slots trade at differing values and implying an efficient market.

We also see that slots trade at higher prices where there is more scarcity of slots, again implying that the value is a scarcity value not rent to monopoly.

Additionally, the findings of this study are also relevant to predictions of future airline industry structure. The emergence of LCCs has initiated a large body of literature assessing the real threat posed by no-frills airlines on traditional carriers, and access to airports under the grandfather rule might stagnate the growth of low cost carriers' market penetration levels. However, in many cases LCCs do not wish to access airports which have slot constraints or congestion; both work against the underlying LCC business model. Although opinions vary on the LCCs ability to gain valuable market power in as much as to create a change in industry characteristics, established carriers have experienced a reduction in market share over the last decade, while LCCs operating from London airports achieved a substantial increase in short-haul routes. Reflective of the success of the newly adopted low cost business model is Jetstar in Australia and Tiger in Singapore. These industry developments could be steered in the wrong direction should loyalty to the current system be left intact.

References

AFX European Focus (2004, 'Alitalia Buys Bankrupt Carrier Gandalf for 7.1 mln Eur'), 25 March.

AFX News Limited (1999), 'AB Airlines PLC – Re Slot Exchange', 25 May.

AirTran Airways Corporate Communications (2004), *AirTran Airways Agrees to Assume Chicago Midway Gates and Acquire Certain Other Assets of ATA Airlines*, 26 October: http://www.prnewswire.com/cgi-bin/micro_stories.pl?ACCT=932993&TICK=AAI&STORY=/www/story/10-26-2004/0002311472&EDATE=Oct+26,+2004.

Air Transport Group, College of Aeronautics, Cranfield University (2002), *A Study into Low Cost Carriers and Network Quality*, May, Ministry of Transport, Public Works and Water Management, The Netherlands.

Bauer, J. (2005), *Do Airlines use Slots Efficiently?*, July, German Aviation Research Society (GARS).

Black, L. (1991), 'Delta Wins $1.7 Billion Battle over Pan Am', *The Independent* (London), 12 August.

Bonin, J.P and Imai, M. (2005), 'Soft Related Lending: A Tale of Two Korean Banks', Wesleyan Economics Working Papers, #2005-011.

Borenstein, S. (1988), 'On the Efficiency of Competitive Markets for Operating Licenses', *Quarterly Journal of Economics*, 103 (2) (May), pp. 357–85.

Borenstein, S. and Zimmerman, M.B. (1988), 'Market Incentives for Safe Commercial Airline Operation', *American Economic Review*, 78 (5) (December), pp. 913–35.

Boyfield, K. (2003), 'Who Owns Airport Slots? A Market Solution to a Deepening Dilemma', in Boyfield, K. (ed.), *A Market in Airport Slots*, London: Institute of Economic Affairs: http://accessible.iea.org.uk/record.jsp?type=publication&ID=256.

Business Wire (2005), 'Republic Airways Holdings Announces Court Approval of Omnibus Agreement with US Airways', 31 March.

Buyck, C. (2004), Air Transport World Online, 29 March.

Carey, C. (1990), 'TWA Deals to Net $57.6 Million Landing Slots, Jets Being Sold', *St Louis Post-Dispatch*, 18 August.

Carey, C. (1991), 'Britain Clears Way for Sale of TWA Route', *St Louis Post-Dispatch* (Missouri), 6 March.

Case Comp/A.38.477/D2 (British Airways/SN Brussels Airlines): http://europa.eu.int/comm/competition/antitrust/cases/decisions/38477/en.pdf.

Carter, D. and Simkins, B.J. (2004), 'The Market's Reaction to Unexpected, Catastrophic Events: The Case of Airline Stock Returns and the September 11th Attacks', *Quarterly Review of Economics and Finance*, 4, pp. 539–58.

Civil Aviation Authority (CAA) (2001), *The Implementation of Secondary Slot Trading*, November.

Civil Aviation Authority (CAA) (2004), UK Civil Aviation Authority's Response to the European Commission's Staff Working Paper on Slot Reform: *Introducing Commercial Allocation Mechanisms*, November.

Commission of the European Communities (2004), *Commission Staff Working Document*, Brussels, 17 September: http://europa.eu.int/comm/transport/air/rules/competition2/doc/2004_09_17_consultation_paper_en.pdf.

Connolly, N. (2002), 'Aer Lingus Heathrow "Slots" Worth €100 million to Potential', *Airline Business*, 1 August.

Dasgupta, S., Laplante, B. and Mamingi, N. (1998), *Capital Market Responses to Environmental Performance In Developing Countries*,, Development Research Group, The World Bank, April: http://www.worldbank.org/nipr/work_paper/market/MARKETS-htmp4.htm.

Dempsey, S.P. (2001), 'Airport Landing Slots: Barriers to Entry and Impediments to Competition', *Air and Space Law*, XXVI/1 (February).

Department of Justice (2005), Docket No. FAA-2005-20704, Congestion and Delay Reduction at Chicago O'Hare International Airport: http://www.usdoj.gov/atr/public/comments/209455.htm.

DOTECON (2001), *Auctioning Airport Slots*, report for HM Treasury and the Department of the Environment, Transport and the Regions.

Economic Regulation Group, Civil Aviation Authority (2001), *Secondary Trading in Airport Slots: A Consultation Paper*, 20 July: http://www.caa.co.uk/docs/5/ergdocs/slots_secondary_market.pdf.

Eckel, C., Eckel, D. and Singal, V. (1997), 'Privatization and Efficiency: Industry Effects of the Sale of BA', *Journal of Financial Economics*, 43, pp. 275–98.

The Economist (1991), 'Delta Air Lines; Global Reach', 9 November.

Feldman, J. (1998), 'Calling the Slots', *Air Transport World*, 35 (7) (July), p. 154.

The Gannett Company (1991), 'United Lands Pan Am's London Landing Rights', 12 March.

Gillen, D. and Lall, A. (2003), 'International Transmission of Shocks in the Airline Industry', *Journal of Air Transport Management*, 9 (1), pp. 37–49.

Global News Wire (1999), 'AB Airlines Says Avoided Insolvency through 2 mln stg Gatwick Slot Sale', 25 May.

Guardian Newspapers (2003), 'BA Rescue Deal Keeps Swiss Carrier Flying', 23 September.

Hoovers (AD&B Company): http://www.hoovers.com/free/.

House of Commons Northern Ireland Affairs Committee (2004–05), *Air Transport Services in Northern Ireland*, Eighth Report of Session 2004–05, Vol. I: http://www.derrycity.gov.uk/Airport/downloads/Air%20Transport%20-%20Volume%20I.pdf.

http://www.garsonline.de/downloads/slots%20Market/031107-matthews.pdf.

IATA Worldwide Scheduling Guidelines, section 5.3.

The Independent (London) (2000) , 'Continental Airlines bids for DC Air', 10 October.

MacKinlay, A.C. (1997), 'Event Studies in Economics and Finance', *Journal of Economic Literature*, 35 (1) (March), pp. 13–39.

National Economic Research Associates (NERA) (2004), *Study to Assess the Effects of Different Slot Allocation Schemes*, Final Report for the European Commission, January.

Newswire Association Inc. (2001), 'American Airlines Announces Three Transactions that Dramatically Increase the Scope of its Network', 10 January.

Official Journal of the European Union (2004), *Regulation (EC) No. 793/2004 of the European Parliament and of the Council*, 21 April: http://europa.eu.int/eur-lex/pri/en/oj/dat/2004/l_138/l_13820040430en00500060.pdf.

Pagano, M. (1991), 'BA Angered by Heathrow Ruling', *The Independent* (London), 10 March.

Pilling, M. (2003), 'BA Grabs Yet More Slots', *Airline Business*, November.

Popper, M. (1992), 'American Wins TWA O'Hare Slots, Secured Bonds Win Value', *Investment Dealer's Digest*, 20 July.

PR Newswire (1991), 'Northwest Launches Expanded Service at Washington National; Plans Inaugural Ceremony', 1 April.

Sabeva, G. (2002), 'Balkan Air Confirms BA Slot Sale, to Lease Boeings', Reuters News, 26 February.

Salpukas, A. (1991), 'Big Airlines Gain in Eastern Sales', *New York Times*, 6 February.

Secretary's Task Force on Competition in the US Domestic Airline Industry, Airports, Air Traffic Control, and Related Concerns (1990).

Singal, V. (1996), 'Airline Mergers and Competition: An Integration of Stock and Product Price Effects', *The Journal of Business*, 69 (2) (April), pp. 233–68.

Skapinker, M. (1999), 'BA Received Approval for £75 million CityFlyer Takeover', *Financial Times* (London), 21 July.

Southwest Newswire (1991), 'Continental Airlines Continues Financial Restructuring', 15 November.

Spitz, W., *Slot Valuations under Alternative Institutional Arrangements*, Presented to: German Aviation Research Society: February 2005.

The Times (London) (2004), 'Special Report: Soaring Cost of Touching Down', 22 February.

UK Competition Commission, Air Canada and Canadian Airlines Corporation (1999), *A Report on the Merger Situation*, ch. 5: Views of Third Parties other than Regulatory Bodies: http://www.competition-commission.org.uk/rep_pub/reports/1999/fulltext/430c7.pdf.

United States Code, title 49, subtitle VII (49USC41714).

United States Government Accountability Office (GAO), Transatlantic Aviation: Effects of Easing Restrictions on US-European Markets, July (2004).

Virgin Express(2002), *Annual Report*: http://www.virgin-express.com/downloads/annualReport2002.pdf.

Wall Street Journal (Eastern edition) (1990), 15 October, p. B5: http://proquest.umi.com/pqdlink?did=4226563&sid=1&Fmt=3&clientId=6993&RQT=309&VName=PQD.

Wall Street Journal (Eastern edition) (1991), 19 February, p. A15: http://proquest.umi.com/pqdlink?did=4239942&sid=1&Fmt=3&clientId=6993&RQT=309&VName=PQD.

Wall Street Journal (Eastern edition) (2005), 16 March, p. A7: http://proquest.umi.com/pqdlink?did=808161711&sid=3&Fmt=3&clientId=6993&RQT=309&VName=PQD.

Walters, J. (1997), 'BA Accused of Stockpiling Landing Slots', *The Observer*, 7 December.

Walters, J. (2002), 'BA Buys Up More Slots', *The Observer*, 21 April.

Whiteman, L. (2004), 'Southwest: Monsters of the Midway', *Daily Deal*, 17 December.

Zagor, K. (1991), 'Eastern Move to Sell 67 Slots Runs into Trouble', *Financial Times* (London), 15 February.

Zellner, A. (1962), 'An Efficient Method of Estimating Seemingly Unrelated Regressions and Tests for Aggregation Bias', *Journal of the American Statistical Association*, 57 (298) (June), pp. 348–68.

Zhang, A. and Aldridge, D. (1997), 'Effects of Mergers and Foreign Alliances: An Event Study', *Transportation Research, E (Logistics)*, 33, pp. 29–42.

Appendix 12.1   Case Studies

British Airways

United Kingdom's largest international scheduled airline and Europe's number two air carrier, British Airways, is one of the founding members of the oneworld alliance. The airline's primary hubs include London Heathrow and London Gatwick airports, the former hosting some of the world's most profitable routes. As a result of a dominant position at London Heathrow airport, the airline has succeeded in obtaining yearly profits since 1996 with the exception of 2002, while the competition issues entailed in BA's increasingly domineering position among its top competitors (Lufthansa, UAL and Virgin Atlantic Airways) have posed significant difficulties for the regulatory authorities.

As stated earlier, although the current regulatory policies established by the European Commission prohibit any form of monetary transaction of slots between air carriers, British Airways (BA), over the last 15 years, has significantly enlarged its share of slots at London Heathrow airport. Officially, these proceedings take the form of slot swaps wherein the buyer swaps a number of inconvenient landing and take-off slots (sometimes called 'moonlight' slots) with very valuable or peak landing rights; the seller is subsequently compensated in cash. Such an event was recorded on 1 December 1997, wherein BA entered into a monetary slot agreement as the buyer with Air UK, now KLM Royal-Dutch carrier. The transaction, consisting of eight daily slots at London Heathrow airport, was valued at US$16.3 million (Feldman, 1998). Hardly coincidental, parallel to this incident a partnership proposal between BA and American underwent review by the Competition Commission. UK authorities required the parties to relinquish 168 slots at London Heathrow airport to ensure there would be enough room for the new transatlantic competitors should the partnership be approved. Unsurprisingly, one spokesman from a rival airline has described this strategic event as 'stockpiling in advance' (Walters, 1997). It follows these events gain greater value if utilized concomitant with partnerships/mergers.

Similar competition concerns were raised by the Commission in its evaluation of the proposed CityFlyer takeover by British Airways. The total amount of slots held by BA would as a result of this purchase have increased to 46 per cent at Gatwick airport with a yet more significant amount of 60 per cent of the total peak period movements.[1] The approved acquisition by the UK government, valued at £75 million, occurred on 20 July 1999 conditional on BA's acquiescence to various restrictions on slot usage (Skapinker, 1999). The benefits of the deal would amass to greater flexibility and improved service as CityFlyer's short-haul operations would be transferred to BA's long-haul operations. However, the costs were estimated to trickle down on BA's competitors, such as Delta, taking different forms amongst which degradation of service could serve as an example. Fear of increased barriers to entry at Gatwick prompted the Commission to set a 41

---

1   http://www.competition-commission.org.uk/rep_pub/reports/1999/fulltext/430c7.pdf.

per cent cap during daytime operation between 1999 and 2004 for BA as well as its subsidiaries, the percentage being calculated as part of the total amount of landing rights at the airport.

Three years later, on 25 February 2002, it was reported BA has bought four slots at London Heathrow airport for £3.15 million from Balkan Bulgarian Airlines (Sabeva, 2002). The Balkan airlines, in need of cash to cover operating costs, swapped 10 weekly slots at London Heathrow airport for slots at London Gatwick airport. The slot exchange was reported to take place on (Sabeva, 2002) March 30, 2002 while the compensation amount was to be paid in two stages. At the time of the slot swap, airport slots at Gatwick had been valued at approximately £600,000 each.

In the same year, on 12 July 2002, it was reported BA bought seven slots at Heathrow for £25 million from SN Brussels (Connolly, 2002). Shortly following the transaction, the two air carriers submitted a proposed alliance report to the Commission, which has received approval in March 2003. The latter airline is legally registered under the name of Delta Air Transport, although commercially it operates under SN Brussels. It began its operations in 2002 while under the ownership of SN Air Holdings, which should be noted also holds a hundred percent of Virgin Express equity. In 2002, British Airways had begun to openly account for slot acquisitions represented in its financial reports by a new category of 'landing rights' under intangible assets.

On 23 September 2003, BA purchased access to eight more slots owned by Swiss Airlines at Heathrow by granting the latter party a loan of £22 million (Guardian Newspapers, 2003). Swiss airlines developed after the collapse of Swissair in March 2002. Additional to the slot-loan binding agreement between the two carriers, BA and Swiss airlines had also planned to enter into a code-sharing contract on Swiss's Heathrow routes from 26 October onwards.

On 10 October 2003, BA acquired 4 more slots at Heathrow from United Airlines for $20 million through an internet auction. The slot acquisition increased BA's slot share at London Heathrow to more than 41 per cent from 37 per cent in 1999. United airlines had sold these valuable assets partly owing to the weak financial position of the air carrier at the time and also due to the lack of usage of slots during the winter season (Pilling, 2003).

Between 2002 and 2003 BA is reported to have completed at least three other more limited slot exchanges at Heathrow, amongst which number Lithuanian Airlines, Avianca of Colombia and Adria Airways of Slovenia (Walters, 2002).

The most recent transaction reported in the news dates back to 22 February 2004, with a record of four slots (or two pair of slots) at London Heathrow airport being sold for $30 million dollars by BA to Qantas airlines (*The Times* (London), 2004). Some current reports have estimated the value of one pair of peak period slots at London Heathrow to fall around £6–£10 million (House of Commons Northern Ireland Affairs Committee, 2004–50).

Qantas and Virgin Atlantic

In January 2004 both Qantas and Virgin Atlantic Airways bought two pairs and four pairs of slots respectively at London Heathrow airport, each air carrier paying a total of £20 million to Flybe, a regional airline owned by a family trust of the late Jack Walker (*The Times* (London), 2004).

Qantas Airways is one of the leading international airlines and Australia's number one air carrier. One of its top competitors is Sir Richard Branson's Virgin Atlantic Airways, part of the Virgin Group conglomerate of more than 200 media, travel, and entertainment businesses.

The gravity of the transaction is augmented if viewed in the context airlines could easily have obtained slots at Gatwick and Stansted airports. This incident highlights the uncertainty with respect to the substitutability of airports, often brought up in merger/alliance competition analysis. Furthermore, both Qantas and Virgin Atlantic Airways could not utilize these slots immediately owing to the complexity of scheduling, market research, recruitment, etc. (see Kilian, Chapter 14 in this volume). The grandfather rule however still posed a real danger on maintaining a hold on the newly acquired slots. Virgin Atlantic's response to this threat was to lease the slots to Air France, such that the Lyon route previously served by its franchisee, Flybe, was kept alive (ibid.). Qantas, however, entered a contract with British airline, Flightline, where the latter is to provide a two daily flight service on the London/Manchester route.

For Flybe the sale represented a change in its business model, from a traditional regional airline to a low-cost low-fare airline. However, since Flybe air carrier is not publicly traded we were not able to observe the effects of the slot sale on its stock price.

American Airlines

American Airlines, a wholly owned subsidiary of AMR Corporation, is ranked the world's first air carrier. Some of the major episodes in the history of this airline include the initiation of the Oneworld global marketing alliance in August 1994 as well as a major expansion in service represented by the acquisition of Trans World Airlines (TWA). The latter event consisted of a transfer of slots and it was thus recorded as one of the three slot transactions between American airlines and other air carriers (Midway Airlines, TWA, and US Airways) in the course of 1990 and 1991.

On 15 October 1990 the AA bought 14 slots at LaGuardia (NY) airport and National airport from Midway airlines for a price between $7 and $14 million (*Wall Street Journal* (Eastern edition), 1990). In 1990s, the value[2] of a slot varied between $500,000 and $1 million depending on whether the slot fell during off-peak or peak periods respectively. The transaction was characterized by a combination of sale and leaseback of slots, although the leaseback was estimated to be short

---

2   The parties refused to disclose the amount of the transaction.

term. The business deal represented Midway's financial trouble as it was further accompanied by a reduction of 9.1 per cent in the company's work force.

In 1991, the AA paid a record of £18 million per slot pair at Heathrow airport to Trans World Airlines (*The Times* (London), 2004). Prior to the deal being approved on March 5, 1991, some commentators observed the transaction, although it would insert a much needed $515 million into struggling TWA, would lead to the company's disintegration, owing to the elimination of a valuable source of revenue (Carey, 1991).

In a fight for dominance, at Chicago O'Hare International Airport, between American and United airlines on 10 July 1992, the former won 40 slots and three gates from Trans World Airlines for $150 million. TWA had filed for bankruptcy earlier that year. At the time of the event, based on other similar transactions, the going rate per slot at O'Hare airport was approximately equal to $2.725 million (Popper, 1992).

American Airlines, announced on 10 January 2001 it will engage in three main transactions meant to improve American's domestic network and create broader customer choice (Newswire Association Inc., 2001). First, the airline agreed to purchase all the assets of Trans World Airlines for a total of $500 million. The transaction also included the acquisition of 173 slots helping to create a new hub for the air carrier in St Louis. Secondly, American airlines decided to enter into a business deal with one of its top competitors, US Airways, in purchasing 14 gates, 36 slots, 66 owned aircraft and an additional 20 leased aircraft for a total of $1.2 billion. Thirdly, the air carrier also acquired a 49 per cent stake in DC Air for a price of $82 million. With respect to this last event, Donald J. Carty, chairman and CEO of American Airlines announced the company will 'generate a level of customer loyalty and achieve a level of growth for American Airlines, American Eagle, and DC Air that would otherwise take years to achieve' (ibid.).

United Airlines

The United Airlines carrier engaged in two major slot transactions in the period falling between years 1990 and 1991. On 18 August 1990, the United Airline paid $19.3 million for 12 slots at Ronald Reagan National airport to Trans World Airlines (Carey, 1990). The payment was divided in two sections: $6 million dollars was paid in cash and the rest through a negotiable, interest-bearing promissory note. At the time of event disclosure the transaction was still subject to approval by the federal regulators.

The acquisition of slots at Heathrow airport by United Airlines in 1991 had dramatically changed the meaning of competition for the British air carriers. On 11 March 1991, United airlines gained approval to purchase Pan Am's Heathrow landing rights, with a record of £18 million per slot (The Gannett Company, 1991). The transaction lead to a lifesaving total cash injection for the Pan Am Corporation of $290 million. Prior to the event the only two US airlines allowed to serve Heathrow Airport were Pan Am and Trans World Airlines. A British Airways spokesman's reaction to the business deal is representative of the strong impact of United's, American's, and Virgin Atlantic's entry at Heathrow: 'We can cope with the Gulf war and the recession, but this could do more damage to BA's profitability than those two together' (Pagano, 1991). The comment was part of a campaign, launched by BA's chairman, Lord King, to persuade the government to reconsider its decision.

Alitalia Airlines

On 25 March 2004, it was reported Alitalia airlines paid €7.1 million for Gandalf company and 224 slots, some of which are located at Paris CDG, a severely slot constrained airport (AFX European Focus, 2004). The transaction took place as a result of the collapse of the Italian carrier Gandolf Airlines (Buyck, 2004).

Continental Airlines

On 10 October 2000 it was reported Continental airlines made a $215 million bid for the assets of DC air, namely 119 jet slots and 103 commuter slots at Heathrow airport (*The Independent* (London), 2000).

Eastern Airlines

The deregulation measures enacted during the 1970s and 1980s in the aviation industry, as embodied by the Airline Deregulation Act of 1978, enabled and promoted greater competition. The effects on the pricing structures were particularly difficult to overlook and many airlines found it problematic to avoid takeovers or mergers, amongst which the sale of the majority of shareholdings by Eastern airlines to the owner of Texas airlines, Frank Lorenzo, serves as an example. Although the restructuring of Eastern Airlines under the new ownership was a success it was short lived. The airline was able to make additional cost cuts as well as acquire a number of its competitors (i.e., People Express, Continental and New York Air) under a non-union regime. However, a further attempt to reduce operating costs by Frank Lorenzo across the machinist labour sector initiated a strike, causing the company to loose millions in daily revenues. The company's inability to recover from the fiscal downfall lead to its demise on 18 January 1991 after striving for nearly two years to reorganize under Chapter 11 of the federal bankruptcy code (Zagor, 1991). As part of the breakup of Eastern Airlines, we recorded four major transactions with four other air carriers.

Firstly, competition between United and Northwest airlines combined with a rarity of events of this nature, more specifically where international routes, gates and landing and take-off slots at congested airport like National, LaGuardia and O'Hare are auctioned, initiated active bidding between the two parties. On 19 February 1991, it was reported Northwest Airlines bought 67 slots at Washington national for $35.5 million (ibid.). The assets bought by Northwest airlines were of particular importance in constructing a hub at Hartsfield Atlanta airport as well as enabling a doubling of flights at Washington National airport (placing Northwest carrier as the second largest carrier at this airport). As a highlight of its flight expansion

Northwest lowered business fares, which feature no cancellation penalties and require just three-day advance purchase. The airline set one-way fares at $129 to Newark and Hartford, and $189 to Atlanta and Boston. The fares represent a $170 savings off one-way coach fares to Atlanta, a $94 savings to Boston, a $93 savings to Hartford, and a $59 savings to Newark. (PR Newswire, 1991)

Secondly, United Airlines also acquired 21 take-off and landing slots as well as three gates at Chicago O'Hare International Airport for a total of $54 million (Salpukas, 1991).

Thirdly, on 19 February 1991, Delta Air Lines purchased nine slots at National for $5.4 million and seven slots at LaGuardia (NY) for $3.5 million (ibid.). The assets gained by Delta Air Lines improved its control over approximately 87 per cent of the passenger traffic at Atlanta's Hartsfield International Airport, thereby making it difficult for its major competitors (American Airlines, Southwest Airlines, United Airlines) to service certain routes. Currently, Delta Air Lines is ranked third in the US and the world with hubs in Atlanta, Cincinnati, New York City (Kennedy), and Salt Lake City.

Fourthly, the result of the liquidation sale of Eastern airlines also consisted of the transfer of 64 slots and six gates at LaGuardia (NY) airport as well as six A-300s to Continental Airlines for a total of $85 million on 15 February 1991 (Wall Street Journal (Eastern edition), 1991). However, the transaction did not involve cash but rather a reduction in debt.

Republic

Republic Airways Holdings owns both Chautauqua and Republic Airlines, with the former being its principal operating subsidiary. The company, based in Indianapolis, offers through Chautaugua Airlines alone more than 700 daily flights to 76 cities in 32 states, Canada and the Bahamas, through code sharing agreements with four[3] major US airlines.

On 16 March 2005, Republic Airways Holdings signed into a $110 million sale-lease deal for 113 Ronald Reagan slots and 24 LaGuardia slots with US airways although the disclosure of the deal took place on 14 March 2005 (Wall Street Journal (Eastern edition), 2005). US Airways had sold its assets in order to cover its operating costs and avoid bankruptcy. The agreement between the two parties was approved by the US bankruptcy court on 31 March 2005 and upon the cash injection from Republic airlines, the court had extended the deadline for the submission of a reorganization plan by US airways until 15 April 2005 (Business Wire, 2005). On 1 July 2005 the company announced it will offer for sale 7 million share of common stock for public sale and the proceeds will be used to buy US airways jets and slots; the asset sale closed 31 July 2005.

---

3   AmericanConnection, Delta Connection, United Express and US Airways Express.

AirTran Airlines

In 2004, AirTran Airways assumed lease on 14 gates at Chicago Midway airport and some slots at Washington National and LaGuardia (NY) for $87.5 million from ATA airlines. AirTran Airways is one of the largest low-fare air carriers in the United States, servicing over 500 flights a day to more than 40 destinations. AirTran Airways is a subsidiary of AirTran Holdings, occupying a status as the second largest carrier at its hub – Hartsfield-Jackson Atlanta International airport. The slot acquisition would represent an additional hub for AirTran Airways at Chicago Midway airport, complementing its strong East Coast network. However, on December 16, 2004, Southwest outbid AirTran securing six gates at Chicago's secondary airport and slots at both Ronald Reagan and LaGuardia airport for a total of $89.9 million (Whiteman, 2004). The CEO of AirTran Airways, Leonard Joseph, commented 'We saw Midway Airport expansion as a good opportunity, but we were not going to overpay for those assets ... Before this opportunity came along, the airline had a successful growth blueprint in place, and we will continue to move forward with that plan' (ibid.). The procurement of assets is estimated to increase Southwest's flights in 2006 by as much as 40 per cent from a current level of 145 flights.

Virgin Atlantic

On 25 May 1999, Virgin Atlantic Airways purchased seven slots at Gatwick airport for £2 million from AB airlines (Global News Wire, 2999). AB airlines, wholly owned subsidiary of AB Airlines PLC, prior to the transaction had employed the exchanged slots on both Gatwick/Shannon and Gatwick/Nice routes. In 1998, the Gatwick/Shannon route produced £6 million in gross revenue while the Gatwick/Nice route generated £1.2 million in gross revenue (AFX News Limited, 1999). The ability of the struggling AB airlines to cover its operating costs and successfully renegotiate loan agreements with its credit facilities heavily depended on the definitive aspect of this transaction. The purchase of slots for Virgin Atlantic translates into a portfolio expansion at Gatwick airport which might entail additional services to Miami, Florida.

Early in April 2001, Virgin Atlantic Airways purchased two more slots at London Gatwick for €2.4 million from Virgin Express Holdings. We were only able to observe the effects of the event on the latter carrier only, since Virgin Atlantic Airways is not publicly traded.

US Airways

On 15 November 1991 it was reported US Airways bought 62 jet slots and 46 commuter slots at LaGuardia airport as well as 6 slots at Washington airport for a total for $61 million from Continental Airlines (Southwest Newswire, 1991). Most of the assets Continental sold to US Airways were acquired through the liquidation sale of Eastern Air Lines earlier that year. It has been commented Continental's operations at LaGuardia (NY) airport were non-strategic and largely duplicative of the service provided at its hub, located at Newark airport (ibid.). For US Airways the purchase of assets represents a key strategic movement in improving service in existing markets as well and establishing a tough competitor reputation in new markets. More specifically, the transaction enabled US Airways to add three daily flights at Washington National Airport on top of the 75 daily departures it held at the time. In addition, USAir increased its gate holdings at LaGuardia to 12, all of them equipped to handle wide-body aircraft. At the time of the press release the sale was subject to Bankruptcy Court approval.

Delta Airlines

One of the world's leading air carriers, Delta Air Lines, expanded its ownership rights over some of Pan Am's assets in 1991, estimated to add $1 billion per year to its total revenues (*The Economist*, 1991). The acquisition, which took place on 12 August 1991, (included Pan Am's transatlantic services as well as some European operations) allowed the air carrier to open hubs at both New York and Frankfurt locations. The transatlantic expansion, placing the air carrier in direct competition with SwissAir, enabled it to provide service to the US, India and the Middle east, from its principle Frankfurt hub. The transaction was valued at $1.7 billion (Black, 1991).

Chapter 13

Flight and Slot Valuations Under Alternative Market Arrangements

William Spitz

Introduction

What is the value to a commercial carrier of being able to fly into and out of a congested airport? How do congestion and the related delay costs affect this value? The answers of course depend on a myriad of variables – the airport in question, the time of day, the amount of and nature of congestion, use restrictions, who the airline competitors are, etc. More fundamentally, the 'value' of a given operation may well depend on the purpose for which the valuation is being made.

The value of a flight takes on a special consideration when one is operating at a slot-constrained airport. In simplified terms, a slot essentially represents the right to take off or land at a particular time of the day. Here one can think in terms of 'operational' value (which may be closely tied to how much a slot contributes to an airlines' short-run bottom line) or 'asset' value (which should reflect how much a daily flight could be bought or sold for on the open market). In principle, of course, these alternative valuations should be equivalent (at least in the long run). But the estimate of a slot's value may well be quite sensitive to how it is measured – for example, measurement via a profit contribution metric may lead to quite a different value than observed lease or sale prices of slots at a given airport.

In this chapter I attempt to address the question of how much an operation is worth by actually estimating the incremental profit contribution of flights at LaGuardia Airport in New York during a recent temporary period of very high congestion *without* slots, and comparing these estimates with conventional slot valuations that are representative of periods with controlled (lower) congestion. The incremental profit estimates are based on a bottoms-up approach where the 'value' of a flight is measured by its effect on the carrier's net profits across its entire network; this approach accounts for network impacts by reflecting the net contribution of the flight to the carrier's network profitability. In contrast, conventional slot valuations are typically based on income studies that project a future income stream created when slots are leased to another airline on a long-term basis.

This chapter is organized into four sections. In the first, I present a brief history of slots in the United States at four major airports and an overview of recent slot transactions. The next section contains a discussion of factors affecting conventional slot valuations, provides an economic analysis of how slot controls may affect carrier profitability, and presents estimates of valuations from recent sales and valuation studies. I then present a new method to value flights at congested airports based on network opportunity costs. The results from applying this method at LaGuardia Airport in New York are compared to conventional slot valuations presented in the preceding section. The chapter ends with a summary and closing thoughts.

Background – A Brief History of Slots in the US

Slot valuations can be affected significantly by the institutional arrangements that are in place. Traditionally, the slot programs at the four slot-controlled airports in the US – LaGuardia, JFK, O'Hare-Chicago and Reagan (National) in Washington – have been subject to administrative allocations where the US Department of Transportation (DOT) has set rules for assigning slots to specific carriers (or other entities). Alternative methods for dealing with congestion have been considered, but not implemented. These include congestion pricing, where slots per se would be eliminated entirely and time-of-day fees set in such a way to encourage carriers to shift flights from highly congested times to less congested ones; and slot auctions, where a predetermined number of slots would be defined and then auctioned off to the highest bidders.

The High Density Rule (HDR) was passed into law in 1968, which established slot controls at five congested airports – Washington National (DCA), LaGuardia (LGA), JFK, Newark (EWR) and O'Hare (ORD). Newark was later exempted from the rule. Three types of slots were defined under the HDR – Commuter slots (jet aircraft < 56 seats and propeller aircraft < 75 seats), Air Carrier slots (no restrictions on aircraft size), and Other (meant to accommodate GA and other non-scheduled operations). Slots at LGA were specified as takeoffs or landings; the slots at the other three airports could be utilized for either type of operation at the discretion of the slot-holder.

Table 13.1 shows the distribution of slots per hour at each airport after passage of the HDR. The slot totals remained virtually unchanged until the mid-1990s. Slot allocations were determined by scheduling committees made up of the participating airlines at each airport.

After the US airline industry was deregulated in 1978, demand for access to the slot-constrained airports continued to expand. The 'Buy-Sell Rule' was implemented in 1986; this created a secondary market for buyers and sellers of slots. Existing slots were grandfathered to incumbents, and a minimum usage requirement (the so-called 'use/lose' rule) was instituted – initially set at 65 per cent (averaged over two months), this was later raised to 80 per cent.

In the mid-1990s, the demand for access intensified and the Department of Transportation created new slot 'exemptions' at LGA and ORD for three types of operations – new international flights, flights by new entrants in 'extraordinary' circumstances, and 'essential air service' (EAS) to certain small communities.

These exemptions could not be bought or sold. During 1997-98, 53 new entrant exemptions were approved at ORD, along with 30 at LGA.

Table 13.1 Hourly slot distributions at US high density airports through the mid-1990s

	DCA	ORD	JFK	LGA
Air carrier slots	37	120	63–80	48
Commuter slots	11	25	10–15	14
Other slots	12	10	0–2	6
Total	60	155	73–97	68

In April 2000, the US Congress passed the so-called 'Air 21' legislation, which mandated that slot controls be eliminated at ORD by July 2002, and at JFK and LGA by January 2007; controls were to remain in effect at DCA. Additional exemptions were provided at LGA for service by new entrants and for flights using aircraft less than 70 seats from defined small or non-hub airports; no limits were placed on the number of such exemptions. At DCA, 24 additional exemptions were provided to eight different airlines (primarily low-cost carriers (LCCs)), and equally divided between destinations within 1250 miles and beyond 1250 miles. In all cases, Air 21 exemptions could not be bought, sold or transferred.

The impact at LaGuardia was swift and dramatic. Over 600 exemption requests were filed during the summer of 2000, and over 200 additional flights were operating by the Fall. This represented an operational increase of over 20 per cent, leading to a huge increase in delays at the airport, as many of these additional flights were scheduled during peak morning and evening times. As a result, a moratorium on additional flights was declared by DOT in September. In January 2002, DOT issued new rules limiting Air 21 exemptions at LGA to 159 per day, and these were allocated via a slot lottery system. This resulted in a total of 75 commercial operations and six non-commercial operations per hour during most hours of the day. As of December 2005, LaGuardia averages about 1,076 daily scheduled operations, with the largest slot holders being US Airways, Delta and American Airlines. Currently, DOT is working on a set of rules to address operations at LGA after the HDR expires at the end of 2006.

At O'Hare, the HDR was formally eliminated in June 2002. There was a muted impact initially due to the terrorist events of 9/11. However, by the end of 2003, more than 100 daily operations had been added by carriers. In January 2004, the FAA negotiated a 5 per cent reduction in operations by American and United Airlines, the two major hub carriers at O'Hare. However, these operational reductions were quickly replaced, mostly by additional flights by Northwest Airlines and the new startup Independence Air. In June 2004 an additional 2.5 per cent reduction was agreed to. In 2005, the FAA imposed further restrictions on total operations by American and United, and as of December 2005, O'Hare averages about 2,470 daily scheduled operations.

At JFK, traffic declined dramatically following 9/11. Recovery was slow, and significant capacity was finally added in 2004 by Delta and Jet Blue. Service is now back to the levels seen in 2000, with about 844 daily scheduled operations. American and Delta are the largest slot holders at JFK.

Immediately after 9/11, GA operations were completely prohibited at DCA. Since late 2005 a small number of GA flights operating under very tight restrictions have been allowed. The largest slot holders are US Airways, American and Delta, who between them hold about 75 per cent of the 742 average daily scheduled operations at the airport.

Recent Slot Transactions

Ever since the Buy-Sell Rule was instituted in 1986, slot holders have had the right to sell or lease slots to others. Although the FAA legally 'owns' the slots and in principle could withdraw them at any of the affected airports, in practice the slot holders treat them as quasi-permanent assets, and their value is recognized explicitly on carriers' balance sheets.

In addition, incumbent slot holders behave strategically based on potential network opportunity costs and knowledge of who current and potential competitors are. This leads to 'hoarding' or 'babysitting' of slots, whereby carriers may sit on some of their slots and fly just enough operations to satisfy the use/lose rule. One possible explanation for this behavior is that the value of slots may be higher as a package than individually. This is particularly true for the legacy hub-and-spoke carriers, who depend on a network of carefully timed flight banks into and out of their hub airports. In this setting, the value to such carriers of any given slot (which can affect their entire network operation) may well exceed the value that a potential buyer would place on that single slot.

In practice, there are few outright sales to current or potential competitors; those that do occur are typically distress sales that take place when a carrier is financially troubled and have a need for short-term cash in order to keep operating. Table 13.2 shows slot sales that took place at LaGuardia between 2001 and 2005. In general, the market for slots in the US has been quite thin, and through the 1990s the major slot holders consolidated their assets, as shown in Table 13.3.[1]

In lieu of outright sales, swaps and leasing of slots are relatively more common. Carriers will often swap one slot for another for scheduling and logistic reasons; these are often for short-term or seasonal operations, but can also be employed for longer-term plans.

Although the carriers must notify the FAA of such swaps, in most cases the official holder of each slot does not change.

Leasing also tends to be a more attractive option than an outright sale because the holder retains control for possible future use. Many such agreements are short-term and spell out specific provisions for early termination. Leases between

---

1   The only exception is at JFK, where additional slots were created specifically to allow new service by low-cost carrier JetBlue starting in late 1999.

Table 13.2 Slot transactions at LaGuardia since 2001

Type of transaction	Year	Seller	Buyer	Number of slots
AA buyout of TWA	2001–02	TWA	American	52
Investment or distress sales	2002	Mitsubishi Bank	Northwest	25
	2003	Northwest	Pension Benefit Guaranty Corp.	49
	2003	Air Canada	Wells Fargo Bank Northwest	45
	2005	Wells Fargo Bank Northwest	Air Canada	42
	2005	Pennsylvania Commuter Airlines (US commuter)	Piedmont Airlines, Inc. (US commuter)	19
	2005	Piedmont Airlines, Inc. (US commuter)	Republic Airways Slot Holding (investor in HP/ US merger)	15
	2005	US Airways	Republic Airways Slot Holding (investor in HP/ US merger)	9
Sales to competitor	2003	Allegheny Commuter (US commuter)	American Eagle (AA commuter)	10
	2003–04	US Airways	American Eagle (AA commuter)	8

Table 13.3 Domestic air carrier slots held by major airlines at US high density airports

Airports	Holding entities	1986	1991	1996	1999	2006
Chicago O'Hare	American, United	66%	83%	87%	84%	na
New York JFK	American (inc. TWA), Delta	43%	60%	75%	84%	61%
Washington National						
New York LaGuardia	American, Delta, US Airways	27%	43%	64%	70%	69%
Washington National	American, Delta, US Airways	25%	43%	59%	65%	66%

Sources: US General Accounting Office (1986–1999 data); FAA (2006 data).

incumbent carriers are fairly common; however, incumbents rarely lease to new entrants because the latter are seen as potential low-cost competitors.

Factors Affecting Conventional Slot Valuations

There are many factors that may affect how much any given slot is worth. The most obvious one is time of day. In general, there is a higher demand for service during morning and early evening hours relative to other times of the day in order to satisfy business travelers. Airport-specific features and restrictions are also important. For example, both LGA and DCA are subject to perimeter rules that limit or ban the operation of flights beyond a certain distance.

The perimeter rule at LGA bans flights beyond 1,500 miles (flights to Denver are exempted); ostensibly the rule is designed to protect service at JFK. At DCA, the perimeter rule bans flights beyond 1,250 miles, although 12 daily exemptions have been carved out to allow service to certain Western cities. These perimeter rules have the effect of reducing the value of slots at the affected airports.

The use restrictions identified earlier (Air Carrier vs. Commuter) also affect valuations. In general, Air Carrier slots are more valuable because both commuter and air carrier operations (as defined above) are allowed; this is in contrast to commuter slots which limit the size of aircraft that may be used.[2]

The availability of related airport infrastructure may also affect slot values. The availability and location of gates, hold rooms, baggage facilities, etc., also can have important impacts on how valuable any given slot is. For example, at LGA, although gate capacity far outstrips runway capacity, the issue of who controls which gates has a large effect on which carriers operate there and how much any potential competitor might be willing to pay for a slot.

As noted earlier, the value of a slot may be higher in the context of a carrier network relative to its individual stand-alone value. In a hub-and-spoke operation, a given flight may affect the service offers available in many origin-destination markets. At an airport such as LGA, however, this effect may be quite moderate since it is not a major connection hub for any carrier.

A potentially more important factor is that carriers operating at an airport that is highly congested are likely to be operating in an environment of increasing marginal costs, i.e., each additional flight costs more and more to operate due to the cumulative effects of congestion from other flights. In such an environment, the level of fares at the airport may well allow higher-than-normal profits to be earned by carriers. This is shown in Figure 13.1, where the equilibrium price level P_e allows carriers to earn congestion rents (profits) shown by the area A+C above the private marginal cost curve. The kink in the cost curve indicates the point at which private marginal costs begin to increase due to congestion as operations expand beyond the 'normal' capacity of the airport. (Note that carrier profits measured relative to the private marginal cost curve do not equate to social marginal benefits because carriers will consider only the congestion delay impact

---

2   'Other' slots (for GA, military and other non-scheduled uses) have no market value. These are allocated on a first-come, first-serve basis and cannot be traded, sold or leased.

that an incremental flight will have on their own operations even though other flights at the airport may also be incrementally delayed by this extra flight.)

The effects of imposing a slot constraint also can be assessed using Figure 13.1. A slot constraint behaves like a quota on the market. An effective quota limits the quantity available in the market to some value less than Q_e. The quota-constrained quantity is labeled as Q_s in Figure 13.1. The price in a quota-constrained market is determined by the marginal buyer's willingness-to-pay at Q_s – this price is indicated by P_s. Depending on the slope and shape of the demand and cost curves, carrier profits under a slot regime (A+B) may well exceed those earned with no cap imposed (A+C).

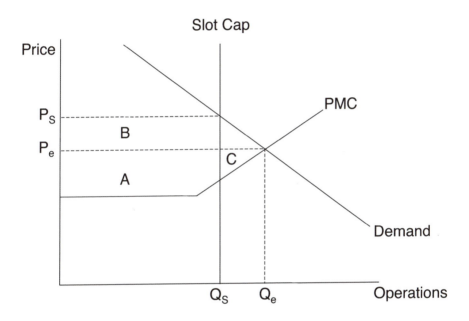

Figure 13.1 Economic profits with congestion and slot constraints

Evidence from Slot Sales and Valuation Studies

There is limited evidence about the value of slots from actual transactions. Between 1990 and 1997 there were 12 transactions at DCA, and the average slot sold for about $1 million. In 2000, US Airways proposed a sale of 199 air carrier slots and 103 commuter slots at DCA to a new entity, DC Air, for $141 million. Industry analyses of the proposed transaction (which was ultimately rejected by DOT) suggest that the implied values were approximately $950,000 per air carrier slot and $300,000 per commuter slot.

Many valuation studies have been completed for air carriers seeking loans or loan guarantees from the Air Transportation Stabilization Board (ATSB), which was formed in the aftermath of the 9/11 attacks to help financially distressed carriers. These studies typically use an income-based approach, projecting a future income stream created when slots are leased to another airline on a long-term basis. Table 13.4 depicts typical slot valuations using the income approach for slots at LGA and DCA. It is important to point out that the lease terms shown below directly reflect the impact of the Air 21 legislation which mandates the end of slot controls at LGA in 2007. In contrast, since slots will continue to be utilized at DCA, the term length used there (about 4.5 years) reflects industry views of how to appropriately measure the long-term value of slots. Despite the differences in lease length, however, the daily 'rental' value for slots is similar at both LGA and DCA, on the order of $150–$800 dollars per day. In addition, there is a fairly significant difference in lease values based on time-of-day; the late afternoon-early evening period generates higher lease rates.

Table 13.4   Typical slot valuations in 2004–05 using income approach at LGA and DCA

	LGA			DCA		
	0600–1430	1430–1930	1930–2130	0700–1400	1400–1900	1900–2100
Monthly lease rate	$10.000	$20.000	$5.000	$17.000	$25.000	$17.000
Implied slot value	$333	$667	$167	$567	$833	$567
Term (months)	24	24	24	53	53	53
Total lease payments	$240.000	$480.000	$120.000	$901.000	$1.325.000	$901.000
NPV @5%	$227.939	$455.878	$113.969	$806.949	$1.186.690	$806.949
NPV @15%	$206.242	$412.485	$103.121	$655.952	$964.635	$655.952

One important caveat is that such valuation studies have typically been used to reflect asset-backing for loans from the ATSB or a private lender, and so there may be some incentive to inflate the estimates. Additionally, the valuations may be impacted by liquidation issues and the financial weakness of large slot holders. These factors suggest that there may be a wedge between observed lease rates and incremental profit opportunities due to congestion and/or slot constraints that may be available to potential buyers, although *a priori* it is not clear in which direction the wedge goes.

Valuing Flights Based on Network Opportunity Costs

As described above, there was a brief period just after the Air 21 legislation was passed when hundreds of additional flights were scheduled into LGA, which resulted in a large increase in congestion at the airport. This period provides an interesting contrast to the slot-controlled environment that was in effect both before and soon after the legislation was implemented. In the context of the profitability analysis depicted earlier, the results from this period should be roughly reflective of the equilibrium shown at P_e and Q_e in Figure 13.1.

The method presented below is a 'bottoms-up' approach where the 'value' of a flight is measured by its effect on the carrier's net profits across its entire network. Thus we are trying to measure the incremental (marginal) network opportunity value of a flight. This means that the estimates should: 1) account for network impacts (i.e., the fact that any given flight operation may have effects in multiple origin-destination markets depending on the entire flight network of the carrier operating the flight; and 2) reflect the net contribution of the flight to the carrier's network (i.e., network profitability with and without the flight), and not just on-board own-segment profitability.

As hinted at earlier, one caveat to be aware of is that the net marginal contribution as valued by the carrier generally will not equate to the 'social value' of the flight. There are at least two primary reasons for this. First, carriers will consider only the congestion delay impact that an incremental flight will have on their own operations even though other flights at the airport may also be incrementally delayed by this extra flight. Second, passenger welfare (which may well include valuations of time and delay in consuming air travel) is not accounted for in carriers' profit-maximizing output decisions, which are based on money fare revenues.

The approach behind valuing the marginal network contribution of each flight at an airport is straightforward:

1) for each carrier, compute network origin-destination (O-D) revenues and segment costs (including own-aircraft delay costs) for all flights to and from the airport;
2) remove a given flight from the schedule;
3) reallocate traffic to remaining flights in each relevant O-D pair;
4) re-compute carrier revenues and costs;
5) compute the network contribution of the specified flight as the difference between network revenues minus costs with and without the flight.

The basic tool employed is a QSI (Quality of Service Index) model that evaluates service levels in individual O-D markets. The model utilizes a given set of flight schedules and specified time and circuity criteria to find all possible nonstop and one-stop services in the selected markets. For each O-D market, projections of market shares for each service are computed based on that service's share of assigned QSI points. The QSI values vary by the frequency of service, equipment type and the type of service offered (non-stop or one-stop).[3] The market shares then are applied to market size and fare data from the DB1B ticket sample to

---

3   The QSI of the smaller plane is used for one-stop connections involving two different equipment types.

generate estimates of carrier revenue from each service offer. An estimate of cargo-related revenue is also made based on a percentage of passenger revenue.

Carrier costs are computed on a flight-by-flight segment basis. These costs are built up from the planned block hours shown in the flight schedules. Overall profitability for a given carrier then is a result of combining the market-based revenues with the segment-based costs across the carrier's network. When removing a single flight from the schedule to compute net contributions, it is important to recognize that this may affect service offers in many different O-D markets. The model utilized here simulates what would happen to carrier profitability on a network basis if a given flight were removed. The carrier would then re-evaluate its network profitability with that one flight removed from its schedule.

For the present analysis, OAG flight schedules at LaGuardia (LGA) were utilized for the chosen sample day 15 November 2000. Note that this is after the Air 21 legislation was passed but before operating restrictions were imposed by the FAA. Thus it represents a high-demand scenario (with about 1,350 daily operations) with very high average delays (about 100 minutes per flight). Delays are estimated via a statistical equation that computes minutes of delay per flight as a function of total hourly flights at the airport. Hourly operations and estimated delays per flight are shown in Figure 13.2.

To facilitate the analysis, all O-D markets involving at least one nonstop or one-stop service offer to, from or through LGA were identified. This effectively identifies the relevant portions of each carrier's network that may be affected by schedule changes at LaGuardia. (Only online service was considered; interline connections were ignored.) Through this process, there were a total of 738 different O-D markets identified.

The DB1B ticket sample from the fourth quarter of 2000 was used to obtain estimates of market sizes and average fare levels. Fares for individual flights were scaled up or down based on time-of-day criteria.

To see how the model works on a specific flight, consider US Airways Flight #3639 from LaGuardia to Providence (PVD) departing at 2200. This is a late-night flight that connects with many incoming flights to LGA. The model identifies all US Airways online connections involving this flight; via these one-stop connections, there are 20 O-D markets (including the nonstop LGA-PVD market) that are potentially affected by this flight. The model also identifies all other nonstop and one-stop online services offered by other carriers in these O-D markets as well. A summary of the markets and carriers involved is shown in Table 13.5.[4]

Thus, from this single flight, there are potentially 20 O-D markets that would be affected if the flight were to be removed from the schedule. Looking, for example, at GRR-PVD, the model computes market shares for each service based on the QSI values described above. This results in the following share estimates: US Airways 22.3 per cent, Continental 15.0 per cent, Northwest 62.7 per cent.

---

4   Note that all of the affected markets involve Providence as a destination. This is because the specified flight is a late-night operation that connects with many inbound flights arriving at LGA; but it does not connect to any outbound flights departing at PVD because of the lateness of its arrival there.

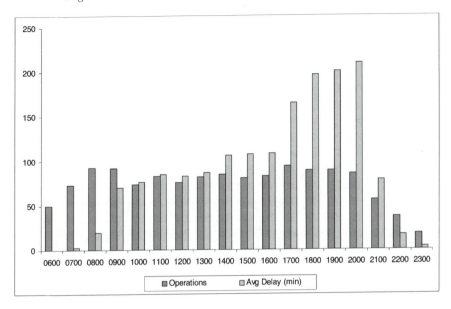

Figure 13.2 LGA operations and estimated delay following Air 21

The shares are then applied to the daily traffic and average fare in the market obtained from the DB1B ticket sample (13.19 passengers per day at an average fare of $210.31). The actual fare used for each individual service offer depends on its departure time – fares are assumed higher than average during peak times and lower than average during off-peak times. This results in the following daily revenue estimates from the GRR-PVD market for the flight: US $614, CO $425, NW $1,735. The process then is repeated for all 19 remaining market pairs, and the total revenue accruing to US Airways in these markets is calculated.

Now what would happen if the flight were to be removed from the schedule? As shown in Table 13.5, the US would still have service in all 20 markets, so most of the revenue would not be lost. In each market, the model assumes that the passengers on the cancelled service would be reallocated to any remaining services in that O-D market proportionate to their QSI values. So, for example, in the Buffalo-Providence (BUF-PVD) market, if the LGA-PVD flight mentioned above were removed, passengers could still choose from among five remaining US Airways services, and three services each from Amercian, Continental and Delta.[5]

When passengers move to another service, the fare that they pay is assumed to be the maximum of the fare paid on their original service and the service they are moving to. If the fare paid rises as a result, there is also assumed to be an

5   The model assumes an 85 per cent load factor constraint for any given segment flight; so if the number of reallocated passengers would cause the load factor to reach 85 per cent on either segment of a specific service offer, then passengers (over 85) for that service are assumed to reallocate again to the remaining services that still have space available.

Table 13.5   O-D markets served on a single flight segment from LGA to PVD operated on 15 November 2000

Flight Offers in O-D Markets Served by US Airways #3639 LGA-PVD							
MKTPAIR	US	AA	CO	DL	NW	UA	WN	Grand Total
BHM-PVD	10		1	8	2		3	24
BUF-PVD	6	3	3	3				15
BWI-PVD	22	2	2				14	40
CHO-PVD	6					4		10
CHS-PVD	11			2		4		17
CLT-PVD	28		3					31
DCA-PVD	15	3	3	5				26
GRR-PVD	4		3		5			12
GSO-PVD	12	3	2	2		4		23
GSP-PVD	9	3	2	1		4		19
IAD-PVD	8		2			4		14
ITH-PVD	5							5
JAX-PVD	12		1	5			3	21
LGA-PVD	7	6		5				18
PIT-PVD	26	3	2			4		35
RDU-PVD	14	2	2	2		4	4	28
RIC-PVD	14	2	2	3		4		25
ROC-PVD	7	5	3	3				18
SDF-PVD	8		3	2	3	4	4	24
TPA-PVD	17	1	3	6		2	5	34
Grand Total	241	33	37	47	10	38	33	439

elasticity effect, so there is an overall reduction in passengers if they are moved to higher-fare services.

After removing the one flight and reallocating the affected passengers in all affected markets, the total revenue accruing to US Airways in these markets is again calculated. The difference between this revenue and the revenue collected before the cancellation represents the current network contribution of the flight—i.e., revenue lost if the flight were removed, all else remaining the same. In this example, the estimated network contribution is $1,817.

This network revenue contribution must now be compared to the cost savings that would accrue from not operating the flight. Estimates of carrier operating costs per block hour for specific aircraft types were obtained from FAA Form 41 data. The relevant cost is applied to the scheduled block time of the flight to arrive at a total operating cost, including aircraft ownership costs.

After removing the one flight and reallocating the affected passengers in all affected markets, the total revenue accruing to US Airways in these markets is again calculated. The difference between this revenue and the revenue collected before the cancellation represents the current network contribution of the flight—i.e., revenue lost if the flight were removed, all else remaining the same. In this example, the estimated network contribution is $1,817.

This network revenue contribution must now be compared to the cost savings that would accrue from not operating the flight. Estimates of carrier operating costs per block hour for specific aircraft types were obtained from FAA Form 41 data. The relevant cost is applied to the scheduled block time of the flight to arrive at a total operating cost, including aircraft ownership costs. The carrier will also benefit from reduced airport delay on its remaining flights at LaGuardia. The model assumes that each carrier considers only its own delay savings in computing this delay benefit (even though all carriers are affected).

The delay equation that is used to estimate these savings generates bigger effects (larger time savings) for flights that are removed from peak periods relative to off-peak periods.[6] In the present example, the estimated cost savings from all three sources are $1,262. The net contribution of the flight, and therefore its value, is the difference between the network revenue contribution and the full costs of operation – in this case, $555.

This process is repeated for each flight at LGA. The primary net contribution drivers in this analysis are time-of-day delay (flights later in the day are subject to more accumulated delay), equipment type (which affects both operating costs and the number of passengers that can be carried from different O-D markets), and the degree to which carriers internalize delay costs (which is a function of their share of operations at the airport).

Table 13.5 displays the estimated average net contributions per flight by time-of-day at LGA during the brief period of very high congestion and no slot controls after the Air 21 legislation was implemented. The estimates range from a low of $62 for operations during the most congested period of the day to a high of $1574; the weighted average across the entire day is $1,350. The smaller value during the late afternoon-early evening period is the opposite of what was found for observed lease rates for slots. One possible explanation for this is that the model does not adequately reflect the higher fares that may be charged during peak times to offset the very large increases in costs associated with the congestion estimated for this period of the day; this may be the case even though carriers internalize only a portion of the overall congestion costs incurred at the airport. Or the congestion costs themselves may be somewhat overstated based on the delay estimates shown in Table 13.5.

---

6   Carriers are assumed to internalize only a portion of the delay savings during peak times due to the increased likelihood that another carrier would come in to replace the cancelled flight, thus resulting in no delay savings.

Table 13.5 Net profit contributions per flight by time of day based on 15 November 2000 sample day analysis

Time of day		
0600–1430	1430–1930	1930–21.30
$1,574	§62	§762

Overall, the flight valuations estimated for the Air 21 period with no slot controls are somewhat higher than the lease rates observed for actual slot transactions previously shown in Table 13.5. One could conclude that this argues against the notion that imposing slot controls leads to higher carrier profitability (recall Figure 13.1).

Summary

In this chapter I have attempted to address slot valuations by comparing conventional slot valuations from lease agreements with incremental profit contributions at LaGuardia Airport in New York during a recent temporary period of very high congestion without slots. While the two methods yield somewhat different results, there are many murky issues that cloud any hard and fast conclusions. For example, it may be the case that observed lease rates for slots do not fully reflect the profit opportunities available to carriers. As noted earlier, this may be the case if the valuations are depressed by the financial weakness of large slot holders who do most of the observed selling. A second confounding factor is the recent rapid growth of low-cost carriers at airports like LaGuardia, leading to downward pressure on air fares and carrier profits. Third, the bottoms-up network profit analysis described above relies heavily on operational and financial data that are subject to considerable variability and uncertainty. Finally, it may well be the case that the estimates generated do not fully reflect market equilibria at LGA during the short period considered when slot constraints were removed. Further research into the effects of airport congestion and slot constraints on flight valuations is clearly warranted.

PART D
International Experiences

Chapter 14

The Development of the Regulatory Regime of Slot Allocation in the EU

Matthias Kilian

Introduction

Anecdotal evidence suggests that a market for slots – scheduled times of arrival or departure available or allocated to an aircraft movement on a specific date at an airport[1] – exists in the European Union. When the subject is discussed with EU officials, however, it is stressed that under the current regulatory regime slot trading is illegal in the European Union, therefore a market should not exist (see, e.g., Paylor, 2005). This is undoubtedly true as far as primary trading is concerned. Primary trading involves buying slots at source – a one-shot purchase in the original allocation of slots at any given airport. With respect to to secondary trading – the sale of slots between airlines after the initial allocation – the picture is much more obscure. There is no denying that secondary slot trading exists (see CAA, 2001a, p. 33; CAA, 2004, p. 3) and some allege that the EU 'resolutely closes its eyes to the slot trading that is taking place' (Buyck, 2004). To what extent it is lawful and covered by the current EU regulatory regime is a question open to discussion. This chapter will, after an introduction into the practical dimensions of the problem, assess the legal aspects of slot trading by outlining the EU regulatory regime introduced in 1993. The chapter will then turn to the challenges EU lawmakers have faced since then, look at the 2004 reform and finally address the ongoing reform discussions and the legal challenges a market based approach to slot allocation could create. The chapter will demonstrate that from a legal point of view, many questions remain to be answered.

The Factual Background

The Flybe Deal

Although academics have been discussing the problem of slot allocation and slot trading for many years, an official market for slots has not developed in Europe.

---

1   EC regulation 95/93. Other, slightly different definitions exist, see CAA, 2001a, paras 1.9 to 1.11. See also Brachmann, 1994, p. 9; Hüschelrath, 1998, pp. 51ff. For a discussion of the economical characteristics see Knieps, 1996, p. 5.

However, every once in a while developments on the grey market stress that it is worthwhile to keep discussing the issue. In January 2004, the trading of slots once again caught the attention of the public when a number of British broadsheets reported a transaction between three airlines involving slots (see O'Connell, 2004): According to these reports, Britain's number two long-haul airline, Virgin Atlantic Airways, and Australia's flag carrier Qantas each paid an alleged £20 million to the small British regional airline Flybe for six slots at Europe's biggest airport, London Heathrow. Heathrow is known as one of the most – if not the most – slot constrained airports in the world. Airlines queue for years to gain access to the airport. The attractiveness of the airport cannot be underestimated, given that two other airports in the London area – Gatwick in Sussex and Stansted in Essex can also offer slots for long-haul operations to interested airlines. Both Virgin Atlantic and Qantas were obviously willing to pay £20 million each to obtain slots at Heathrow rather than using an alternative gateway. A report by the United Kingdom's Civil Aviation Authority demonstrated in 2001 that a typical short-haul service at Gatwick by British Airways that was just breaking even would make a profit of £1.8m a year (or generate an additional £21 per passenger) if shifted to Heathrow (CAA, 2001c, para. 2.35). Unsurprisingly, one commentator said that access to Heathrow's runways 'brings profit, prestige and a seat alongside the big boys' (O'Connell, 2004).

The Seller

Flybe had made good use of the slots in question even before selling them. The slots were used since 1996 for services to Lyon and Toulouse franchised by Air France, i.e., they were operated under the Air France brand and designator by Flybe, allowing the French carrier to use its own slots at Heathrow for services on its trunk route between Paris and London. Flybe, in the midst of a transition from a traditional regional airline to a low-cost low-fare airline concentrating on services in the regions, decided to dispose of its slots at Heathrow in mid-2003. Obviously management had realized that more money could be made from disposing of the slots than from continuing franchise services or services in its own right. The airline, which was privately-owned and had to finance an expensive fleet renewal programme, thus – as it has been described – 'auctioned the slots off'. The revenue from the disposal of the six slots earned Flybe the equivalent of three brand-new Bombardier DHC8-Q400 turboprop aircraft, of which it was acquiring 17 examples at the time. The Flybe-deal was not the first transaction in which a small airline holding slots at Heathrow for many years left the airport with a golden hand-shake. Over the past decade a number of smaller airlines with slots at Heathrow have gradually been bought out by the 'big boys'. Nowadays, not many slots are left which are not in hands of long-term players that have the resources to make investments into landing rights.

The Buyer

Virgin Atlantic, the airline that acquired two of the six slots offered by Flybe, is a typical example for a new entrant struggling to get a foot in the door at slot-constrained airports. The airline, headed by British entrepreneur Richard Branson, began operations in 1984 with a transatlantic service from London's secondary airport at Gatwick. Since then it has grown dramatically with long-haul services despite strong competition by Britain's flag carrier British Airways. Ever since its launch, the airline has been extremely popular with travellers, but while trying to meet market demands it has again and again faced problems obtaining the necessary traffic rights and slots at its bases Heathrow and Gatwick. In recent years, Virgin Atlantic has invested a considerable amount of its revenues in the acquisition of slots from other airlines which, for whatever reason, were willing to surrender their slots.[2] Virgin Atlantic's main competitor British Airways, which already holds more than 40 per cent of slots at Heathrow, quite frankly admits keeping a 'hit-list' of carriers that it judges might be willing to sell slots sooner or later. According to its annual reports, over a three-year period British Airways has spent some £72 million on the acquisition of slots at Heathrow. Airlines involved in such dealings with British Airways in the past were, for example, Air UK, SN Brussels Airlines, Swiss and United Airlines (Gow, 2004).

Probably the most famous example for these grey market dealings was a slot trade struck between American Airlines and United Airlines and the ailing US airline Pan American shortly before the latter's demise in 1991. American and United, the largest domestic carriers in the United States, had desperately but unsuccessfully tried to launch transatlantic services from the States to London for a couple of years when incumbent Pan American, the dominant player on that market for four decades, got into serious financial trouble and was forced to sell off some of its 'crown jewels' in order to avoid bankruptcy. Traffic rights to and slots at Heathrow changed hands at an estimated £18 million per slot (CAA, 2001c, para. 2.39). The 'going rate' for a slot at Heathrow currently is an estimated £4 to £6 million per slot.[3]

The European Landscape

Slot availability and slot trading is a problem that exists at airports worldwide and has repeatedly led to legal disputes. As a US Court held in 1985, 'at high density airports slots are scarce and, hence, quite valuable'.[4] In Europe, in addition to Heathrow and Gatwick, the German airports in Frankfurt and Duesseldorf are said to be the most overcrowded airports on the European landscape, with demand for slots outnumbering availability by tens of thousands each scheduling period. In January 2004, small Danish airline Sun Air inaugurated a new weekday service between Duesseldorf and Billund and somewhat bitterly remarked that it had

---

2   For example, in spring 1999 struggling AB Airlines, a small British low cost airline on the brink of bankruptcy, sold a number of the slots it held at Gatwick for £2 million to Virgin Atlantic, see Kilian, 2000, p. 159.

3   The CAA has valued a pair of slots at Heathrow held by British Airways in 2001 at 9.1m GBP (see CAA, 2001b, para. 1.4.

4   See *Eastern Airlines v. FAA*, 772 F.2d 1508, 1510 (D.C. Cir. 1988) (D.C. Cir. 1988)); also *Air Canada v. United States Deparment of Transportation*, 843 F.2d 1483 (D.C. Cir. 1988);

taken the airline almost ten years to obtain a pair of economically useful slots at Duesseldorf. In the late 1990s in a bid to expand, home-based carrier LTU International Airways took the unusual step to acquire a small regional airline (RAS Airlines) with a number of attractive slots at Duesseldorf. Such a disguised slot transfer through the aquisition of smaller or defunct airlines seems to be an approach taken predominantly on the continent where there is a bigger reluctance to get involved into straightforward slot sales.

How serious the problem is in general has been illustrated by a study carried out by the British Civil Aviation authority. It has demonstrated that 27 of the 44 busiest routes in Europe depart either at Heathrow, Gatwick, Frankfurt or Duesseldorf. Even more striking is that according to this study in the foreseeable future 27 European airports will suffer from slot constraints (CAA, 1998, para. 180). With very little prospect of additional runway capacity in the EU becoming available over the next decades, the pressure for a market for slots will mount.

EU Regulation 95/93

The process of slot allocation started in the early 1960s when airlines, on a voluntary basis, agreed to respect specific take-off and landing times at an airport as allocated by a' slot coordinator'. Over time the process spread to a growing number of airports and agreed guidelines were adopted by airlines attending the bi-annual IATA scheduling conferences. These guidelines were first set down in 1967 in a Scheduling Procedures Guide (SPG).[5] As IATA is an airline organization with voluntary membership, the rules were non-binding for airlines or regulators. Although slot allocation evolved as a system administered by airlines, government involvement has increased over the past decades as congestion worsened and market access became increasingly more difficult. In Germany, the first slot-coordinator was appointed in 1971 by the Secretary of Transport (see Kilian, 2000, p. 159). Today, a number of countries have their own rules for the allocation of slots, one prominent example being the member states of the European Union. Before EU regulation 95/93 came into force in 1993,[6] slot allocation in the EU followed the rules of the Scheduling Procedures Guide (SPG). EU regulation 95/93 is modelled after the SPG and has led to harmonization of rules governing slot allocation in the EU.

Slot allocation follows the rules prescribed in Regulation 95/93 at so-called 'slot-coordinated' airports. If an airport suffers from slot-constraints, it has to be designated as 'slot-coordinated'. Additionally, member states can designate any other airport as 'slot-coordinated' at their discretion. Consequently, not all 'slot-coordinated' airports in the EU actually suffer from slot-constraints. The Regulations require the member states to create the office of a national 'slot-coordinator' that allocates slots and monitors the use of allocated slots. In Germany, the slot-coordinator is a natural person, while in other member states it

---

5   For a detailed discussion of the IATA SPG, see Brachmann, 1994, p. 34.
6   Official Journal L14, 22 January 1993.

is a body corporate, usually set up and funded by national airlines.[7] The airports have to notify the slot-coordinator twice yearly about the number of available slots for the upcoming seasons. These slots are then allocated by the slot-coordinator after a consultation process with a consultation committee created at each airport subject to Regulation 8(1a). Priority is given to airlines that have used slots in the preceding scheduling period for at least 80 per cent of the time. In this case, the airline can claim the same slot for the next equivalent scheduling period. This right has been described as some sort of a 'squatters' right'. If the slot has not been used at least 80 per cent of the time, it will be lost and returned to a pool of unused slots ('use-it-or-lose-it' rule).[8] The allocation mechanism therefore is usually referred to as 'grandfathering of slots'. What is left in the slot pool after the allocation of grandfathered slots – newly created, returned or lost slots – is allocated to applicants. If 50 per cent or more of the requests have been made by new entrants, 50 per cent of those remaining slots are to be allocated to new entrants. The remainder of slot capacity is preferentially allocated based on a range of factors including the introduction of a year-round service, a general preference for commercial air services, the size and type of market and curfew restrictions.

The allocation mechanism described above differs from the system the EU Commission had proposed back in the early 1990s. At the time it was suggested to pool a limited number of grandfather rights, an approach somewhat akin to the early system of 'scheduling committees' known from the United States where slots were allocated by mutual consent after each season by a committee of all airlines serving the airport in question.[9] The suggestion of pooling grandfather rights even in a limited way was met with resistance by the incumbent carriers and – these mainly being flag carriers back then – consequently the member states. This resulted in the actual system whereby only slots which are unused are pooled and re-distributed to new entrants. Unsurprisingly, a main criticism against the current system is that it is beneficial for incumbent carriers and makes it difficult for new entrants to gain access to slot-constrained airports.

Slot Trading under Regulation 95/93

Introduction

Regulation 95/93 does not explicitly mention slot trading. As slots can only be allocated by the national slot-coordinator, it follows from the allocation mechanisms of Regulation 95/93 that a market for primary slot trading cannot

---

7   E.g., in the UK ACL Aviation Co-Ordination Ltd, in France COHORS, in the Netherlands ACN Airport Coordination Netherlands. A full list can be found at http://www.euaca.org/, the website of the European Union Airport Coordinators Union.
8   For the economics of the use-it-or-lose-it rule, see in detail Hüschelrath, 1998, p. 140.
9   For the operation of 'scheduling committees', see in detail Grether et al., 1989, p. 13; Wolf, 1995, p. 27.

evolve: newly-created slots are allocated through the national slot-coordinator and are not traded on a market. Primary slot trading in the EU, therefore, is an issue primarily discussed by economists who debate whether the allocation of slots by market forces rather than by an administrative procedure would lead to a more effective and competitive market. As primary slot trading cannot take place at the moment in the EU, it is not a legal issue. The picture is different, however, as far as secondary slot trading is concerned, i.e., the transfer of a slot already held by one airline to another airline.

IATA Scheduling Guide

The aforementioned IATA Scheduling Guide – after which Regulation 95/93 is modelled – only allows the exchange of slots for operational reasons, but does not treat slots as tradable rights. The United States departed from the Scheduling Guide's approach in 1986 when the so-called 'buy-sell-rule' was introduced (see Langner, 1996, p. 146). Part 93, Subpart S § 93.221 Code of Federal Regulations 14 since April 1986 provides that 'slots may be bought, sold or leased for any consideration and any time period and they may be traded in any combination for slots at the same airport or any other high-density traffic airport'.

Art. 8(4) Regulation 95/93

Despite these overseas developments, the European Commission followed the principles laid down in the IATA Scheduling Guide when it drafted Regulation 95/93. The wording of Art. 8(4) Regulation 95/93 closely follows the Scheduling Guide as it states that 'slots may be freely exchanged between air carriers or transferred by an air carrier from one route, or type of service to another, by mutual agreement or as a result of a total or partial takeover or unilaterally.'

Since Regulation 95/93 came into force on 21 February 1993, EU officials have repeatedly stressed that despite of the terms 'freely exchanged' and 'freely transferred', Art. 8(4) Regulation 95/93 prohibits any form of slot trading. Among other EU officials, then EC Commissioner for Transport Neil Kinnock expressed his view on slot trading under Regulation 95/93 in 1997: '[T]he regulation in no way provides for slots trading, but simply for the exchanges foreseen in Article 8(4). In addition, there can be no reason related to Art. 85(3) EC[10] to envisage monetary considerations for transfers. On the contrary, such requirements would make it more difficult for a competitor to gain access to a route' (see Balfour, 1997, p. 109). A year later the Commission's position was reiterated by Commissioner Karel van Miert at a hearing at the House of Lords.[11] He stressed that when Regulation 95/93 was discussed in the early 1990s, there was mutual consent among all member states that slot trading was illegal: '[I]t was crystal clear that

10   Numbering has changed since: Art. 85 EC is Art. 81 EC of the treaty of Amsterdam.

11   See, e.g., the remarks of the former EU Commissioner Karel van Miert in House of Lords on 8 July 1998, *Hansard*, 282.

slot trading, according to the Code, is illegal because no one accepted that at the time and that has not even been discussed. One delegation raised that question and all the others said no, so it was crystal clear.'

Despite these comments, the wording of Art. 8(4) has given rise to much discussion as to whether it indeed prohibits any form of slot trading from a legal point of view. Art. 8 certainly is no masterpiece of law-making as it merely states that slots may be exchanged, but does not set up rules under which conditions and particularly fails to address the question whether the exchange of slots may be accompanied by financial considerations.[12]

The Underlying Problem

Considering the problem superficially, the case against slot trading could be based on the argument that Art. 8(4), which requires an 'exchange of slots', undoubtedly forbids a straightforward sale of a slot by one airline to another independent airline. Such an exchange of slots is only possible if both airlines in question already hold slots at the same airport. How can there be slot trading under these circumstances?

To illustrate the underlying problem, it is useful to understand that slots come in many 'forms and shapes'. The problem for an airline at any airport in the world, even the most slot-constrained airport, is not to obtain a slot, but to obtain a commercially useful slot. The requirement to hold a slot to be able to exchange it with another airline therefore is not a real obstacle to enter the market of slot trading: So-called 'moonlight slots' are readily available at any airport, even at London-Heathrow. 'Moonlight slots' are slots that are commercially useless because they grant a right to take-off or land at times where there is no demand for the air services in question, i.e., during off-peak periods late at night (at 'moonlight') or on weekends. An airline therefore can always apply for a moonlight slot with the intention to merely use it as a 'bait' for a slot trade with another airline. Literally, then a slot exchange according to Regulation 8(4) takes a place. A legal problem arises from such a slot exchange if it amounts to an uneven trade because the airline holding moonlight slots is willing to pay the other airline holding the more attractive slots for the exchange to take place (see Kilian, 2000, p. 159). Under such circumstances, it is usually evident that the airline being paid for the exchange has no intent to use the moonlight slots it receives, but will exit the relevant market after having cashed in or received any other form of non-monetary compensation.[13]

---

12   For a detailed discussion, see Kilian, 2000, p. 159.

13   Non-monetary considerations in the past have included code-sharing arrangements and slots at foreign airports, CAA, op. cit. para 2.30.

The Guernsey Case

The Factual Situation

Such a scenario was the background of a landmark decision of the British High Court published in 1999.[14] The States of Guernsey Transport Board sued the British slot-allocator, Airport Co-Ordination Limited (ACL), because ACL had approved a slot exchange between Air UK and British Airways. Air UK, a British regional airline later bought by Dutch flag carrier KLM and re-named KLM uk, had maintained Guernsey's most important air link to London Heathrow for more than 15 years. For Guernsey, a tax haven for investors und business-people, this route, which guaranteed access to international air links through a major airline hub, was a lifeline. Because of the development of London's third airport Stansted into Air UK's main base, the airline decided in 1997 to withdraw its sole remaining route into Heathrow, its service to Guernsey, effective from the summer schedule 1998. No carrier already operating into Heathrow was interested in taking over the route and any possible new entrants that had come forward had no real chance to obtain slots from the slot-pool allowing a four-times daily, one aircraft operation on the route. Loss-making Air UK approached British Airways and offered to exchange the four pairs of daily slots it held for the Guernsey route with moonlight slots British Airways held. The deal went through with the approval of ACL for the winter and summer scheduling periods. Unsurprisingly, British Airways was uninterested in maintaining the Guernsey route (which in fact it had surrendered to Air UK in 1980) and used the former Air UK slots for services on more attractive international routes. As a result, Guernsey lost its important Heathrow link (and since then has been served only through London's secondary and tertiary airports), while Air UK earned an alleged £15.6 million from the deal with British Airways (CAA, 2001c, para. 2.38). Air UK then quietly returned the moonlight slots it had received from British Airways to the slot pool and disappeared from Heathrow.

The Legal Dimension

When the dealings between British Airways and Air UK became known, the Guernsey government[15] decided to sue ACL. Guernsey's objection was that ACL had given its approval to a poorly disguised slot transfer rather than a slot exchange and therefore to a blatant violation of Art. 8(4) Regulation 95/93. Guernsey argued that both British Airways and Air UK had not exchanged slots in a way envisaged by Regulation 95/93 which only allowed for economically or numerically even trades. British Airways, so the argument went, thus had simply bought slots illegally from Air UK which was acting as British Airways' servant

---

14   *Regina v. Airport Co-Ordination Ltd* (ex parte the States of Guernsey Transport Board) = [1999] European Law Reports S. 745ff. = Business Law Europe 1999/14, S.2f. For an in-depth discussion of the ruling see Kilian, 2000, p. 159.

15   Guernsey is a 'Crown dependency'.

when returning BA's moonlight slots. Under Art. 8(4), between two independent airlines only a slot exchange was lawful, not a slot transfer. As it was impossible at the time of the court proceedings to prove that money had actually been paid by British Airways to Air UK, the legal arguments focused on the distinction between an 'exchange' and a 'transfer' of slots, as the latter was only allowed subject to much more stringent conditions than an exchange. The resulting court proceedings were of an additional delicacy because of a rift between the EU commission and the British Department of Transport and Industry. While the EU commission's point of view was strictly against any form of slot trading (see above), the DTI was very much in favour of slot trading and had made its different point of view known in an other context before (Balfour, 1997, p. 109).[16]

The Court's Decision

The High Court, certainly much to the surprise of the Commission, decided in favour of ACL and did not object to an exchange through an uneven trade. The court held that an 'exchange' takes place whenever the loss of any slot previously held is compensated by the acquisition of another slot through the same transaction ('reciprocal transfer'), regardless of the unevenness of the exchange from an economical or numerical point of view. Having arrived at this point, the Court came to the conclusion that the wording of Art. 8(4) left little, if any room for further interpretation. Why the court did not double-check its findings with the help of a teleological analysis is probably only understandable by taking different methodological approaches of the common law and civil law systems into consideration. The court declined to submit the case to the European Court of Justice (ECJ) because it believed that its interpretation was beyond reasonable doubt. In this authors opinion, the methodology the ECJ applies when interpreting statutory law could have led to a different result in the Guernsey case. The most important methodological principle for common law courts is 'fidelity to the written word', while the ECJ (and civil law courts) put much more emphasis on teleological and historical aspects of a law by analysing preparatory material. Consequently, and unlike common law courts, the ECJ takes the written word usually only as a starting point of its interpretation of a statute. A teleological interpretation would have taken the genesis of Regulation 95/93 – it was derived from the IATA Scheduling Guide – and the point of view of the people who had drafted it into consideration. It certainly would have been much more difficult to refute the case against slot trading if those aspects had been properly addressed by the High Court.

---

16   The problem was raised between the EU and the DTI when British Airways and American Airlines entered into negotiations to form an alliance. For anti-trust-reasons, both were required to surrender more than 200 slots at Heathrow, resulting in a dispute whether those slots could be sold or simply had to returned to the slot pool (in the end, the alliance project was abandoned before this problem was sorted out).

Regulatory Reform Part 1: Regulation 793/2004

The Aftermath of the Guernsey Case

The findings of the High Court in the Guernsey case are not binding for courts in other EU member states. Also, the European Court of Justice has not yet had an opportunity to decide on the exact meaning of Art.8(4) and thus end the legal uncertainty by issuing a decision binding for all member states.[17] Unsurprisingly, since the High Court decision the discussion about slot trading has continued.

After its defeat before the British courts, Guernsey and its neighbour island Jersey – Jersey lost its Heathrow link in 2001 – have been actively lobbying the issue of slot trading. In 2001, Jersey, guided by the US experience, suggested the creation of a market for the trade of slots in which not only airlines, but also interested regional authorities and local bodies are allowed to acquire slots. In its proposal, Jersey indicated its willingness to acquire slots at Heathrow and to delegate its power to exercise the rights to an airline that undertakes to serve the route in question (*Jersey Evening Post*, 8 and 16 September, 2000; Guernsey Press Report, 13 October 2000).

The Civil Aviation Authority went a step further in its discussion paper 'The Implementation of Secondary Slot Trading' published in 2001 (CAA, 2001a, paras 2.1–2.5). In it, the CAA not only advocates the legitimisation of secondary slot trading, but also the admission of non-airline entities to such a secondary market. The Civil Aviation Authority's belief is that the admission of financial intermediaries will lead to a more sophisticated market and allow airlines to use slots as collateral in raising capital. Thus, the Civil Aviation Authority strongly supported the legitimisation of secondary slot trading. It based its findings also on discussions with respondents which were asked to give their views on the subject and were all but one in favour of slot trading. It has to be noted, however, that among the CAA's ten respondents were only airlines which would obviously benefit from slot trading. No smallish, newly established or regional airlines were asked their opinion. Surprisingly enough, Manchester airport was the (only) respondent that expressed scepticism whether mechanisms designed to maximise economic efficiency would be able to take fully into account social and economic externalities necessary to maximise social welfare.

The European Commission's Point of View

The European Commission has been considering the issue of slot trading ever since Regulation 95/93 came into force in 1993. The Regulation placed a duty on the European Commission to report to the European Parliament and the European Council on the effects of the regulation three years after its inception, and to place a proposal for the continuation or revision of the regulation before the European Council by 1 January 1996. Although the European Commission

17   For a discussion of how the ECJ's methodology would be applied to the regulation in question, see Kilian, 2000, p. 159.

had research carried out by Coopers & Lybrand in 1995, the 1 January deadline was missed. In a 1996 discussion paper the Commission, however, stated that it had 'opted to facilitate exchanges of slots ... by allowing monetized trading. ... At the same time, safeguards will be needed to ensure that secondary trading ... does not enable dominant carriers to strengthen their position unduly' (Hüschelrath, 1998, p. 281). This statement – which also indicates that under the current regulatory regime slot trading is unlawful – was not followed up because of resistance from the member states. Instead, the consultation process continued and eventually led to another report, this time prepared by PriceWaterhouseCoopers and published in May 2000. It did not specifically address the problem of slot trading, but convinced the Commission that the provisions of Regulation 95/93 needed to be significantly amended (see European Commission, 2000).

Following a meeting with member states and representatives of the airline industry in July 2000, the Commission entered into yet another consultation process which finally resulted in a proposal for a considerably amended Regulation 95/93. In the Commission's view, the existing rules have proven to be insufficient to provide for clear definitions and appropriate enforcement mechanisms, making it necessary to clarify the legal nature of slots, ensure the transparent, neutral, and non-discriminatory airport capacity determination and set up allocation procedures by legally and factually independent coordinators. The problem of slot trading was another major concern of the Commission and one of the aspects it focused on during the consultation process.

In June 2001, the European Commission finally tabled a first proposal amending Regulation 95/93. On 10 June 2002, the European Parliament approved the proposal subject to a number of amendments, resulting in changes made to the proposal by the European Commission in November 2002. The European Council approved the proposal in December 2003 after some more changes. The draft went further through the parliamentary process in 2004 and was discussed in mid-March 2004 by the European Parliament's RETT-committee. On 21 April the European Parliament and the European Council adopted a new Regulation 793/2004 that amended Regulation 95/93.[18] It came into force on 30 July 2004.

Regulation 793/2004

a) expressed only in factual terms, is clarified in Regulation 793/2004. Under the previous regulatory regime, some airlines regarded slots as their assets, while on the other hand airports believed that, as slots are inextricably linked to their infrastructure, slots constituted their property. To clarify the legal problem, the term 'slot' is given a new definition. Pursuant to Art. 2, slot shall mean the entitlement of an air carrier to use the full range of airport infrastructure necessary to operate an air service at a coordinated airport on a specific date and time for the purpose of landing and take-off. The term 'entitlement' clarifies that a slot is

---

18   Regulation (EC) 793/2004 of the EP and of the Council of April 21, 2004 amending Council Regulation (EEC) 95/93 on common rules for the allocation of slots at community airports (Official Journal L138 0f April 30, 2004).

not an airline's or an airport's property, but merely a public good allocated to an airline that can be renewed under certain conditions prescribed by the Regulation. As a public good, the entitlement can also be withdrawn if it is not renewed without giving the concerned airline a legal claim. Art. 8(1) further stresses the legal nature of slots as it states that at the expiry of the relevant scheduling period all slots are to be returned into the slot-pool. Technically, the grandfathering of slots is made an exception to this general rule in Art. 8(2) in order to avoid the impression that slots are owned by the airlines.

Allocation criteria The general framework of the amended Regulation reconfirms the principle of 'use-it-or-lose-it' and the 80 per cent threshold is kept for scheduled air services. Analysis of the current system has convinced the Commission that the continuation of slot allocation based on historical precedence in the usage is justified from both the passenger and the airline side. However, the distinction between scheduled and non-scheduled air services is not upheld. Therefore, non-scheduled air services have to meet the 80 per cent threshold as well (previously, it was set at 70 per cent).

The amended Regulation, according to Art. 8(4), allows the re-timing of grandfathered slots, i.e., the return of a grandfathered slot to the slot-pool in exchange for a new slot at a more favourable time. The previous Regulation, if taken by its literal meaning, did not allow such a move as (1) all slots returned to the coordinator had to go into the slot-pool for general redistribution and (2) newly allocated slots could not be grandfathered immediately.

In order to more effectively use scarce commodities, Art. 8(2) allows member states to allocate slots only to use for minimum sized aircraft. Through this provision, smaller aircraft can be kept out of slot-constrained airports allowing for the most effective use of airport capacity.

New entrants The amended Regulation aims to provide a better protection of new entrants. Only genuine new entrants will benefit from the preferential allocation of unused slots. An airline is not considered a new entrant if it (a) holds 5 per cent per cent of the slots at the airport in question, or (b) if it has a joint operation, code sharing or franchise agreement with another airline which is not a new entrant, or (c) if it is a subsidiary or parent company of an airline not considered a new entrant.

Enforcement The amended Regulation strengthens the position of the national slot co-ordinator, but also gives him additional responsibilities. In particular, the coordinator will have extended rights to withdraw slots immediately if they are used in violation of the Regulation, especially in cases of abuse. Examples for such an abuse given by the Commission are the regular and intentional operation of slots at times different from those cleared by the coordinator and the illegal transfer of slots dressed up as exchanges (see below). In addition to the right to withdraw slots, fines can be imposed for the abuse of slots.

To counterbalance the additional powers of the coordinator, his decisions must be subject to judicial review before national courts. Finally, in order to enable slot-coordinators to exercise these new powers, they have to be independent from those affected by their decisions. Therefore, the proposal aims to strengthen the position of the slot-coordinator. His/her office is to be financed independently rather than through an airport or a group of airlines (an 'interested party').

Other aspects The Commission has refrained from adding environmental criteria to the allocation mechanism, but has already announced that it will look into bringing an environmental dimension into the allocation system at a later stage. Another idea brought forward, but not contained in the final proposal was to prioritise slot allocation by assessing to what extent alternative modes of transportation are available between two city pairs.

Slot Mobility

The Commission was also interested to address the problem of slot trading in the amended Regulation. To stress the importance of the problem, aspects of 'slot mobility' – the transfer or the exchange of slots – are addressed in a new Art. 8a and no longer in a sub-section of Art. 8.

Consultation process With the exception of the United Kingdom and – to some extent – Luxemburg and the Netherlands, all member states were against legitimizing slot trading before an in-depth analysis of the impact on market access has been carried out. The Commission summarized the results of the consultation process with regard to slot trading as follows:

> [T]he large majority of industry stakeholders ... warned strongly against any market access measures that would deviate from established industry practice such as the IATA scheduling guidelines. The majority of industry stakeholders have not, at this stage, expressed support for unilateral slot transfers in form of slot trading as a means to introduce more slot mobility and incentives for the efficient use of slots. In areas of the world, like the United States, where slot trading has been a method of market entry, past experiences have not led to the reinforcement of the competitive situation at the most saturated airports. This may be due to the fact that slot trading favored carriers with significant financial strength and large slot portfolios of grandfather rights. It is also difficult for air carriers to know when slots are available and to bid for them. Unrestricted slot trading therefore risks simply reinforcing the dominant positions of incumbent carriers at congested European airports. (European Commission, 2001, p. 3)

As a result of the consultation process, the European Commission was unwilling to introduce market access measures in the amended regulations. In contrast, the European Commission decided to suggest much stricter rules to seal any loopholes that had emerged in the past and were actively used on the grey market as illustrated by the Guernsey case.

The European Commission's proposal The European Commission suggested Art. 8a to read:

(1) Slots may be ...

(b) transferred
- between parent and subsidiary companies, and between subsidiaries of the same parent company
- as part of the acquisition of the majority of the capital of an air carrier
- in the case of a total or partial take-over when the slots are directly related to the business taken over.

(c) exchanged, one for one, between two air carriers where both air carriers involved undertake to use the slots received in the exchange.

(2) Slots cannot be transferred in any way between air carriers or between air carriers and other entities with or without monetary compensation other between those air carriers referred to in Art. 8a(1)(b). (Ibid., p. 31)

To effectuate this provision, the Commission additionally suggested that a slot-coordinator should be able to withdraw slots if a '*fake*' exchange had taken place (ibid., p. 11). The Commission's definition of a fake exchange was exactly the scenario from which the Guernsey case had resulted – when practically unusable slots are exchanged for peak hour slots along with monetary compensation and thereafter returned to the pool. Under the Commission's proposal, the coordinator would have had the right to refuse the exchange if he is satisfied that one airline was actually planning not to use the slots (ibid., p. 18). Pursuant to Art. 14(3), he would additionally have had the right to withdraw the slots that had been exchanged if these were not operated by the carriers as intended.

The political agreement of the European Council The Commission's proposal, backed by the European Parliament in its first reading of the draft, was met with resistance by the European Council in December 2003. Particularly the United Kingdom opposed the suggested explicit prohibition of secondary slot trading – the declared key issue for the United Kingdom. To reach a political agreement on the amending Regulation, a compromise text was finally agreed upon.[19] While Art. 8a(2) was completely deleted, Art. 8a(1)(c) now simply reads: 'Slots may be exchanged, one for one, between air carriers.'

Surprisingly, Art. 14(3), which only makes sense in conjunction with the original wording of Art. 8a(1)(c), was not deleted from the proposal. Slightly reworded, it still reads: 'The coordinator shall withdraw and place in the pool the series of slots of an air carrier which it has received following an exchange pursuant to Art. 8a(3)(c) if they have not been used as intended.'

It is difficult to understand what the purpose of Art. 14(3) could be as the 'intention to use' can only be defined if an 'undertaking to use' has been given when the exchange took place. An 'undertaking to use', however, is no longer required as a result of the last minute changes by the European Council. The situation is even more confusing as from the United Kingdom's point of view,

---

19 See Council of the European Union, 16257/03 AVIATION 265 CODEC 1856 (18 December 2003).

these last-minute changes to Art. 8a safeguard the possibility of secondary slot trading in accordance with the High Court ruling in the Guernsey case.

Assessment The new regulatory regime has not led to changes as far as secondary slot trading is concerned. Neither is secondary slot trading explicitly allowed nor is it explicitly forbidden. It will be up to a thorough interpretation of Art. 8a by the national courts or eventually the ECJ whether secondary slot trading is prohibited by the Regulation.

Both the arguments of the High Court of England and Wales and of those who derived a prohibition of secondary slot trading from the old Art. 8(4) need to be reassessed. It might be argued that the last minute changes to the new Art. 8a simply left the traditional system untouched, allowing an interpretation in both ways. On the other hand, one has to take into consideration that the changes were made, at least according to the United Kingdom's understanding, to safeguard the possibility of secondary slot trading. From a methodological point of view, this could make for a somewhat different approach to a genetic interpretation of the statute necessary.

The European Commission, however, is of the opinion that Regulation 793/2004 has not changed the status quo. i.e., that exchanges of slots for financial consideration is not permitted under Community Law (European Commission, 2004, para. 2.3). While this opinion is not binding for the European Court Of Justice, it appears unlikely that the ECJ would rule to the contrary. The European commission also admits that the current system is unable to prevent a grey market for slots in which uneven exchanges of slots occur (ibid., para. 2.6).

Regulatory Reform Part II: The 2004 Consultation

Consultation

It is probably fair to state that the outcome of the regulatory reform has caused more problems than it has solved – at least as far as slot trading is concerned. Whatsoever, the discussion has not come to a halt since Regulation 793/2004 entered into force at the end of July 2005. As a result of the failure to bring its proposal through the Council, the Commission published a consultation paper on 17 September 2004 which initiated a consultation period addressing the problem of slot mobility again (ibid., para. 2.3) and will eventually lead to yet another amendment of the current regulatory regime (see Paylor, 2005, p. 52). The Commission is of the opinion that the current administrative system is fundamentally flawed as it lacks incentives for carriers to make unused slots available (European Commission, 2004, para 2.3).

In its consultation paper, the Commission has submitted a proposal that would represent a radical departure from the current system of slot mobility.

The alternatives (which could eventually be combined) proposed for discussion all differ in detail, but would all lead to the creation of a market for slots:[20]

1) slots will be subject to primary trading: airlines would be forced to pay for slots on a seasonal basis at rates set by the airport operator depending on demand and supply. As a result, the market for slots would be regulated through additional charges set by the airport operators;

2) slots will be subject to secondary trading, albeit not directly between airlines but via the airport's slot coordinator who would publish all relevant details and seek the best bidder for slots an airline is willing to surrender (with the airline not necessarily receiving 100 per cent of the sale price). Intrinsically linked to this problem is the question of ownership of slots under civil law, and, consequently, if non-airlines, e.g., financial institutions, should be able to own slots and lease them to airlines;

3) slots will be auctioned off: those slots that become available on an ad hoc basis have to be auctioned off rather than allocated. As this would not guarantee the availability of sufficient slots, additionally all slots at an airport have to be auctioned off over a ten year period, i.e., 10 per cent of all slots at an airport have to be made available every year;

4) a certain percentage of grandfathered slots will be redistributed at airports where requests have neither been satisfied through slot allocation nor through slot trading over a three year period.

Support for this departure from the current regime will most likely come from the United Kingdom which has long advocated a market-based approach to slot allocation. The United Kingdom calls for an immediate introduction of formalised secondary trading across the EU[21] and plans to push ahead with the submission of legislative proposals during the UK Presidency of the EU in the second half of 2005.[22] It has, however, acknowledged that it expects resistance from many of the member states.[23]

---

20   For reactions to the proposal on secondary slot trading, see IATA (2004), p. 6; CAA (2004), p. 5.

21   See Letter from the Secretary Of State For Transport to the House Of Lord's Internal Market Sub-Committee B, dated 7 December 2004.

22   See the Operational Programme of the Council for 2005, submitted by the incoming Luxembourg and United Kingdom Presidencies, para 44 (http://www.fco.gov.uk/Files/kfile/UK-Lux%20prog.pdf).

23   See Letter from the Secretary Of State For Transport to the House Of Lord's Internal Market Sub-Committee B, dated 7 December 2004 : 'Support for the Commission's suggestions among other Member States is likely to be limited. No other Member State has the same extremity of capacity and slot constraint as the UK, and many are strongly opposed to changing the present system.'

Legal Challenges in a World of Slot Trading

Market control through competition law Although the outcome of a renewed reform process cannot be predicted at this point, it seems likely that a market based approach will eventually prevail. While the resulting creation of a market for secondary slot trading may solve some problems for economists, it will create a plethora of new problems for lawyers. In a system where slots are allocated under an administrative mechanism, the control of market forces is not necessary. If a free market is created, market forces need to be controlled by competition law to avoid the emergence of a dominant market position by one airline or alliance – or, as competition law puts it, by one 'undertaking or a group of undertakings' – and the possible abuse of this dominant position. Such an abuse is prohibited by both EU and national competition law.

Accumulation of slots Competition law does not prohibit dominance as such, but only its abuse. It could not stop the accumulation of a dominant slot position by a single airline, e.g., British Airways at Heathrow or Lufthansa at Frankfurt, even to the point where an airline holds all slots at a congested airport because it has the market force to foreclose a market by acquiring all available slots. It is open to discussion whether the pure accumulation of slots as such is undesirable and will eventually lead to abuse, as some airports in the US are dominated by a single airline – e.g., Atlanta Hartsfield by Delta Air Lines, Dallas/Fort Worth by American Airlines or Houston Intercontinental by Continental Airlines – without a resulting market failure. Unless such abuse can de demonstrated, competition law cannot control the accumulation of a slots (unless it is achieved through a merger of some size). If desired, the accumulation of slots has to be controlled through a regulatory solution, e.g., by imposing restrictions on the buying of slots above a defined 'cap'.

Hoarding of slots Another problem to be considered is the possible hoarding of slots by an airline. In a competitive market, a dominant airline might be unwilling to sell surplus slots (or ask for monopolistic prices) knowing that the competitor would use it for new or additional services competing with those of the seller. This might be an incentive to hoard slots rather than to make them available to the market as the revenue gained through the sale might be offset by the additional competition involuntarily created by the sale in the long-term. Competition law can address this problem only in exceptional circumstances where the essential facilities doctrine applies. Again, only a regulatory mechanism can provide a solution below this threshold. Currently the 'use it or lose it' rule to some extent controls the denial of spare capacity, although it cannot address the problem of clearly inefficiently used capacity.

Summary

- Since 1993, slot allocation in the EU member states follows EU Regulation 95/93 which is modelled after the IATA scheduling guide. Slots are not subject to primary slot trading, but allocated by a national slot-coordinator.
- EU regulation 95/93 does not explicitly mention secondary slot trading, but only addresses 'slot exchanges'.
- Anecdotal evidence shows that a grey market for secondary slot trading exists in the European Common Market.
- The European Commission has expressed its view that under the current regulatory regime secondary slot trading is illegal, while an exchange of slots without financial considerations is lawful.
- When interpreting the Regulation from a legal point of view, legal methodology supports the view that any deal that includes a slot exchange accompanied by a financial consideration is in violation of European Law.
- The British High Court, however, has approved of a secondary slot trading deal in a decision from 1999 which was disguised as an exchange of slots by the airlines involved.
- Under an amended Regulation 95/93 that came into force in July 2005 secondary slot trading remains unlawful.
- A proposal by the European Commission calling for stricter rules to seal any loopholes that have emerged in the past and which are actively used on the grey market, was approved by the European Parliament but did not pass the European Council.
- In September 2004, the European Commission has initiated a consultation process and submitted a proposal that would represent a radical departure from the current system of slot mobility by allowing primary and secondary slot trading.
- If secondary slot trading becomes lawful, resulting legal problems need to be addressed by the regulator from the outset. These include market control through dominant carriers, the accumulation of slots and the hoarding of slots.

References

Balfour, J. (1997),' Slots For Sale', *Air and Space Law*, 22, p. 109.

Brachmann, U. (1994), *Die rechtlichen und tatsächlichen Möglichkeiten zugunsten eines effektiven Wettbewerbs trotz überlasteter Flughäfen im europäischen Luftverkehr*, Konstanz.

Buyck, C. (2004), 'A Difficult Delivery', *Air Transport World*, February.

Civil Aviation Authority (1998), *The Single European Aviation Market – The First Five Years*, London: CAA.

Civil Aviation Authority (2001a), *The Implementation of Secondary Slot Trading*, London: CAA.

Civil Aviation Authority (2001b), *Estimating Demand Valuation*, London: CAA.

Civil Aviation Authority (2001c), *Heathrow, Gatwick, Stansted and Manchester Airports' Price Chaos 2003–2008*, London: CAA.

Civil Aviation Authority (2004), *Introducing Commercial Allocation Mechanisms: The UK Civil Aviation Authority's Response to the European Commission's Staff Working Paper on Slot Reform*, London: CAA.

European Commission (2000), *Study of Certain Aspects of the Council Regulation 95/93 on Common Rules for the Allocation of Slots at Community Airports* (report prepared for the European Commission by PricewaterhouseCoopers), Brussels.

European Commission (2001), *Proposal for a Regulation of the European Parliament and of the Council Amending Council Regulation (EEC) No 95/93 of 18 January 1993 on Common Rules for the Allocation of Slots at Community Airports*, COM (2001) 335 final, Brussels.

European Commission (2004), *Commercial Slot Allocation Mechanisms in the Context of a Further Revision of Council Regulation (EEC) 95/93 on Common Rules for the Allocation of Slots at Community Airports* (Staff Working Document), Brussels.

Gow, D. (2004), 'BA Outbid for Heathrow Slots', *The Guardian*, 21 January.

Grether, D., Isaac, M. and Plott, C. (1989), *The Allocation of Scarce Resources*, Boulder, CO: Westview Press.

Hüschelrath, K. (1998), *Infrastrukturengpässe im Luftverkehr*, Wiesbaden.

IATA (2004), *Comments On The European Commission's Staff Working Document Regarding Commercial Slot Allocation Mechanisms*.

Kilian, M. (2000), *Der Handel mit Slots*, Transportrecht 2000, p. 159.

Knieps, G. (1996), *Slothandel als marktwirtschaftliches Instrument bei Knappheitsproblemen an Flughäfen*, Freiburg.

Langner, S. (1996), *The Allocation Of Slots In The Airline Industry*, Baden-Baden.

O'Connell, D. (2004), 'Soaring Costs of Touching Down', *Sunday Times*, 22 February.

Paylor, A. (2005), 'The Slots Game', *Air Transport World*, April.

Ulrich, C. (1996), ,Die Flugplankoordination der Bundesrepublik Deutschland', in DVWG (eds), *Drittes Luftverkehrsforum*, Bergisch-Gladbach, p. 17.

Wolf, H. (1995), *Möglichkeiten und Grenzen marktwirtschaftlicher Verfahren zur Vergabe von Start- und Landerechten auf Flughäfen*, Kiel.

Chapter 15

Slots, Property Rights and Secondary Markets: Efficiency and Slot Allocation at US Airports

Daniel M. Kasper

Introduction and Overview

Despite the fact that air travel has increased faster than US airport capacity for many years, congestion-related delay does not yet appear to be pervasive problem for the nation's aviation system. Rather, the vast majority of delays are caused by weather, not congestion, and thus, as explained below, are unlikely be ameliorated by the adoption of market based methods for allocating airport/runway access. As shown in the following exhibit based on data compiled by the FAA, volume-related delays accounted for less than 15 per cent of all delays in each of the four years preceding September 11, 2001, and the percentage of delays related to volume has declined steadily over the past decade. In contrast, weather-related delays consistently account for over 70 per cent of all delays.[1]

Nonetheless, there was a significant increase in delays in 2000 that contributed to the perception of a delay 'crisis.' This increase, however, was due in large part to the increase in flights at New York's LaGuardia Airport (LGA) precipitated by the passage of legislation (known as 'AIR-21') that removed the limits on flights operated by small commercial aircraft at LGA that had previously been imposed under the High Density Rule (HDR). Thus, in September of 2000 (i.e., immediately following the implementation of AIR-21), delays at LGA alone accounted for 25 per cent of flight delays for the entire country (US DOT and FAA, 2001, p. 10). Following the FAA's subsequent re-imposition of flight limits at LGA, delays both at LGA and nationwide fell significantly.

In addition, delays during the summer of 2000 were inflated as the result of a labour dispute involving pilots at United. This labour dispute caused a substantial increase in reported flight delays during the summer months of 2000 and resulted in delays of more than 15 minutes for 57.8 per cent of United's arrivals during July of 2000. As a result, United alone accounted for 30 per cent of all reported

---

1   Moreover, existing delays appear to be concentrated at only a handful of major airports. See US Department of Transportation and Federal Aviation Administration, p. 2.

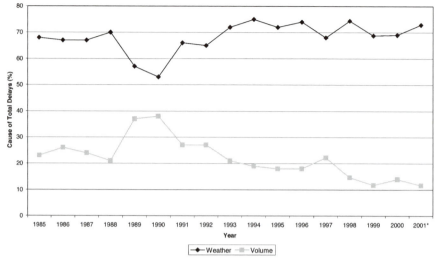

Notes: Volume Delays include Airport Terminal Volume and Air Route Traffic Control Center Volume. 2001 data is for Jan-Aug.
Source: *Aviation Capacity Enhancement Plans*, U.S. DOT and FAA

Figure 15.1 Source of air carrier delays, 1985–2001

delays. Nearly a third (31.5 per cent) of United's scheduled flights were late 70 per cent or more of the time during July of 2000 compared to 0.5 per cent for American and 0.2 per cent for Delta during that same period, and 0.7 per cent for United in July of 2001 US Department of Transportation, 2000, Table 6). With the elimination of these significant (and non-recurring) events, the total number of delays declined substantially in 2001 – even prior to September 11.[2]

As shown in the chart below, on-time arrival rates since at least 1989 have generally ranged between 75 per cent and 85 per cent. Notwithstanding the well-publicized delays of 2000 and the fact that the number of aircraft operations grew roughly 15 per cent from 1989 to 2001,[3] on-time arrival rates have remained within that same range in recent years, including 2000.

With the exception of the airports subject to the 'high density rule' (HDR),[4] runway access at US airports has been allocated historically on a 'first come, first

---

2   '[B]eginning in March 2001 the number of delays declined in every month except August. From April – June 2001, delays declined by 11.21 per cent compared to the same period in the previous years. During June, July and August, when convective weather disrupts many operations, delays were down by 7.99 per cent from the previous summer' (US DOT and FAA, 2001, p. 17).

3   Air carrier and commuter operations at airports with FAA traffic control service grew from 20.80 million in 1989 to 23.84 million in 2001 (*FAA Aerospace Forecasts*).

4   Airports subject to the HDR are LaGuardia (LGA), Kennedy (JFK) and Reagan Washington National (DCA). O'Hare (ORD) was exempted from the HDR as of 1 July 2002.

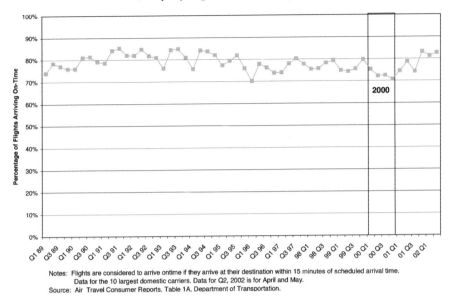

Notes: Flights are considered to arrive ontime if they arrive at their destination within 15 minutes of scheduled arrival time.
Data for the 10 largest domestic carriers. Data for Q2, 2002 is for April and May.
Source: Air Travel Consumer Reports, Table 1A, Department of Transportation.

Figure 15.2 On-time arrival rates 1989–2002, Q2

served' basis.[5] Airport runway (and other airfield) costs are typically recovered
via weight-based landing fees set at the level required to recoup the actual airfield
costs incurred by the airport to provide, maintain and operate the airfield.

When available runway capacity is sufficient to accommodate all existing
demand – as is typically the case at most US commercial service airports – 'first
come, first served' is widely accepted as the least costly way of attaining an efficient
allocation of runway resources.[6]

When demand for access to an airport exceeds the existing runway capacity for
a significant period of time, however, it can lead to congestion-related delay. But
not all congestion or delay is inefficient: As with any product, there is an efficient
level of delay.[7] Congestion-related delay becomes an economic problem only when
the full 'social' costs imposed by the use of a congested runway (including delays

---

5   Federally-imposed administrative limits on operations at certain other airports
were temporarily imposed following a strike by air traffic controllers, but these restrictions
were subsequently lifted.

6   For the vast majority of US airports, which are not chronically congested, the
prevailing 'first come, first served' system thus remains the most efficient method for
allocating runway access.

7   These are delays that would be too expensive to eliminate; that is, delays that users
would be willing to tolerate rather than pay enough to eliminate them. For example, airlines
(and their customers) would be unwilling to eliminate the possibility of weather-related
delays because the cost – limiting the number of flights year round to the level that could be
operated in bad weather conditions – would be too high for airlines and their customers.

imposed on other aircraft and passengers) exceed the private costs (i.e., the costs borne by those seeking to use the runways when they are congested).[8]

Since adopting and implementing measures to reduce delays would be costly, the elimination of some delays is not economically worthwhile – i.e., users would be willing to tolerate such delays when that would be less costly than eliminating them. For example, airlines (and their customers) presumably would be unwilling to eliminate the possibility of weather-related delays because the cost – limiting the number of flights year round to the level that could be operated in bad weather conditions – would be too high for airlines and their customers. Likewise, users might be willing to tolerate a few congested hours per week if the costs of eliminating those delays exceeded the benefits of delay reduction to users (see, e.g., Brueckner, 2002; Meyer and Sinai, 2001).

For the vast majority of US airports (which are not chronically congested) the prevailing 'first come, first served' system thus remains the most efficient method for allocating runway access.

At a handful of high-demand commercial service airports in the United States, however, airport authorities have been unable or unwilling to add runway capacity. As a result, unrestricted demand would regularly exceed available runway capacity for many hours each day, thereby producing on a regular basis an economically inefficient level of congestion-related delays. Under these circumstances, policymakers have three basic options:

1) adopt measures that limit demand (demand management) to levels consistent with available capacity;
2) build new runways and/or adopt new technologies that increase capacity; or
3) do nothing (except what is required for air safety) and permit the 'first come, first served' system to determine the delay equilibrium.[9]

It should be noted that the most efficient solution to the problem of congestion delays may well involve capacity expansion (i.e., option 2, above).[10] The focus of my analysis, however, will be on option 1, specifically on the efficiency implications of alternative approaches to allocating runway access when demand (runway access) has been restricted.

---

8   See Coase (1960) for the seminal treatment of social costs.

9   With the addition of some occasional 'jaw boning,' the delay equilibrium option could be used to describes the FAA's approach to congestion at airports such as Boston Logan and San Francisco International which, while busy, may not yet face the same level of 'excess demand'(under the first come-first served system) as the HDR airports.

10   Although 'demand management' techniques can be used to suppress demand, failure to add runway capacity can be expected to lead to reduced economic efficiency and societal economic when the revenues generated by the new runway capacity would be sufficient to cover the full cost of those runways. In short, the economically efficient response to congestion-related delays may well involve expanding runway capacity. Thus, policies that focus almost exclusively on restricting demand rather than on facilitating capacity expansion can be expected to result in economically inefficient approaches to congestion-related delay.

Once operational restrictions are imposed to limit demand, issues arise as to how the limited runway access rights should be allocated. The allocation options can be categorized as either market-based or non-market based. Non-market approaches administratively allocate access rights[11] In contrast, market-based approaches rely on prices and markets to allocate scarce resources. Examples of market-based approaches include the use of congestion-based landing fees, auctions and, importantly, secondary markets.[12]

Using Market Based Systems To Allocate Runway Access

The Case in Favour of Market Based Slot Allocation

When operational restrictions are imposed to limit demand, questions arise as to how the limited access rights should be allocated. The allocation options can be categorized as either market-based or non-market based. Non-market approaches include 'grandfathering' (which allocates access rights to existing users),[13] administratively awarding such rights (e.g., by DOT's exemption authority or distributing them via lotteries). In contrast, market-based approaches rely on prices and markets to allocate scarce resources. Examples of market-based

---

11   When runway access restrictions were first imposed at HDR airports, access rights (called 'slots') were initially awarded to the carriers then serving each airport (i.e., 'grandfathered') based on each carrier's level of service at that airport when the HDR was adopted. As additional operations became possible, new access rights have been allocated by DOT, typically via its discretionary (exemption) authority as well as by lottery. Although slots at HDR airports were initially allocated to the carriers then operating them, most of the slots currently held by airlines serving HDR airports have been acquired by their present owners via purchase – often at substantial prices. Thus, to the extent that any scarcity rents that accrued to airlines as a result of HDR, they were captured by the original slot holders when they sold those slots. For subsequent acquirers, the cost of slots represents a significant capital investment.

12   Regardless of how the scarce access rights are allocated *initially*, however, a secondary market – i.e., one where slot holders are free to buy, sell and/or lease their slots – is essential to ensure that scarce runway access rights continue to be allocated efficiently. Indeed, as discussed further below, a secondary market is sufficient *by itself* to ensure the efficient allocation of slots.

13   Runway access rights at HDR airports were initially allocated to the carriers then serving each airport (i.e., 'grandfathered') based on each carrier's level of service at that airport when the buy/sell rule was adopted. As additional operations became possible, new access rights have been allocated by DOT, typically via its discretionary (exemption) authority and by lottery. Although slots at HDR airports were initially allocated to the carriers then operating them, most of the slots currently held by airlines serving HDR airports have been acquired – often at substantial prices – by their present owners. Hence, any scarcity rents that accrued to airlines as a result of HDR were captured long ago by the original slot holders when they sold those slots. For subsequent acquirers, the cost of slots represents a significant capital investment.

approaches include the use of congestion-based landing fees, auctions and, importantly, secondary markets.

Economists typically favour market-based over non-market allocation methods because experience has shown that markets tend to allocate resources more efficiently than non-market methods and because rising prices induce competitive firms to enter or expand existing capacity to meet market demand.

Perhaps the main benefit of market-based allocation methods is the expectation that their adoption would alleviate the volume-related delays that can occur when demand exceeds available capacity.[14] The anticipated delay reductions would result from increasing prices to a level where the demand for runway access equals the number of hourly flights that can be accommodated by the existing runways. In the short run, well-designed market mechanisms allocate scarce resources to those who value them the most, and hence, could be expected to put scarce resources to their most economically valuable uses.

Likewise, market-based allocation methods tend to encourage the efficient allocation of resources in the long run because, as capacity becomes scarce, rising prices are likely to attract the capacity-increasing investments needed to drive prices down to long run competitive levels. Thus, market-based allocation methods can provide both a signal that new investment is needed and an economic incentive for investors to make such investments.[15]

Some Important Caveats

Nonetheless, there are a number of circumstances where it is generally recognized that markets cannot be relied upon to produce an efficient allocation of resources. For example, where – as is common in the case of airports – sellers enjoy monopoly or substantial market power, unregulated prices charged for access would likely be higher – and output lower – than required for an economically efficient allocation of resources.

Likewise, where the cost of establishing and operating a market allocation system exceeds the expected benefits – as would be the case at airports where existing supply of runway capacity consistently exceeds demand for access to those runways – using auctions or congestion-based landing fees would only add costs without improving the allocation of runway resources. As discussed above, volume-related delay does not appear to be a widespread or significant problem for US commercial aviation.[16] In the few cases where inefficient levels

---

14   See Docket DOT-2001-9849 at 4397; also Docket FAA-2001-9854-1 at 31741.

15   As discussed below, however, even this theoretical virtue is open to considerable doubt in the case of airports since both the need for new capacity and the inability (or unwillingness) to provide it have been apparent at the nation's most congested airports for many years.

16   Even at Boston Logan, an airport identified as having unacceptable delays, the FAA has determined that 'airline over scheduling does not represent a significant cause of recent delays a Logan Airport' (Logan International Airport, June 2002, Final Environmental Impact Statement, at ES-38). Likewise, scheduled traffic at SFO and

of congestion-related delay have arisen because of chronic excess demand (e.g. at LGA), the FAA has dealt with the problem directly by capping operations at levels consistent with the available capacity.

A similar caution is in order regarding the ability of some market based allocation systems to provide information about the need for – and thereby to attract – new investments that would increase capacity. [17] There is little evidence, however, that an inability to raise the capital (rather than political or environmental opposition) has prevented the few seriously congested US airports from expanding their runway capacity. On the contrary, the added revenues that could be generated by adding runway capacity at these high-demand airports would likely be more than sufficient to fund the runway development.[18] And for airports where no new runway capacity is likely to be added regardless of whether any of the proposed options is adopted, it would be disingenuous to suggest that the adoption of a market-based allocation system would be necessary – or even helpful – in adding runway capacity. Thus, to a considerable extent, proposals to use auctions or congestion pricing to reduce delays at US airports appear to be a 'solution in search of a problem.'

In addition, the 'market based' allocation methods often espoused by policymakers and some economists have frequently failed in some important respects to satisfy the conditions necessary for market-based systems to allocate resources efficiently. For example, where the expected revenues generated by the new runway capacity are sufficient to cover its cost (circumstances that clearly exist at the few US airports facing significant congestion problems), adding runway capacity would be the best way to enhance economic efficiency and increase society's economic welfare. Yet policymakers often fail even to consider this efficiency and welfare-enhancing option.

Likewise, policymakers have often failed to recognize (or even to acknowledge) the critically important role of secondary slot markets in ensuring the efficient allocation of runway access rights.[19] Although well-designed auctions or landing fees might, in theory, produce an efficient initial resource allocation, on-going (i.e., 'dynamic') efficiency requires a cost-effective means of reallocating resources as market conditions and participants change. Hence, the ability to exchange resources in a secondary market is essential to maintaining overall efficiency.

PHL 'can be handled effectively during good weather conditions' (FAA 2001 Capacity Enhancement Plan).

17   Since that information is already available (or could easily be obtained) from the existing slot market for slot-controlled airports, the need for such information cannot justify the adoption of auctions, congestion-based landing fees for HDR airports.

18   For example, Chicago is proposing $6 billion for runway development and capacity enhancement at ORD that would be funded from existing revenue sources.

19   For example, the USDOT RFC failed to mention secondary markets although it specifically included auctions, congestion pricing, and peak-period pricing as 'market based approaches' (RFC, at 43948). Nonetheless, there are numerous examples of secondary markets throughout the economy, including major stock, futures and commodities exchanges, as well as more recent creations like eBay.

Finally, the adoption of either auctions or congestion pricing runs the risk of creating economically perverse incentives for airport proprietors, a problem that has been recognized by numerous governmental authorities that have considered permitting airports to levy 'market-based' access charges.

> [A]irports are often monopoly providers of service within their particular regions, and there is a risk that they will have an incentive to under-invest in capacity in order to exploit their monopoly position .... If all revenue were collected by airport operators it would generate large financial surpluses which may blunt [their] incentives for efficient behaviour, including the investment needed to relieve the constraint. (Castles, 1997, p. 11)

Similarly, authorities in Australia concluded 'while higher airport charges may signal clearly the need for new investment, that investment may not be forthcoming if the airport retains the scarcity rents'.[20] And it was for these reasons, among others, that the USDOT promulgated its policy governing airport rates and charges.[21]

Issues Raised by Alternative Market Based Allocation Systems

Congestion-Based Landing Fees

Establishing efficient, market-clearing levels for congestion fees would not be a simple matter. The rate setting authority first would have to estimate the full cost of congestion-related delays to users – and, hence, how much they would be willing to pay to eliminate those delays (see Brueckner, 2002). Starting with that estimated price, authorities would then have to iteratively set and reset prices until they determined the level of landing fees required to produce the efficient number of flights. While this experimentation is underway, prices will almost certainly be set at inefficient levels, resulting either in a significant waste of scarce slot resources (caused by setting prices above efficient levels) or excessive congestion (when landing fees are set below efficient levels). As noted by the Federal Trade Commission's Bureau of Economics, 'whether or not a regulatory body or airport administrator would (or could) choose the appropriate levels of peak and off-peak prices both to avoid congestion and to utilize capacity optimally is not clear' (US Federal Trade Commission, 1994, p. 49). Moreover, since the FAA

---

20   Productivity Commission, 2002, p. 442: '[W]hile market clearing prices may provide good signals to airports of where new capacity is desired, they may not provide good incentives to actually deliver it at the socially desirable time' (quoting from UK CAA, 2001, p. xiv). It should also be noted that any scarcity rents accruing to airlines as a result of HDR were captured long ago by the original slot holders when they sold those slots. For subsequent acquirers, the cost of slots represents a significant capital investment.

21   Because 'ownership' of runway access rights would vest in users rather than with airports, the use of secondary market would also obviate the problem of airport market power that would result if slot were controlled by airports.

has already determined the effective runway capacities for US airports (as part of its responsibility for aviation safety), there is no reason to risk the significant efficiency losses likely to result from such experimentation. Instead of trying to experimentally determine both the efficient levels of capacity and price, it seems far more sensible to treat the capacity (e.g., defined as the number of slots perhour) as given and then to let the market determine the appropriate price.

Moreover, there is little or no economic advantage to be gained from congestion fees when the effective capacity of the runways has been already determined. As noted by the FTC Bureau of Economics:

> If that capacity is known and reflected in the hourly slot quotas, the prices at which slots trade will fluctuate as demand fluctuates: higher prices during high demand periods, and lower prices during low-demand off-peak periods. Thus, given the level of capacity, market forces will lead naturally to peak and off-peak prices. [And] the only information that regulators or airport administrators need is an estimate of the capacity of the airport. Since [that] capacity does not generally change from day to day or hour to hour, the degree of regulatory oversight necessary to administer slot-based regulation is modest.[22]

Moreover, since the demand for access rights (slots) can vary widely depending on the time of day, time of year and changing economic circumstances, congestion fees would have to be changed constantly in order to continually clear the market.[23] Thus, '[t]he amount of information required to implement peak and off-peak pricing may be formidable' and those responsible for setting such fees would likely find it extremely difficult to keep up with constantly changing demand conditions (Castles, 1997, p. 11). As a result, the administrative costs required to operate a system of congestion-based landing fees would be substantial. And, like most other airport costs, these administrative costs would be borne by airport users – including airlines, passengers and shippers. It therefore seems unlikely

---

22   Although '[s]ome critics promote the use of price rather than quantity [i.e., slot] regulation as the appropriate regulatory instrument, [t]he underlying reason for this preference is not always clear. While *market* prices are superior economizers of information compared to administratively set output levels, *administratively set* prices do not necessarily possess the same advantage [citation omitted]' (US Federal Trade Commission, 1994, p. 49).

23   '[T]he demand for air travel (both at peak and off-peak times) will fluctuate with changes in the business cycle, seasonally, and as a result of purely random events such as terrorist threats and the weather. If administratively determined prices cannot respond quickly to cyclical and random changes in peak and off-peak demand, the welfare of air travelers may be reduced. If prices are set too low, travelers may be faced with congestion and congestion related delays; if prices are set too high, airport capacity may be underutilized as the number of operations falls below levels necessary to control congestion. ... In practice, there may be difficulties in estimating and setting accurately the structure and level of airport charges which would match supply and demand for airport capacity taking into account the peak profile of demand' (US Federal Trade Commission, 1994, p. 49).

that – as a purely practical matter – congestion pricing would enhance economic efficiency at US airports.

Thus, even if congestion fees could, in principle, produce allocations as efficient obtainable from secondary market slot trading, the latter approach – defining enforceable and marketable property rights in slots – 'has a significant advantage … in terms of the level of knowledge required of the government.' So that 'it appears likely that such a [congestion] pricing structure is not superior to slot-based regulation' (US FTC, 1994, pp. 49–50).

Auctions[24]

Because auctions would entail the sale of a defined number of runway access rights (slots), they could avoid some the iterative efficiency losses inherent in an allocation system based on congestion fees. And like the existing secondary market for slots at HDR airports, any system for auctioning slots would require the clear definition of the legal rights that would be acquired by the winning bidders (i.e., making it clear what property rights are being transferred). Also like the existing secondary market system, *it is the act of limiting operations by creating slots – and not auctions – that would reduce congestion-based delays.* In short, auctions are merely a mechanism for allocating the slots.

Even if an auction system were adopted to allocate slots, however, a secondary slot market would still be necessary to ensure that the allocation of slots remained efficient between auctions.[25]

Nor is an auction system is necessary for the efficient allocation of slots since *regardless of how those slots were allocated initially*, a properly functioning secondary market would permit the reallocation of slots to their highest valued uses. Moreover, so long as slots can be freely traded in a secondary market, the initial allocation will affect wealth distribution rather than efficiency (cf. Coase, 1960).

In addition, if auctions are contemplated for the reallocation of existing slots (rather than simply the allocation of newly created slots), it would raise potentially thorny legal issues – unless slot holders were fairly compensated for the loss of their investments existing slots at HDR airports.[26]

Moreover, the design and management of an efficiency enhancing slot auction for multiple airports would require considerable expertise and expense as well as sophisticated software and bidding facilities. The complexities in designing an economically sound slot auction arise principally from the interdependencies inherent in the airline business. That is, 'there are strong interdependencies between the values of different slots, which are only useful to an airline as a component of

---

24   My analysis of auction theory and practice has benefited from the assistance of my LECG colleague, Darin Lee, an expert in the fields of auctions and game theory.

25   As previously discussed, first come, first served would remain the most efficient allocation method until demand for access/slots consistently exceeds the available runway capacity.

26   FN re takings issues.

a viable schedule. Airline's schedules establish the relationships between connecting services and these relationships determine the value of slots to individual airlines' (Castles, 1997, p. 10). For example, these schedule interdependencies mean that the value an airline serving LGA-DCA places on a 7.00 a.m. departure slot from LGA will depend on whether or not it also holds an 8.00 a.m. arrival slot at DCA. Similarly, there are important complementarities between slots at the *same airport* for different times of the day. That is, the value an airline places on an arrival slot will depend on whether or not it is able to secure a corresponding departure slot later in the day.

It is well established – both from a theoretical perspective and from the experience of numerous radio-spectrum auctions worldwide – that simultaneous auctioning of related resources is efficiency enhancing because it allows firms to aggregate portfolios of complementary resources (see, for example, Cramton, 1998 or Ausubel et al., 1997). Thus, to produce efficient outcomes, an auction for slots would require that all slots at an airport be auctioned simultaneously: 'To allow firms to establish meaningful service patterns, any auction should be a simultaneous one.'[27] Moreover, in order for an auction process to be able to allocate access rights efficiently throughout the national (or international) aviation system, the simultaneous auctioning of access rights at all slot-constrained airports would be required, thus adding considerably to the cost and complexity of the resulting auction.

To be fully efficient, moreover, an auction should also permit airlines to submit package bids (i.e., to make bids conditional on winning matched 'pairs' of take-off and landing slots at different airports or sets of slots at the same airport).[28] Indeed, three leading auction experts noted that 'the auction must allow package bids, in which bidders can express preferences for packages of items, rather than just individual items'.[29] But package bidding would require an extremely complex auction.

> The rules for a [slot] auction that took account of this interdependence would therefore [have to] be highly complex in order to allow, for example, for multiple contingent bids with different values bid depending on the complementary services under different outcomes from the overall auction. In practice, no such auctions have ever been implemented. (Castles, 1997, p. 10)[30]

------

27   Comments of the United States Department of Justice, Docket No. FAA-2001-9854-76.

28   For example, airlines should be able to bid for a matched slot pair involving a departure at LGA and an arrival at DCA. Likewise, they should be able bid on a portfolio of arrival and departure slots at any constrained airport which best fit their overall network schedule.

29   See Comments of Peter Cramton, Lawrence Ausubel and Paul Milgrom, Docket FAA-2001-9852, FAA-2001-9854.

30   The need for 'package' bidding in auctions with synergistic resources arises from what auction practitioners refer to as the 'exposure' problem. The exposure problem arises when bidders trying to assemble portfolios of complementary resources bid above their standalone value for individual objects, with the expectation of winning all desired resources, thus capturing economic 'synergies.'

Although there has been much research on the potential efficiency gains of package bids,[31] they have yet to be implemented in any major auction. The FCC is currently proposing to use package bidding in its upcoming auction for spectrum in the upper 700 MHz band. But implementing even this relatively simple package bidding system has proven to be exceptionally difficult in practice, and the upper 700 MHz auction has already been postponed five times while the FCC attempts to work through these (and other) difficulties.[32]

An auction involving package bidding would impose significant costs on both the FAA/DOT – to design the rules, software and bidding facilities necessary to administer such an auction – and on airlines that would be forced to devote considerable resources in preparing for the auction. By way of illustration, the FCC's proposed auction for spectrum in the upper 700 MHz band, which involves only 12 licences and 4,095 potential packages, is much simpler than any potential slot auction involving package bidding. Even though the prospective bidders in that auction had considerable experience in spectrum auctions, however, they expressed grave doubts regarding the cost and complexity required by the use of package bidding.[33] A slot auction utilizing package bids would involve millions if not billions of potential packages, a substantial portion of which airlines would be forced to evaluate. 'Compared to most of the other costs involved in conducting combinatorial auctions, bidder valuation costs are relatively less affected by advancing technologies, particularly when the asset valuation process requires substantial human inputs' (Ausubel and Milgrom, 2002).

---

31   For a survey of recent developments in package bidding, see de Vries and Vohra, 2005.

32   Developing an efficiency enhancing slot auction that allowed for package bids would be considerably more complex than for the FCC's upper 700 MHz auction. The current proposal for the upper 700 MHz auction involves a total of only 12 individual licences, which in turn generate 4,095 possible licence packages. But a slot auction involving even a single airport (i.e. ignoring for the time being the interdependencies between airports) would involve significantly more slots, and thus, substantially more potential packages. For example, auctioning only one arrival and one departure slot at a single airport for each of the 16 hours between 6.00 a.m. and 9.00 p.m. (i.e., 32 slots in total) creates over *600 million* potential packages even after restricting packages to have an equal number of arrival and departure slots. But the FAA's proposed auction option for LGA would require auctioning substantially more (i.e. 246) slots (see FAA-2001-9854-1 at 31747-31748) and the DOJ has asserted that 'To be effective in promoting competition, an auction should have a large number of slots available for bidding at regular intervals ...' (see Comments of the United States Department of Justice, Docket No. FAA-2001-9854-76).

33   For example, Verizon Wireless stated 'The level of complexity that combinatorial bidding brings to this auction greatly increases the amount of analysis by bidders and has a major impact on bidding strategy' (Comments of Verizon Wireless, DA 00-1075, p. 4). Similarly, Voicestream Wireless stated '... in spite of its substantial expertise, VoiceStream believes that the rules for combinatorial bidding are unusually complicated and that it will be extraordinarily difficult to create software to track the auction and develop bidding strategies' (Comments of VoiceStream Wireless, DA 00-1075, p. 4).

In sum, while auctions could provide a more efficient initial allocation of newly created slots than the administrative approaches heretofore favoured by US DOT, the development and implementation of an economically sound auction methodology for large scale application (i.e., involving existing slots and multiple airports) would be both highly complex and costly for both the auctioneer and the prospective bidders. Some of these complexities might be avoided by using a simpler auction format and then permitting slots to be reallocated in a secondary market. But if ultimate efficiency depends on the secondary market, that lays bare the primary wealth distribution/redistribution motivation for utilizing auctions for the initial allocation of slots. At least in terms of existing slots (as opposed to newly created slots), it is difficult to see an efficiency justification for this type of 'robbing Peter to pay Paul' approach to slot allocation.[34]

Secondary Slot Markets

For secondary slot markets to function properly, slot holders must enjoy cognizable legal ('property') rights in the slots they hold.[35] Although these ownership rights need not be full and complete (e.g., 'fee simple' title under English and American common law), they must be substantial enough in extent and duration to justify the expense of exchange. In the case of airport slots, this requires the recognition of private property rights in resources that many consider to be 'public property.'

Nonetheless, it is well established that conferring certain private rights to use a public 'commons' can lead to a more efficient use of the commons' resources. And it is also reasonably clear that while the distribution of those rights may affect the distribution of wealth (as between recipients and non-recipients), how wealth is initially distributed does not necessarily have significant efficiency implications.

In any event, this 'public v. private' debate has been resolved – temporarily, if not definitively – in the United States where some private property rights in slots at HDR airports have already been legally recognized. Thus, US airport slots are already subject to a market-based allocation system – and have been since the adoption of the buy/sell rule in 1985.[36] Indeed, the secondary market that exists in the United States for the purchase, sale and/or leasing of airport slots is the only system in currently in use that approaches a fully market-based system for airport access rights.[37]

---

34 In addition, if the use of auctions requires the taking of slots from existing slot holders, it is likely to result in litigation and considerable uncertainty in the financial markets where slots have been long viewed as airline assets and relied upon by lenders and investors as collateral.

35 This is true in the United States for HDR airport slots and, to a somewhat lesser extent, in the United Kingdom.

36 Since 1985, carriers have been permitted to buy, sell and/or lease slots at HDR airports. See Amendment 93–49, 50 Fed Reg 52195.

37 Certain property rights in slots at UK airports have also be recognized and these slots can be traded – but subject to more restrictive conditions than apply in the US market.

The ability of airlines to acquire HDR access rights in the secondary market provides an on-going mechanism for reallocating these scarce resources to new and more productive uses and users.[38] As explained by the FTC Bureau of Economics: 'An economically efficient solution ... would limit the use of the resource and allocate right of use to those who value them highest. The HDR ... [has] largely accomplished this by creating a slot market ... [that] allowed slots to be transferred to carriers with the most highly valued flights. Other things equal, the value of a given flight rises as consumer demand for the flight rises. Thus, the slot market – that is, the ability of carriers to buy, sell and lease slots freely – helps ensure that the flights offered are those that consumers value the most' (US Federal Trade Commission, 1994, p. 7).[39]

Further, FAA slot data indicate that the market for slots is a reasonably active one. During the six-month period from March to August of 2001, for example, the FAA recorded approximately 1,328 slot transactions (leases and sales).[40] Four hundred and twenty (420) of these transactions involved weekday slots. Based on 1,296 daily slots, this is *equivalent* to approximately one-third of all weekday slots being transacted once during this six-month period. The turnover for weekend slots is equivalent to a turnover of about 30 per cent of available weekend slots. And slot transactions do not appear to be limited to large network carriers.[41]

Nor does the buy/sell system appear to have imposed an unfair or disproportionate burden on new entrants. As the FTC observed: 'HDR [including buy/sell] promotes rather than limits new entry because it creates a market in which potential new entrants can obtain operating privileges' (US Federal Trade Commission, 1994, p. 2). By permitting slot holders to convert their slot holdings into cash, moreover, buy/sell provides incumbent slot holders with strong

---

38   Secondary markets have also been adopted to reallocate scarce resources in other regulatory contexts. E.g., The Clean Air Act permits the purchase and sale of emission credits in order to reduce emissions as cheaply and efficiently as possible.

39   'The HDR was adopted in order to allocate existing capacity. If that capacity is known and reflected in the hourly slot quotas, the prices at which slots trade will fluctuate as demand fluctuates: higher prices during high demand periods, and lower prices during low-demand off-peak periods. Thus, given the level of capacity, market forces will lead naturally to peak and off-peak prices. To implement [such] slot-based regulation, the only information that regulators or airport administrators need is an estimate of the capacity of the airport. Since airport capacity does not generally change from day to day or hour to hour, the degree of regulatory oversight necessary to administer slot-based regulation is modest' (US Federal Trade Commission, 1994, p. 7).

40   The FAA does not include one-for-one slot trades when counting transactions. If those transactions are included, the total number of slot transactions would be approximately 3,100 (rather than 1,328).

41   Source: LECG analysis of FAA slots data. Based on 1,296 daily slots. See Docket FAA-2001-9854-1 at 31747. Note that because some slots have been leased or sold more than once during this six-month period, the actual number of unique slots involved in transactions will be somewhat lower.

incentives to sell slots to those – including new entrants – who are willing to pay more for the slots than incumbents expect to earn by retaining them.[42]

Although it is often assumed that the combined effect of HDR and buy/sell has caused HDR airports to become more concentrated than airports not subject to HDR, that assumption does not appear to be correct.[43] Concentration levels at LGA and Washington Reagan National (DCA), for example, are lower than the average for other comparable, non-hub airports. This is shown in Table 15.1.

Table 15.1 HHIs at the largest 15 non-hub airports, 2001

Airport	HHI	O&D Rank
OAK	4.43	14
MDW	3.06	11
SJC	2.26	13
BWI	2.2	6
HNL	2.13	12
DCA	1.88	10
SAN	1.87	9
SEA	1.83	5
JFK	1.7	15
LGA	1.68	3
BOS	1.67	4
TPA	1.41	8
MCO	1.41	2
FLL	1.38	7
LAX	1.35	1
Mean	2.016	
Median	1.832	

Note: HHIs based on domestic O&D passengers.

Source: US DOT DB1A Database.

---

42   Indeed, the increase in new entry – particularly with small jet aircraft – has been so great as to prompt the the inclusion of an option in the RFC that would permit the imposition of minimum aircraft size limits at LGA. In addition, all of the recent exemptions from HDR at LGA have been awarded to new entrants and small carriers. As a result, such carriers currently operate a significant share of slots at LGA.

43   And even if HDR had produced increases in concentration, there is little reason to believe that the outcome would have been any different under any other market based allocation system – including auctions and congestion fees.

In addition, concentration at the most severely constrained HDR airport has been declining – whether measured by actual slot usage (arrivals and departures) or slot ownership. In terms of LGA slot usage, the HHI in 2001 was at its lowest level in a decade, as shown in Figure 15.3 below, while the HHI based on slot ownership fell from 1,747 in 1993 to 1,661 by April of 2001.[44] Likewise, concentration measured in terms of O&D traffic at LGA for 2001 was 1,680, a level that was lower than the 1996 HHI level of 1,720.[45]

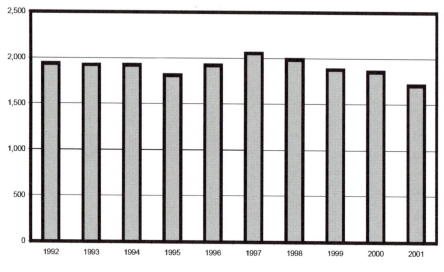

Notes: Aircraft Departures and Arrivals at LGA Source: U.S. DOT T-100 Database.

Figure 15.3 HHI of LGA flight operations, 1992–2001

Summary and Conclusions

Not all market-based allocation systems are created equal, at least when it comes to the allocation of runway access. In order for market-based approaches to the allocation of runway access rights to enhance efficiency, it is first necessary that demand for access to those runways exceed available runway capacity to the point of causing economically inefficient levels of delay. In short, runway capacity must be scarce. But according to FAA data, volume-related delays a problem only at a relative handful of US airports and are minimal on a national basis. For this reason, there appears to be little justification for most US airports to consider the use of auctions, congestion fees or other 'demand management' options.

----

44   1993 HHI is from US Federal Trade Commission, 1994; 2001 HHI was calculated by LECG based on FAA slot data for April 2001.
45   The average HHI for 1994–2000 using O&D traffic at LGA was 1,674 compared to 1,680 for 2001. Source: LECG analysis of US DOT data.

Where access rights are already subject to a market-based allocation/reallocation system – as is the case at HDR airports in the United States – adoption of other demand management measures cannot be expected to reduce delays or otherwise improve efficiency for reasons set out above. Rather, the primary effect of shifting from reliance on secondary markets to a system of auctions or congestion pricing would thus be to transfer wealth from airlines to airport proprietors – without adding runway capacity or otherwise alleviating delays and at the risk of causing considerable economic harm. Yet policymakers do not appear to have given much consideration to evidence that a secondary market is likely to be more effective than auctions or congestion fees in ensuring efficient allocation of scarce runway access rights.

Moreover, it is not clear why – or how – efficiency would be enhanced by replacing the existing secondary market system with a system of auctions or congestion pricing. And even if a system of auctions or congestion fees were adopted, secondary markets would still be required to ensure dynamic efficiency by continually reallocating access rights to their highest valued uses. Thus, whatever means are used in the initial allocation of runway access rights, a secondary market will be needed to ensure continued efficient allocation of those slots.

Congestion fees (including peak pricing) and auctions are also plagued by serious practical problems that raise grave doubts not only as to their effectiveness but also as to their very workability.[46] In this context, it is perhaps worth noting that at no commercial airport is runway access allocated solely – or even principally – by means of auctions or landing fees (including peak period pricing).[47] In part, this may be due to the fact that there are only limited airports where such methods might be able – even in principle – to improve efficiency. But the absence of auctions and congestion pricing may also be due to the fact that these allocation methods suffer from problems that appear serious enough to cast substantial doubt on their ability to allocate scarce runway access resources efficiently.

And in those relatively few cases where demand exceeds capacity and the potential for capacity expansion and other technical 'fixes' has been exhausted, policymakers should consider extending the existing system used at HDR airports to ensure that scarce runway access rights continue to be allocated as efficiently as possible over time. But regardless of the method used to allocate scarce runway

---

46   In addition, such approaches are likely to diminish the incentives for airport proprietors to develop new capacity in response to increasing consumer demand for runway access: if airport proprietors cannot profit from the scarcity value of slots, they are more likely to expand capacity in order to increase their revenues and profits.

47   Thus, even London's Heathrow and Gatwick airports, where landing fees substantially exceed the airports' actual (historic) costs, runway access requires airlines to hold time-specific slots. These slots were initially allocated based on historical usage. Subsequent users have acquired slots via acquisition (e.g., AA and UA by acquiring TW and PA's US-LHR rights) or by administrative allocation as new slots have been created.

access rights initially, a secondary market will is necessary to ensure that slots continue to be efficiently allocated over time. [48]

References

Ausubel, L.M. and Milgrom, P. (2002), 'Ascending Auctions with Package Bidding', unpublished working paper, University of Maryland and Stanford University.

Ausubel, L.M., Cramton, P., McAfee, P. and McMillan, J. (1997), 'Synergies in Wireless Telephony: Evidence from the Broadband PCS Auctions', *Journal of Economics and Management Strategy*, 6 (3), pp. 497–527.

Brueckner, J. (2002), 'Internalization of Airport Congestion', *Journal of Air Transport Management*, 8, pp. 141–47.

Castles, C. (Coopers & Lybrand) (1997), 'The Role of Market Mechanisms in Airport Slot Allocation', report prepared for the European Commission.

Coase, Ronald (1960), 'The Problem of Social Cost', *Journal of Law & Economics*, October, pp. 1-44

Cramton, P. (1998), 'The Efficiency of the FCC Auctions', *Journal of Law and Economics,* 41, October, pp. 727–36.

Meyer, C. and Sinai, T. (2001), 'Network Effects, Congestion Externalities, and Air Traffic Delays: Or Why All Delays are Not Evil', Unpublished Manuscript, University of Pennsylvania, Wharton School.

Productivity Commission (2002), *Review of Price Regulation of Airport Services*, Inquiry Report of the Australian Productivity Commission, January

UK CAA (2001), *Heathrow, Gatwick, Stansted and Manchester Airports' Price Caps*, November.

US Department of Transportation (2000), *Air Travel Consumer Reports.*

US Department of Transportation and Federal Aviation Administration (2003), *Operational Evolution Plan*, Version 3.0.

US DOT and FAA (2001), *Aviation Capacity Enhancement Plan.*

US Federal Trade Commission (1994), *Comment of the Staff of the Bureau of Economics*, Study of the High Density Rule, Docket No. 27664, 23 November.

Vries, S. de and Vohra, R. (2005), 'Combinatorial Auctions: A Survey', *INFORMS Journal on Computing* , 17 (4), January, pp. 475–89.

---

48   In short, as long as slot holders can freely engage in exchange (i.e., buy, sell or lease) their slot holdings in a functioning secondary, it makes little difference in terms of efficiency how newly-created slots (e.g., from an increase in runway capacity) are allocated to airlines.

PART E
Auctions and Alternatives

Chapter 16

Auctions – What Can We Learn from Auction Theory for Slot Allocation?

Kenneth Button

Introduction

Many major airports are already seriously congested. The Boeing Commercial Airplane Group (2004) are also very much in line with other sectors of informed opinion when forecasting that air traffic will grown, on average, at about 3 per cent per annum into the foreseeable future. It is unlikely that runway capacity at airports will increase at this rate, and certainly there will be many facilities where capacity will remain stagnant or grow more slowly. These are often airports where demand is high but, mainly because of environmental constraints, additional runways will not be constructed because of local opposition. There are also few 'quick-fix' technologies on the horizon, although enhanced air navigation equipment and air traffic control practices are likely to offer some relief. The challenge in these circumstances is to make the best use of the capacity that is available and to stimulate the adoption of efficient management practices.

The existing mechanisms for allocating take-off and landing slots[1] generally rely upon administrative procedures embracing a combination of formal and informal arrangements often overseen by a scheduling committee that includes incumbent users.[2] It is normal to have some 'grand-fathering' element in the institutional structure, and often there are limitations on the ability of incumbents to trade their slots with other carriers. There is no reason to assume that this type of arrangements, and variants of it, will result in anything like optimal use of slots (Jones et al., 1993). There are few incentives for reallocation between carriers as the demands for their services fluctuates even when this is permitted. It is difficult for new carriers to enter specific markets, and especially those dominated by a

---

1   A slot is defined throughout as the scheduled time of arrival or departure available or allocated to an aircraft movement for a specific date at a given airport. Most countries – or groups of countries as with the European Union (EU) – have regulations governing the ways slots are allocated. For details of the EU regulations see National Economic Research Associates (2004).

2   There are variants on this; for example in the UK, Airport Coordination Limited, a company owned by 11 airlines, allocates slots at 12 of the largest airports. It operates independently of its owners and allocates slots neutrally.

major airline.[3] Equally, there are less than maximum incentives for incumbents to make efficient use of slots they retain in terms of the congestion that is imposed on other operators. The problems are compounded by slot charging regimes based an accounting view of cost recovery as its pricing principle, and that seldom involve such features as full peak-load premiums.

Considerable intellectual attention has been expended on defining appropriate pricing principles for slots and in seeking ways to estimate the subsequent charges. Some of this work has been centered on placing monetary values on the various environmental costs of runway use (UK Department of the Environment, Transport and the Regions, 2000a) and on ensuring that the authorities operating the system have incentives to minimize allocative and X-inefficiency. This is not, however a topic that is at the center of the discussion here; it is assumed that more direct fiscal or command-and-control instruments will be deployed to deal with environmental issues and that the institutional structure underlying the operation of airports and air traffic control is economically efficient.

Other work has been aimed at looking at the ways administrative fees could be modified to make better use of slots. This has included some work on more efficient use of pricing for cost recovery when there is no congestion (Morrison, 1982) but has largely been focused on situations where there are at least some periods where demand under the current charging regime exceeds the fees being levied (Carlin and Park, 1970; Morrison, 1983; Morrison and Winston, 1989; Martín-Cejas, 1997). While the approach examined here to slot allocation is essentially about situations where there is excess demand under current regimes, it is not concerned with the estimation of optimal slot charges per se but rather with adopting an institutional structure that automatically reveals these fees and imposes them on users.[4]

One suggestion for significantly improving the slot allocation situation so that they are used more effectively is the adopting of some form of auctioning system. Unlike concepts such as Pigouvian congestion pricing of slots, auctions do not involve any external estimates of demand elasticities and marginal congestion costs; the prices are determined 'within the market'. The objective here is not to provide a literature survey of all the theoretical work on slot auctions – quite rightly this is running well ahead of any practical or politically acceptable system[5]

---

3   Humphreys (2003) provides data on entry levels at some European airports.

4   There are also situations where in the absence of congestion an airport cannot recover its full costs because of the existence of decreasing costs and competition from other airports. In these cases, there may be sound reasons for providing the capacity but a need to adopt differential pricing based on the 'willingness-to-pay' by airlines. In these situations, auctions can, if the airport is genuinely economically viable, extract rent from the airlines to cover the costs of slot provision (Button, 2005). As far as this paper goes, there are no distinctions made as to the nature of airport competition.

5   Textbooks on auction theory abound but Krishna (2002) provides a useful overview and Milgrom (2004) provides a more practical guide to their uses. McAfee and McMillan (1987) is a good survey of the basic literature from the economic perspective and Milgrom and Weber (1982) provides some idea of the complexities that can emerge in devising an optimal auction.

– but rather to look more generally at how auctions may be useful in slot allocation without completely disrupting current institutional structures. It, thus looks at the merits of auctions and how they may, in broad terms, help ameliorate some of the current inefficiencies in the airport slot allocation process. In doing this there are some references to the use of auctions in other markets where there are experiences of these types of mechanism.

The Slot Problem

The traditional way of allocating slots at airports in Europe and many other countries has been through some form of management procedure, and has generally involved a scheduling committee of interested parties and most notably the airport and incumbent airlines. To assist in co-ordination across airports – every flight requires a take-off and landing slot – there are various inter-airport and airline meetings; the International Air Transport Association (IATA), for example, has, since 1947, had twice yearly Schedule Conferences that now involve some 300 airline and 200 airport representatives. Grandfathering of rights has typically been honored in this process, in part to retain simplicity of a highly complex activity but also because of the desire, at least in the past, for continuity of service to passengers and cargo interests. Technically, however, it may also be seen in economic terms as a way to ensure that existing participants retain their economic rents (Button, 2005). The IATA rules do allow for new entrants; defined as airlines with negligible or no presence at the airport. The EU, for example, in its regulations is effectively encouraging grandfathering provided that the airlines holding slots do not hoard them (i.e., they must use them at least 80 per cent of the time or they go into a pool for reallocation – the 'use-it-or-lose-it' rule).[6]

Ignoring the fact that many individual scheduling committees have incumbent slot users as members and the biases that this may cause, this approach can be criticized generically in that the scheduling committee has far from perfect information regarding the economic benefits derived by the various carriers from holding any slot. Simply using a slot does not mean that a slot is being used to maximum efficiency. Without the threat of competition for the market there may arise demonstrable inefficiencies as incumbents pursue strategic policies to protect their overall networks through *de facto* cross-subsidization of services; essentially using a slot to keep out competition in other parts of its network.

Second, new market entrants often have a threshold of viability; simply gaining a few slots does not always allow them to enjoy network economies on the cost side and makes it difficult to build up economies of market presence on the demand and revenue side. Simply being able to gain a few slots when the vast majority is being grandfathered, as is the case in the EU, can stifle competitive market entry and limit pressures for incumbents to improve their efficiency. Added

---

6   The allocation procedure falls within the EU Single Market under European Council Regulation N0. 95/93.

to this, incumbent airlines have an incentive to 'over-bid' for slots that they will little use to limit potential competition.

The situation for slot allocation is somewhat different in the US. Here anti-trust laws have meant that airlines in general simply schedule flights taking into account expected air traffic control and airport delays. An exception to was initiated in 1968 under the High Density Rule that desgnated five 'slot-controlled airports' (JFK and La Guardia in New York, O'Hare in Chicago, Newark, and Washington Reagan National) where, because of extremely high demand for runways, the federal government limited the number of aircraft movements during certain hours although the airlines from 1986 could buy and sell, or lease, the slots designated for domestic use amongst themselves.[7] At La Guardia, because of demand, there was a freeze on slot numbers and from 2000 a lottery system adopted to allocate new slots. Lack of use of slots at these airports meant them being returned to the Federal Aviation Administration for reallocation.

Linked with the rights to use slots is the matter of the fees that are levied. These are largely cost related in a strict accountancy sense; for example, the International Civil Aviation Organization (ICAO) (2001) talks of

> The cost to be shared is the full cost of providing the airport and its essential ancillary services, including appropriate amounts for costs of capital and depreciation of assets, as well as the cost of maintenance and operation and management and administrative expenses, but allowing for all aeronautical revenues plus contributions from non-aeronautical revenues accruing from the operation of the airport to its operators.

Similar wording appears in many bilateral air service agreements. While this type of approach may have had some practical rationale when it was first devised – basically, in the international case, to prevent countries from manipulating rates to favor national carriers – it runs counter to the Anglo-Saxon economic philosophy that now forms the back-bone of many other air transportation policies.

In practical terms, while charging regimes vary between airports, this has meant that the majority of slot fees have been related to the weight of the aircraft; there are also fees for such things as parking and passenger arrivals/departures to reflect other costs to the airport. There are very few airports that vary their landing and take-off fees according to demand conditions; in Europe in 2004 only Amsterdam, Dublin, Helsinki, London Gatwick, London Heathrow, and Stansted did so and the peak-off-peak differentials were not strictly tied to temporal variations in demand even in these cases. This means that many airports inevitably suffer from severe peak-load problems, and in addition the revenues from the charging regimes are not providing useful insights into investment needs.

Added to this, many countries have interpreted these rules to apply across an airport system rather than to individual airports. The result is that in countries such as Spain, smaller airports have effectively been support by the two main

---

7   Newark was removed from the list in the 1970s, O'Hare was taken from it in 2002 but, after large scale entry, controls were reintroduced in 2004, and the others are due to be taken from the list in 2007.

ones (Madrid and Barcelona) and this, *ipso facto*, means cross subsidization of the airlines and passengers that use the smaller facilities.

The ICAO principles also raise issues concerning the overall level of fees. Whilst they do allow for capital cost recovery, they also require non-aeronautical revenues to be taken into account in setting fees. This has posed problems in cases where airports earn significant incomes from commercial, non-aviation activities such as retailing and car parks (Zhang, and Zhang, 1997). In the particular context of London Heathrow, for example, where concession revenues are particularly large, there have been extensive debates about how to regulate the pricing policy of the BAA, the company owning the facility, when it has income form two 'tills' (Starkie and Yarrow, 2000). This type of issue is not a major consideration here. It is assumed that any auctioning process would be designed to make the best use of runway capacity and that how the resultant revenues are then used is a matter of public policy. But having said that, it would seem logical that more rational use and pricing of slots would, at the very least, make cross-subsidies between commercial and runway activities more transparent.

Auction Theory

Auctions

'An auction is an market institution with an explicit set of rules determining resource allocation and prices on the basis of bids from the market participants' (McAfee and McMillan, 1987). In the context of airport slots the auctions involve buyers bidding for slots, although theoretically, and occasionally in practice, there are markets where it are the sellers that do the bidding.

The underlying economic theory of auctions is that the willingness-to-pay by a user reflects the welfare enjoyed by that user and that this in turn is a reflection of the social welfare created. Existing slots would thus be used to their maximum value because the airline gaining the greatest benefit would make the highest bid. This payment, when compared to costs of investment, also provides signals as to the social desirability of additional capacity or changing the operating practices of the airport. No-one claims that all the links underlying these short- and long-term mechanisms would ever work perfectly even if a fully efficient slot auction regime were established. Competition between airlines, even in 'deregulated markets' is not perfect and thus profits, and hence willingness-to-pay is an imperfect guide to social welfare. There may be wider social reasons, for example ensuring services to particular location, that out-weight private benefits. Also, there are indivisibilities in the capital employed by airports that make capacity decisions difficult.

Despite these potential imperfections, auctions are often seen as providing a more effective way of meeting social welfare criteria than alternatives and have attracted recent attention for their potential regarding airport slots.[8] In particular, since there is often limited competition between airports and the demand for slots at particular times of the day they create 'competition for the market' along the lines Demsetz (1968) and others have advocated. They recognize the monopolistic nature of the slot market but force the optimal use of slots through the auction mechanism that extracts the maximum economic rent from those holding them by selling short-term property rights.

Demonstration effects from the use of auctions in other quasi-monopoly markets have added to this recent interest. Auctions exist in a many spheres and are adopted in preference to other allocation devices for a variety of reasons. In particular, they are used in many countries in government procurement, in the issuing of government securities, in marketing new share issues, to allocate radio band width and television broadcasting rights, and to sell primary products ranging from flowers, fish and live-stock to minerals. While there is a temptation to devote excessive space to looking at the transferability of the experiences in these markets, this is avoided because of the specifics of slots. Suffice it to say that their effects have been mixed, in part this can be explained by the novelty and experimental nature of many of the applications of auctions to newly deregulated or privatized industries, but there has also been a learning experience on both sides of the 'market' that has not only led to better practices but also to quite a considerable expansion of our theoretical understanding of the issues involved.

The reasons for preferring auctions, or particular forms of auction, stem from the nature of the good or service being marketed. Demand and supply conditions, for example, may fluctuate rapidly rendering any notion of a standard value meaningless, or the product appears on the market so rarely that there is a need to rediscover its 'value' every time it does. There has been something of a global up-surge of interest in the actual use of auctions as an allocation device in recent years. The move towards more liberal markets in the 1970s and 1980s, with wide-spread deregulations and privatizations, required mechanisms that allowed effective competition 'for' natural monopoly markets in addition to competition 'in' markets that are more naturally competitive (Williamson, 1976). The subsequent advent of the worldwide web and other forms of advanced telecommunications has stimulated auctions at a much more disaggregate level by facilitating much larger participation rates. Technology has also changed the role of the auctioneer and, by initiating electronic auctions, removed any potential bias inherent in many traditional auction rooms.

There are four broad types of auction:[9]

---

8   For example, 'in order to make the most efficient use of capacity, a market in slots might be created. In the absence of time limits on grandfather rights, this would essentially mean newly created slots. But if there was a recycling process, the pool would be larger and would hold more peak-period slots' (UK Department of the Environment, Transport and the Regions, 2000b).

9   It is assumed throughout that the aim of the auction is to get the highest bid but in some cases, as for example tendering for subsidies to support bus transport in the UK, the aim is to get the lowest bid to provide a service. This has some relevance for providing

- English or ascending-bid auction. This is the most common form of auction for the sale of goods with an auctioneer allowing successively higher bids until only a single bidder is left. At each point in the auction, the various potential bidders know the prevailing highest bid;
- Dutch or descending-bid auction. In this model, the auctioneer gradually lowers the price until a bidder accepts it;
- first-price sealed bid auction. This approach involves bidders submitting sealed bids and the 'winner' is the one with the highest bid when they are opened. In this case the bidders have no idea what the competing bids are when they make their offer;
- Vickrey or second-price sealed-bid auction. This variant on the sealed bid model awards the 'prize' to the highest bidder but the winner only pays the amount of the second highest bid (Vickrey, 1961).

There are numerous variants of each of these (e.g., there may be reserve prices, charging an entry fee to bid, multi-rounds of bidding, auctioning bundles of goods rather than single items, etc), and analysts looking at airport slot auction models, as with those interested in other goods, have nuanced the institutional structures in seeking to meet the specific requirements of that market and currently available technology.[10]

The type of auction normally envisaged for landing and take-off slots at airports is the simultaneous ascending auction whereby all slots are sold periodically at a single auction – singularly or in combinations – for a given period with repeated rounds to ensure that carriers have viable take-off and landing slots for their services. Most analysis (e.g., National Economic Research Associates, 2004) assume, other say it is necessary, that the auction system will work in conjunction with secondary trading, which may or may not permit buying and selling involving monetary payments, to allow adjustments at the margin and ensure that airlines do not end up carrying slots that they cannot use or are short of slots at particular times between auctions.

The Nature of the Air Transport Market

The removal of specific pricing and market entry controls in the domestic US air market from 1978, the phased deregulation of the intra-European market from 1987 and completed by 1997, and the move international towards "Open Skies" bilateral air service agreements, coupled with the liberalization of many smaller air transportation markets, has transformed the way air transportations services are provided. There are more airlines, offering more diverse choices of service to passengers, a larger selection of routes, and generally lower fares. To provide this, however, the airlines have transformed the ways in which they operate.

---

social air services but is somewhat peripheral to this paper, although the general points made here seem to be applicable.

10   Starkie (1998) contains a discussion of secondary trading of slots.

Economies of scale, scope, and density, coupled with those of market presence, have fostered the creation of hub-and-spoke structures of service that has flights converging on a hub airport with passengers being transshipped to continue their journeys to final destinations (Button and Stough, 2000). Difficulties emerge not only because of concerns that hub dominance can lead to monopoly actions by slot holders – the charging of 'hub-premiums' – but also because there is a fear that congestion at hubs will lead to inefficient use of facilities[11]. While there is clearly some link between these two concerns, the first is more closely tied to the grandfathering of rights, and the second to inappropriate pricing of slots, meaning this may not always be a simple link. As Brueckner (2002) has argued, however, if there is hub dominance, then any congestion costs associated with slot utilization fall mainly on the shoulders of the major airline that will internalize these in its decision-making.

Airports have traditionally been owned and operate by state authorities, sometimes nationally but often regionally, and run as traditional public entities, often with some state financial support. There has also often been extensive cross-subsidization across state owned airports. The move towards deregulation and privatization has changed this in many countries. Whilst some countries still see airports as offer a public service where commercial principles have little place, others have added a different philosophy to reduce capture and X-inefficiency. Privatization has become common, although in a variety of forms. The UK, for example, has privatized its airports but has imposed price-cap regulation on the larger ones and has specific planning laws regarding capital expansion. Canada has adopted a not-for-profit model and the federal government has almost removed itself entirely from the sector beyond taking a broad oversight role. The ownership and the institutional constraints on management influence the way that slots are viewed, A 'legacy airport' with no profit motive has little incentive to look for better ways of suing slots – satisficing tends to be common. The move to greater commercialization, however, tends to favor change.

Air navigation services (ANSs) are as important in determining the slot capacity of an airport as the amount of tarmac on the ground. They are also important in terms of how slots may be allocated because they control the flow of aircraft into the system. Again national ownership has been the norm. Recent changes have seen the UK NATS being privatized, Germany's DFS moving in the same direction and other systems being commercialized through a variety of less dramatic institutional developments; e.g., the Canadian non-profit structure. It is too soon to say whether this will off-itself bring about major changes in the efficiency of ANSs, but, while these systems are neutral in terms of the airlines using them, it would seem wasteful to miss any opportunity for maximizing gains in ANS efficiency by not developing a mechanism that does not allow airlines that gain the most from making the most use of them.

---

11   National Economic Research Associates (2004) find there are seven capacity constrained airports in Europe throughout the day and 14 with capacity constraints during peak hours.

Slot Auctions versus Other Allocative Mechanisms

The prevalent form of slot allocation, essentially allocation by precedent and committee, has endured for a long time. Changing it will result in disruption, not only as new procedures are introduced but as airlines and administrators get used to the auction regime; essentially learning how to 'play the game'. Other alternatives have been considered, and some modifications to the traditional administrative approaches have been suggested (DotEcon, 2002). In weighing up their merits vis-à-vis auctions it is important to consider the objectives of the change. For example, if it is simply to provide more useful capacity then the picture is somewhat different to making better use of existing capacity. The practical problem in assessing different regimes is that while it is simple to say that the aim of any system is to maximize social welfare, the specification of the welfare function is often not explicit in policy setting.

In this context it is helpful to compare slot auctions with other suggested reforms. The comparisons are made in general terms and the implicit straw-man benchmark is the current grandfathering type of model used in Europe.

Modifications to the existing administrative methods of allocation The existing regimes of slot allocation have the advantage of familiarity to the airlines and airports. It has also traditionally been claimed that they have advantages for the ultimate users of air transport services – passengers and cargo companies – in that they inject a high degree of stability into the portfolio of services offered at an airport and, therefore, reduce personal and commercial uncertainties over investments based largely on transportation considerations. Setting aside any academic argument over whether a high degree of certainty has significant utility attached to it, the reduced stability that has accompanied the deregulations of airlines, and the subsequent structuring of the industry, has considerably weakened this position in practical terms; there is already quite a lot of uncertainty. Nonetheless, it is against this sort of background that some argue there may be a case for tweaking the system as has been done in the past to meet particular problems (Humphreys, 2003).

Such changes in recent years in Europe have seen coordination across many national regimes, the introduction of explicit 'use-it-or-lose-it' criteria, and the permitting of more extensive secondary trading and secondary markets. The EU came forward in 2004 with proposals to ensure that coordination/scheduling committees are neural in the ways that they treat airlines and can exercise discretion in meeting local needs together with new penalties for carriers that try to circumvent slot allocation rules; for example practices of over-bidding or repeatedly operating at times significantly different to the allocated slot. This does not exhaust the range of possible administrative reforms, some, for example, have proposed a cap on the number slots any airline may hold, and others that carriers pay a deposit on acquiring a slot that would be retained if use was below a defined threshold, and no doubt whatever reforms are introduced they will include an administrative element.

The difficulty with simply modifying the status quo is that it would seem to offer little guarantee of significantly more efficient use of scarce capacity. For example, even if there were more slots released for new entry the administrative approach provides little guidance on how an optimal selection process between the airlines is to be developed, and the overall process begs the question anyway about whether efficiency might not be better used by transferring more little used slots to another large operator at the facility.

Expand capacity A rather dated notion when there is a perceived 'shortage' of slots is that there is an automatic need to expand capacity to meet demand; basically build one's way out of any congestion problem. There may be good grounds for capacity enhancement in some cases but that does not mean there is an automatic need to add capacity, it all depends on how existing capacity is being allocated. Without adequate pricing signals from the fees charged for landing the estimates of whether additional runway capacity is economically justified amounts to a series of quasi-market calculations; essentially a cost-benefit assessment. These seldom give reliable impressions of how a market would react, although one suspects there is likely to be an aviation equivalent of the road engineering concept of 'Down's Law' – traffic expands to fill the freeway space available.

Congestion charges If the primary problem with the current regime of slot allocation is perceived as one of congestion, either at peaks or, as airports such as Heathrow, throughout the day, then adjustments to the fee structure and the removal of administrative allocation become an option. This approach to infrastructure allocation first suggested by Pigou (1920)[12] in the context of road traffic congestion treats the problem as one of negative externalities that require appropriate mark-ups to prices to reflect the associated inefficiency costs.

The literature on exactly what this charge should be is fairly extensive, and indeed the voluminous nature of these writings highlights the problem with congestion charges. They have to be set by an external body that needs to be able to calculate their appropriate level; in contrast auctions actually automatically reveal information about relevant parameters as well as acting as an allocation mechanism.[13] There are also complexities in tying the charges to investment needs and funding (Oum and Zhang, 1990). In conditions of a competitive supply of runway capacity and constant cost the link is a simple one; appropriate charges indicate where capacity is needed and provides the revenues to complete new works. This quickly breaks down where, as in the case of runway capacity, supply

---

12   In fact Pigou removed the discussion of congestion from later editions of *The Economics of Welfare* after being convinced that the underlying problem was, in modern terminology, one of inadequate property right allocation and that if a commercial company owned, in his case, the road in a competitive environment the problem would go away.

13   There is some US evidence that although there may be no premium on slots at most congested airports, there are higher fares at these airports that, through their effects on passenger demand, acts to limit the overall use of these facilities (Kleit and Kobayashi, 1996).

is not competitive and there are obvious technical non-linearities in the long-term cost function.[14]

Free allocation of existing rights and secondary markets If one accepts the Coasian line of reasoning (Coase, 1960) then the initial allocation of property rights to slots, be it by lottery, auction, a hand-over to incumbents or whatever, should, if, subsequent secondary markets are permitted, make no difference to the long-run efficiency with which slots are used. There will inevitably be a short-term transfer of rents between players, and there may be reasons to consider the potential short term disruptions that may accompany the various forms of initial allocation, but subsequent trading will ensure those that can make the most efficiently use of slots will be able to obtain the numbers they require.

In practice, there are important matters of political economy to consider as well as the actual nature of subsequent secondary markets. The initial allocation of slots and the ways rents are thus distributed is not a neutral concern. Windfall gains, for example, accrue if incumbents were give property rights that would not be enjoyed by those wishing to join the market; it is perhaps no accident that many legacy carriers find voice in supporting this give away approach (Sentence, 2003).[15] Lotteries by definition spread the rents randomly but there do arise issues about who can participate and whether there are any payments to be made by winners. There is also the matter of whether the lotteries should be for pairs of slots, combinations of slots, or groups of slots. Any secondary market would also have to be extensive because the initial allocation is likely to be far from optimal.

The matter of secondary trading in slots and the degree to which such a market would conform to economic notions of a competitive market also poses other challenges. As institutional economists frequently point out, markets do not arise and function in a vacuum but rather operate with a structure of formal laws and governances. In practical terms, normally there is a place for trade to take place – in the twenty-first century this can be an electronic exchange – there is the need for oversight of transactions, and methods of recording. These are not problematic and, indeed, exist in embryonic forms in several airport slot markets already and seem to impose minimal transactions costs. A more important concern is the possibility that the market will prove to be imperfect with monopoly power distorting its efficient functioning. In particular there may be concerns that participants will 'bank' certain slots that other airlines need to make efficient use of those slots they already have. The extent this can occur often depends not only on the underlying nature of the airline market and the details of the secondary slot market but also generic nature of competition laws in the country.

---

14   There has also been some work done on equity issues regarding existing charging regimes and possible modifications; e.g., Morrison, 1987).

15   The counter argument oft voiced by airlines is that the slots only have a value because the incumbents have developed them and that any revenue gained is a reflection of the risks and investments they have borne.

Auctions Without at this stage being specific about the particular form of auction to be used, the primary aim of a well-structured auction is make sure that the user who wins is the one that will generate the most utility from the use of a slot or bundle of slots. It is a neutral process whereby exogenous information has no influence. In addition to this, however, an auction also, though the revenues raised, both provides signals as to whether additional capacity is need and also provides resources for capacity expansion (Sentance, 2003). Indeed, the collection of a bid from the auction transfers the value of a slot to the airport and avoids the concerns with arbitrary initial allocations whereby airlines enjoy the wide-fall gains. Incorporating secondary markets into the structure allows for fine-tuning as markets evolve and to make technical adjustments for imperfections in the auction procedure.

A difficulty with this that is often raised is the matter of how auction revenues should be used; they effectively transfer economic rent from carriers to the airport authority (be that a private or public agency). If the airport is itself seen as a rent maximizer then the auction becomes a form of price discrimination with each slot being sold at a price that reduces its value to the airline purchaser to that of yielding a normal profit. The airport in this case would use the revenue to make investments that offer the highest return irrespective of the sector involved. Strictly, if an airport is congested then there are grounds for using the revenue for capacity expansion and the return would justify this. In reality there are often other considerations, and political expedience, that may involve it being used to reduce other air transportation taxes so that the overall impact is fiscally neutral. Of course, if these other taxes were initially distortive then this also produces efficiency gains. This, however, then comes down to a political rather than a decision in positive economics.

Auctions circumvents judgments on the part of any 'committee' as to who should have the rights to a slot although in some forms of auctions the role of the committee is not entirely eliminated. For example, if there are to be periodic re-auctioning of slots then the number and time frame is something determined outside of any market style process. Indeed, the nature of the auction itself, because of a variety of natural and institutional market imperfections, involves judgment and is not value neutral. As with many things, the application is always tempered by the complexity of application.

The Potential Forms of Slot Auctions

The variety of possible forms auctions may take poses challenges in isolating the most suitable for airline slots. There is also the matter of whether secondary markets are to be allowed and, if they are, the forms that they may take. Some of the concerns are to due with the nature of the product being sold, and thus unlikely to vary with time, but others are institutional in nature, and in discussion auctions in more detail we again largely focus on the potential for the European situation to allow comparison with existing allocation regimes.

Designing slot auctions

The focus is on the particulars of auctioning slots but that is really only part of the story. An airline to be able to offer a viable service also needs access to many others elements of the value chain such as ground handling and air traffic control. In particular, there is the need to be able to access gates. These are often let out on long leases or, in the case of the US, often owned by airlines. In true economic fashion this problem is circumvented in much of the literature on slot auctions by assuming that there is no problem in gaining appropriate gates. If this is not the case, any form of auction can be essentially captured, or at the least manipulated, by those holding the complimentary capacity. The monopoly power in the system is at another point in the chain but can still affect the overall size of the air transportation market and the ability of airlines to compete in it. Defining second-best rules for slot allocation in this world and how to modify auctions to cope with the imperfections elsewhere in the system is beyond our scope.

The detailed design of any auction system for slots is important. The four generic types of auction have already been outlined, but it is the simultaneous ascending auction that has attracted the most attention, but its exact form raises some complex problems (DotEcon, 2002; National Economic Research Associates, 2004); a number practical issues emerge, some of which have already been touched upon briefly.

Ownership Property rights, by definition, require a clearly defined legal framework of what is actually owned and by whom.[16] There is an agreed definition of a slot and that is taken as axiomatic. Traditionally slots have been seen as government property with the airlines, via the scheduling committees, having rights of use for a period; a sort of free lease. Where there has been a secondary market, this has in effect been done 'in good faith' with those holding the slots trusting that government would not take them back. This implies a rather vague legal background since sales and trades are technically of property rights owed by a third party. The sales to date also have taken place within a given institutional structure that has implied a fairly fixed supply of slots in the short term and strong indicators of any capacity changes in the future.

An auction system transfers the property right of each slot by law to the winner. This means that ownership has, by contract, been taken from government for a designated period. Whilst this should enhance transparency about who exactly has the property rights, the exact nature of the contract, penalties for violating the contracts of others, etc. may become complicated. Details of this kind need to be clear for an auction to function smoothly.

Coordination of take-off and landing slots and threshold levels Since a single slot is of no practical value, airlines want bundles of slots. At the minimum they want one take-off slot at the origin of the service and one landing slot at the destination.

---

16   Whilst focusing on the particulars of the UK situation, a useful discussion of ownership issue is found in Boyfield (2003).

This, however, because of economies of scale, scope and density is not likely to provide a viable package for a sustained service in most cases. Efficiency from an airlines perspective generally requires a threshold minimum number of flights to provide an attractive service to passenger.

A number of mechanisms have been suggested to meet these related issues. Rassenti et al. (1982) suggested, and tested the possible outcomes using experimental techniques, a combinational sealed bid auction in which the airlines bid for packages of slots at each period. Secondary markets for individual slots provide a mechanism for fine-tuning. An alternative approach is to have conditional bidding for slots that allows the airline to set conditions on its payments – e.g., it will pay $X for slot B provided that it wins the auction for slot A.

Inter-airport coordination As seen, airlines require pairs of slots at different airports to operate a viable service. Ideally this would involve both airports operating similar auctions to reduce the number of constraints imposed within the system. But even if there are similar systems, given that there are gaming issues involved, airlines would likely pursue a number of options given the uncertainty inherent in the system (e.g. purchase several slots to ensure that at least one matches or simply not bid for any slots if matches are thought unlikely). One way around the high potential inefficiencies is to make the game more complex, and reduce the long-run uncertainty of unfavorable outcomes, and the other is to offer non-terminal prizes. The latter involves, for example, the use of the secondary market, but if this course of action in adopted then this also could imply the primary allocation may be modified; for example the use of simultaneous multi-round auctions, combinational bids (a system where bids are for pairs of slots, one for take-off and another for landing), or a staggering of the market to feed more information into each round.

The coordination problem can take on additional dimensions when international jurisdictions are involved. Within large unified markets, such as the US or within the European Economic Area, appropriate standard auctioning procedures would allow slots to be allocated efficiently but there are, in practice, constraints imposed by the to need take into account the requirements of carriers serving airports outside of these markets. Nonetheless, even added flexibility for parts of the network would intuitively seem to offer the scope for wider gains.

Length of time for the auction process There are really two important time issues in the consideration of slot auctions. The first involves the length of time an airline has the right to a slot won in an auction. Technology changes take place over time as do variations in demands for particular slots that affect the flow of potential benefits from any slot. Posner (1972), although looking at the cable television industry, suggests where this type of situation arises, there is a need for repeated auctions at specified times. Such a structure does beg practical questions about the period between auctions and about whether there may be an advantage to incumbents in repeat auctions because of their knowledge of the market. It does, however, also have a practical advantage in that the auctions could be phased so that, for example, only 20 per cent of slots come up for re-auctioning in any one year; this allows more complete information to be generated in the early rounds of auctions as experience is gained and allows new entrants to compete for larger numbers of slots than they may be able to obtain in secondary trading.

The second overall matter concerning time is of an altogether different kind and essentially concerns transactions costs. There are really two underlying issues here. The first concerns the 'product' being auctioned. At present a slot gives its holder a 15 minute window during which a flight may land and or take-off. This allows for any minor problems on the ground (e.g., tarmac congestion) or during the flight (e.g., unusual wind conditions). By changing the duration of these windows it would be possible to create more slots; for example by making the windows longer but giving allow traffic control more flexibility in the way the flow of planes is handled.

The second transactions cost issue concern the efficiency of auctions. Auctions can become extended and complex exercises that take a considerable amount of time before the final contract is made. The existing IATA twice-annual conference system has a major advantage in that agreements are reached fairly expeditiously at meetings. Many of the more complex auctioning structures involving multiple auctions that have been explored by academics, whilst offering the theoretical prospect of an optimal outcome, are often time consuming. The simultaneous multiple round auction[17] used in a number of countries for spectrum auctions, for example, may take many iterations but then the license may be for a number of years.

Coordination with terminal and stand capacity   The focus has been on the particulars of auctioning slots but that is really only part of the story. In particular, there is the synergetic need to be able to access gates at appropriate times. For an airline to offer a viable service it also needs access to many others elements of the value chain including such terminal as ground handling. The possible inability to obtain these complementary facilities introduces additional risk in to the bidding process.

The problem is compounded when those supplying these non-slot services have a vested interest in the slot market. In practice, gates are often let out on long leases or, in the case of the US, often owned by airlines, similar links exist between ground handling and many airlines at a number of airports; a feature that has attracted the attention of the European Commission in the past. Recently a number of airlines have also been involved in vertical integration and invested in other elements of the supply chain that complement the slot market; e.g., the airlines with shares in UK NATS air traffic system. In true economic fashion, this problem is circumvented by assuming that there is no problem in gaining appropriate gates. In this is not the case any form of auction can be essentially

---

17   In these auctions, after an initial round of bids the auctioneer calls for a second round with the highest bids from the first (sometimes modified) are set as floors, and so this goes on until an equilibrium is reached.

captured, or at the least manipulated, by those holding the complimentary capacity.

Secondary Markets

Some form of secondary system of allocation is generally seen to be a necessary concomitant of any auction mechanism. The area between the most obvious of these – secondary slot trading and secondary marketing – has however, changed somewhat after a court ruling in the UK in 1999. Traditionally, although the jargon sometimes gets fuzzy in the literature, trading involves exchange – 'barter' – whereas a market involves a one-way purchase of a slot. Whoever holds the slots has to conform to any original obligations such as a use-it-or-lose it rule. In some cases the slots may be leased rather than sold. From the perspective of efficiency, the market both avoids the issue of 'double-coincidents' of wants that is a feature of any barter system and also allows for new entry. The ICAO does not favor the buying and selling of slots in a market and it is illegal in some countries. The grey area has emerged because under many regimes only trading is allowed but a UK court ruling made it possible to trade with financial side-payments.

The buying and selling of slots is disliked by many authorities because it is often seen to pose a number of problems (European Competition Authorities, 2005; National Economic Research Associates, 2004; Civil Aviation Authority, 2001).

- The holder of slots may hoard some to force up the price of others.
- They attract conditions to the sale; e.g. the purchaser may not use them for competitive routes to the seller (restrictive covenants) or that they also have to use the ground handling facilities provided by the vendor.
- Only slots at unattractive times are put on the market.
- Slots will only be sold to certain airlines not considered a competitive threat.
- There may be imperfect information in the market preventing buyers and sellers coming together.

The list of objections is not, however, powerful. These types of objection suggest inherent imperfections in the nature of a market for slots, but of a kind that are often found in any market. The practical issue is whether they have a major impact and whether, even with imperfections, secondary markets are preferable to other options such as more frequent allocations of primary rights. The evidence that is available is that secondary markets do enhance efficiency under existing regimes where the primary allocation by scheduling committees is far from optimal but that this benefit may be limited.

The most extensive studies have been conducted in the US at four slot-controlled airports. The findings indicate that the size of the secondary slot market was not large between 1986 and 1992, after which it was phased out at Chicago, and that new entry was extremely limited (Starkie, 2003). However, in a comparative study looking at Chicago, O'Hare and Washington National (where slot trading is permitted) and Atlanta and Los Angeles (that do not have slot trading) it was found that at the former slots were used more efficiently (in terms of load factor) and smaller fare increases (Sened and Riker, 1996).

Conclusions

Take-off and landing slots are scarce commodities at many airports and their more efficient use is now becoming of paramount concern. In particular, since the widespread deregulation of the airlines there have been significant changes in the types of service offered to travelers. The deregulation of elements further up the value chain have been much more piecemeal and evidence indicates that this has restricted the full range of benefits being enjoyed by passengers and airfreight service users. Airports, air navigation services, and ground handling are gradually being offered in more rational economic environments, and there is considerable interest in rethinking the way landing and take-off slots are allocated to ensure adequate access for airlines with new models of operation as well as between more traditional carriers.

A variety of options are available for slot allocation to meet the challenges of the new business environment, including more sophisticated planning algorithms and congestion pricing of capacity. Auctions have been used in similar sectors and with a degree of success when carefully structured to meet the inevitable compromises needed to maximize theoretical benefits with the practical realities of the industry in question; it is a world of 'be-spoke' models. The issue in practice is rather less one of whether an optimal auction system could be deployed and more a challenge of defining a practical system that performs better than the various alternatives currently being mooted. The auction option does seem one that does have potential if technical and institutional challenges can be overcome.

References

Boeing Commercial Airplane Group (2004) *Current Market Outlook*, Seattle: Boeing.

Boyfield, K. (2003), 'Who Owns Airport Slots? A Market Solution to a Deepening Dilemma', in Boyfield, K. (ed.), *A Market for Airport Slots*, London: Institute for Economic Affairs.

Brueckner, J.K. (2002), 'Airport Congestion When Carriers Have Market Power', *American Economic Review*, 92, pp. 1357–75.

Button, K.J. and Stough, R.R. (2000), *Air Transport Networks: Theory and Policy Implications*, Cheltenham: Edward Elgar.

Button, K.J. (2005), 'A Simple Analysis of the Rent Seeking of Airlines, Airports and Politicians', *Transport Policy*, 12, pp. 47–56.

Carlin, A. and Park, R.E. (1970), 'Marginal Cost Pricing of Airport Runway Capacity', *American Economic Review*, 60, pp. 310–19.

Civil Aviation Authority (2001), *The Implementation of Secondary Slot Trading*, London: CAA.

Coase, R.H. (1960), 'The Problem of Social Cost', *Journal of Law and Economics*, 3, pp. 1–44.

Demsetz, H. (1968), 'Why Regulate Utilities?', *Journal of Law and Economics*, 11, pp. 55–65.

DotEcon Ltd (2001), *Auctioning Airport Slots: A Report for HM Treasury and the Department of the Environment*, London: Transport and the Regions, DotEcon.

European Competition Authorities (2005), *Progress Report of the Air Traffic Working Group on Slot Trading*, London: ECA.

Humphreys, B. (2003), 'Slot Allocation: A Radical Solution', in Boyfield, K. (ed.), *A Market for Airport Slots*, London: Institute for Economic Affairs.

International Civil Aviation Organization (2001), *ICAO's Policies on Charging for Airports and Air Navigation Services*, Montreal: ICAO.

Jones, I., Viehoff, I. and Marks, P. (1993), 'The Economics of Airport Slots', *Fiscal Studies*, 14, pp. 37–57.

Kleit, A.N. and Kobayashi, B.H. (1996), 'Market Failure or Market Efficiency? Evidence on Airport Slot Use', *Research in Transportation Economics*, 4, pp. 1–32.

Krishna, V. (2002), *Auction Theory*, New York: Academic Press.

Martín-Cejas, R.R. (1997), 'Airport Pricing Systems in Europe and an Application of Ramey Pricing to Spanish Airports', *Transportation Research E*, 33, pp. 321–7.

McAfee, R.P. and McMillan, J. (1987), 'Auctions and Bidding', *Journal of Economic Literature*, 25, pp. 699–738.

Milgrom, P.R. (2004), *Putting Auction Theory to Work*, Cambridge: Cambridge University Press.

Milgrom, P.R. and Weber, R.J. (1982), 'A Theory of Auctions and Competitive Bidding', *Econometrica*, 50, pp. 1089–122.

Morrison, S.A. (1982), 'The Structure of Landing Fees at Uncongested Airports: An Application of Ramsey Pricing', *Journal of Transport Economics and Policy*, 16, pp. 151–9.

Morrison, S.A. (1983), 'Estimation of Long-Run Prices and Investment Levels for Airport Runways', *Research in Transportation Economics*, 1, pp. 103–30.

Morrison, S.A. (1987), 'The Equity and Efficiency of Runway Pricing', *Journal of Public Economics*, 34, pp. 45–60.

Morrison, S.A. and Winston, C. (1989), 'Enhancing the Performance of the Deregulated Air Transportation System', in Bailey, M.N. and Winston, C. (eds), *Brookings Papers on Economic Activity*, Washington, DC: Brooking Institution.

National Economic Research Associates (2004), *Study to Assess the Effects of Different Slot Allocation Schemes: A Final Report for the European Commission*, DR TREN, London: NERA.

Oum, T. and Zhang, Y. (1990), 'Airport Pricing: Congestion Tolls, Lumpy Investment and Cost Recovery', *Journal of Public Economics*, 43, pp. 353–74.

Pigou, A. (1920), *The Economics of Welfare*, London: Macmillan.

Posner, R. (1974), 'The Appropriate Scope for Regulation in the Cable Television Industry', *Bell Journal of Economics and Management Science*, 3, pp. 98–129.

Rassenti, S.J., Smith, V.L. and Bulfin, R.L. (1982), 'A Combinational Auction Mechanism for Airport Time Slot Allocation', *Bell Journal of Economics*, 13, pp. 369–84.

Sened, I. and Riker, W.H. (1996), 'Common Property and Private Property: The Case of Air Slots', *Journal of Theoretical Politics*, 8, pp. 427–47.

Sentance, A. (2003), 'Airport Slot Auctions: Desirable or Feasible?', *Utilities Policy*, 11, pp. 53–7.

Starkie, D. (1998), 'Allocating Airport Slots: A Role for the Market?', *Journal of Air Transport Management*, 4, pp. 111–16.

Starkie, D. (2003), 'The Economics of Secondary Markets for Airport Slots', in Boyfield, K. (ed.), *A Market for Airport Slots*, London: Institute for Economic Affairs.

Starkie, D. and Yarrow, G. (2000), *The Single Till Approach to the Price Regulation of Airports*, London: Civil Aviation Authority.

UK Department of the Environment, Transport and the Regions (2000a), *Valuing the external costs of Aviation*, London: DETR.

UK Department of the Environment, Transport and the Regions (2000b), *The Government's Consultation Document on Air Transport Policy*, London: DETR.

Vickrey, W. (1961), 'Counter-speculation, Auctions, and Competitive Sealed Tenders', *Journal of Finance*, 16, pp. 8–37.

Williamson, O.E. (1976), 'Franchise Bidding for Natural Monopolies: In General and with Respect to CATV', *Bell Journal of Economics*, 7, pp. 73–104.

Zhang, A. and Zhang, Y. (1997), 'Concession Revenue and Optimal Airport Pricing', *Transportation Research E*, 33, pp. 287–96.

Chapter 17

Formal Ownership and Leasing Rules for Slots

Erwin von den Steinen

Introduction

Airport slot policy in the European Union remains controversial. Today's system, despite years of dedicated effort to reform it, may thus still suffer from inherent flaws. In particular, there is a problem of ill-defined property rights in slots. This chapter explores a different formal approach to slot ownership (as well as leasing and sub-leasing) designed to simplify and, arguably, deregulate the commercial position of slots. The following aspects are discussed in this chapter:

- the problem;
- a proposed solution;
- pros and cons from the standpoint of key stakeholders;
- issues of transition from today's system.

The Problem

The problem can be outlined by way of offering six propositions:

1) a slot is a time-specified reservation of airport capacity;
2) for a commercial operator, such a reservation is a means of production;
3) when held recurrently across time slots acquire property attributes (requiring an airline to surrender such slots thus constitutes a 'taking');
4) under current regulation, the slot property value is difficult to realize (EU 'transfer' restrictions hinder if not forbid buying and selling);
5) such a slot market is essentially a *secondary market*; that is, financial transactions tend not to occur between suppliers and users (primary slot rights, particularly in the congested markets, are allocated by a slot coordinator rather than being market-determined by transactions between suppliers [the airports] and users [the aircraft operators]);
6) deregulation (meaning here essential reliance on market forces to stimulate supply rather than using administrative allocations to distribute it) will not succeed without direct participation of producers in buying and selling.

Thus, even if transfer restrictions on users were removed, as long as the EU formally or even tacitly defines a slot as a free entitlement (to which the user retains title if he meets minimum use conditions), reforms are unlikely to work.[1]

A Proposed Solution: Airport Ownership/Airline Leasing (Sub-leasing)

Consistent with other international and European laws (including competition law), the European Union should define slot rights *not* as public entitlements but as private leases.

Capacity *per se* would thus be owned by the airport, and airports would charge for each slot reservation. Slot users would, however, be free to make contracts of varying terms for leasing such reservations and *not* be constrained by regulation, as is currently the case, from subletting slots or transferring leases to other users.

Specific elements or aspects of such a regulation might include:

1) *Unbundling of slot fees.* Consistent with user fee rules and precedents, EU airports would be required to publish for each traffic season an unbundled, non-refundable charge for a reserved slot. Subject to appropriate oversight, each airport would be free to set/adjust its '*reference slot fee(s)*'.[2]
2) *The reference slot fees.* Unless adjusted by agreement (see 3 below), these fees would be levied in full on all operations for which advance reservation was made. As user fee income, slot fees would work to reduce landing fees or overall user charges. Unlike typical landing fees, reservation fees should be determined purely by the value of runway time at the airport and not be based on weight. Such a pricing mechanism would serve to allocate scarce capacity at congested airports to larger aircraft.
3) *Variances from the reference fee.* Airlines and airports would be free in the case of slot series (Repetitive Flight Plans) to negotiate variations from the basic fee *reflecting the economic value to the respective partner of the capacity commitment,* in a manner compliant with governing transparency and non-discrimination rules. (see also 7) below.)
4) *Principles governing the costing of leases.* As user fees, airport slot reservation fee rates would have to be consistent with prevailing pricing rules imposed by national and international law (e.g., bilateral air services agreements) as adjusted from time to time. In principle, however, lease charges could be set to cover:
   a) *Investment and debt-service costs.* An airport needing and willing to execute an expansion of capacity would presumably charge higher slot fees; and

---

1   The Commission's regulations not only protect existing grandfathered rights; they essentially create new grandfathers. Each new slot awarded carries the same entitlement conditions.

2   Since the unbundled slot fee is designed to account for the independent cost and value of time, airports congested during peak hours could and should establish independent peak and off-peak reference fees.

b) Fixed costs (up to a certain point at least).

5) *Terms (length) of leases.* A central policy question is whether and how to regulate the length of leases. Perpetual leases would be contrary to the spirit of the proposed reform. On the other hand, any specific time limitation will tend to be arbitrary. *As a hypothesis for discussion* it is suggested that: Framework regulation at the European level should set a maximum period (say 24–30 years) for new lease contracts.

6) *Deregulated subleasing.* As compensation to airlines for loss of perpetual grandfather rights, subleasing would be deregulated. Operators could sublet slots or even sell leases (unless constrained by the specific terms of individual leases).

7) *Open competition for primary leases.* Regulation would require an open and transparent process for initiating or renewing primary leases, i.e. the grant of recurrent capacity from a supplier (airport) to a user (airline). The regulation should, however, be cautious about prescribing exact forms of competition. Auctions may be an advantageous mode for certain cases but not for others. Thus *it might be left open to the parties (lessors and lessees) themselves to decide how to bid the capacity subject to certain legal standards.* Such standards might include:

   a) Timely transparency. Leases are not final until interested parties have had reasonable opportunity to object or respond with counter offers.

   b) Rights of pre-emption. Where a formal open auction procedure is not optimal (example: a big user wants a large, long term package), a competitor willing to match scope/performance terms and price should have rights to challenge/pre-empt the contract.

   c) Rule of minimum scope. It may be desirable to give oversight authorities the power to exempt small and shorter term leases of capacity, especially at large and/or uncongested airports from bidding rules.

8) *Airport product liability.* Since this proposal would give airports new market power, it may be desirable to countervail ensuing risks by establishing product standards and liabilities. Illustratively, concrete penalties might be associated with selling slots airports do not have. That is, if an airport with the realistic ability to operate say, 40 effective slots per hour, sells 44 hourly reservations, which leads to a cascading of delays, then users have not received the product they paid for. Procedures might then allow for rebates to users in a manner analogous to denied boarding compensation. Similarly, the market framework regulation should enable, even concretely foresee, institutional procedures for recovery of rents from a congested airport that raises slot fees to underwrite expansion but then fails to make needed or timely investments in growth of capacity.[3]

---

3   If Bryan Matthews' thoughtful distinctions between 'congestion' and 'scarcity'; (Menaz and Matthews, Chapter 3) is accepted; that is, a definition of the former being the costs imposed on current operators and consumers by over scheduling – whilst 'scarcity' is defined by the denial of a slot to a would-be operator – then appropriate tools or penalties to deal with separate (even if related) problems can be derived. In the framework of a

9) *Role of the Slot Coordinators.* The institution of the Slot Coordinator would no longer be a point of first contact for a slot seeker, who would first exhaust sublease possibilities and/or compete for available primary capacity. The Slot Coordinator would, however, exercise at least two vital functions: 1) It would oversee slot and runway policy at the airport(s) under its jurisdiction, specifically approving or recommending (to higher authority) disapproval/modification of capacity plans; 2) it would also closely monitor conditions of market access. Operators who perceived denial of fair and equal opportunity to compete would file appeals with the Slot Coordinator as a point of first review.

10) *Overriding Position of Competition Law.* Just as provided in existing EU Slot Regulations, the proposed reform would not constrain application of existing and future rules on competition. That is, consistent with the facts of a specific situation, authorities would still act to require divestitures (of slots) if they found it necessary to preserve competition. Thus, airlines with a dominant position in particular markets could be excluded from auctions or, conversely, auctions might be reserved to airlines that have been denied access. Generic rules of new entrant preference should be removed if possible from slot regulation.

Reaching Political Agreement: The Pros and Cons for Key Stakeholders

Any change proposed will be 'dead on arrival' if it fails to address the hopes, and especially concerns, of stakeholders. In the following the focus is on the potential concerns of three groups:

* regulators (and politicians);
* airports;
* aircraft operators.

Concerns of the Regulators

Regulators must be concerned about implementing the standards of the Chicago Convention, namely safety (as paramount priority) followed by efficiency and fairness as well as (in the context of EC law and policy): freedom of market access, consumer welfare, environmental protection, respect of competitive ground rules (e.g., preventing subsidies) and overall cost-effectiveness (including public sector performance).

---

general legal standard requiring fair and equal opportunity to compete for all qualified operators, airlines suffering demonstrable denial of needed market access could be offered administrative and judicial paths of individual remedy, including civil damage claims against collectors of any scarcity rents.

The issue of market access has arguably been the key driver behind the Commission's slot regulation policies. The Commission correctly foresaw that scarcity allocation would turn into a problematic issue as not only demand grew in major EU metropolitan areas with insufficient runways, but also competitive interests and the ability to serve that demand through the workings of the revolutionary Third Package of air transport liberalization. Rationing and allocation of slots as a pro-competitive measure appeared unavoidable And even forced re-allocations were considered.

Thus serious deregulation would seem a difficult task for European politicians and regulators alike. Yet three broad assumptions could, if shared, lead to a rethinking of the issues:

Belief that evolving public attitudes toward airport expansion can help make the supply of new slots more elastic. Communities in Europe are becoming increasingly concerned about global competitiveness and the need to promote local jobs. Meanwhile levels of objection about airport noise seem to be levelling out. Aircraft manufacturers' ability to deliver Stage 4 aircraft and prospects for reduction in pollutants may further ease resistance to new runways.

Awareness that better use of existing capacity can also be stimulated by market forces. The entry of low cost carriers has made airports notably more competition-oriented, and the public is responding. Travellers seem increasingly willing to accept longer personal trips to the origin airport in exchange for low cost, on-time, point-to-point service.

Recognition of the potential merits of an individual case rule (as opposed to a fixed general standards rule) to justify intervention in market allocation processes. An assumption of current slot regulation is the perceived need to establish blanket 'new entrant' preferences in law (whose individual equity is open to question). Alternatively, when 'scarcity' is not a general condition, the issue of redistribution of assets could be looked at on a case-by-case basis. Disadvantaged aircraft operators – ranging from a corporate jet operator (arguing that airports in key business areas should keep a reserve of 'x' slots open for individual flights) to a network carrier (challenging slot deals of rivals) – would be offered efficient paths of appeal.

Concerns of Airports

Airports should certainly find ownership reform worth exploring. However, they could still raise a range of concerns. First, they might question the utility of new user fees if the result was a 1:1 reduction of landing fee charges.

Second, airport lawyers could be concerned by the liability risk of tailoring arrangements with key users – without simultaneously getting a clear conditioning of the like-treatment requirements in law for all other users (e.g., Article 15 of the Chicago Convention regulating non-discriminatory access to aviation infrastructure).

Overall, however, there seem to be persuasive, if not compelling, arguments for airports to favour the establishment of their formal ownership of airport capacity. These include:

- *Reducing the risks that rents will be inefficiently appropriated by a third party.* It is not in the interest of airports and their owners (typically local communities) that slot series granted by a third party (the Slot Coordinator) become a permanently free entitlement of user(s). Slot sales in a pure secondary market would, moreover, exercise cost-push effects at the congested airports, as buyers of expensive slots push these costs through to consumers (e.g., citizens of the owning community) and probably at a time when the airport itself needs new income for needed capital investment.
- *Reducing the wastage of lost slots.* Under this proposal, every slot reserved would require payment, whether it is used or not. As in the case of ticket-holders failing to show up for the performance, airports, having sold the ticket would be paid and would not have to price no-show costs into their landing fees.
- *Enhanced capabilities for internally generated capital investment.* As demand increases, slot fee income can also increase. It should be made possible to establish capital reserves to fund growth cost-effectively and to avoid or minimize seeking subsidies from the public sector (typical in many countries, notably including the United States).

Concerns of the Aircraft Operators

An important question is to what extent a diverse universe of aircraft operators – ranging from individual pilots to new low-cost carriers, to international system carriers – can possess a commonality of concerns? Or are conflicts among operators so deep that the state must arbitrate to arrive at solutions in the public interest?

This cannot be settled immediately. However, perhaps it is possible to agree to a hierarchy of concerns among operators; that is, the first concern of any operator is its own access to slots. Competitors' problems come second. Moreover, one can make a practical case that if sufficient slots are available – thereby implying a surplus of infrastructure capacity at the margin – operators as a class will be better off. Put another way, the scarcity affecting *you* adds to general congestion pressure and thus affects *me*.

Therefore (even though this may contradict the theology of professional user organizations like IATA), apparent 'over investment' in infrastructure, if it results from competitive efforts to attract market share, can be positive from the users' perspective. Margins of capacity are preconditions for both competition and putting market pressure on costs.

Next come considerations regarding the once-and-future grandfather rights holders – the slot 'owner' or custodian – as presently foreseen in regulations: why should a user accept transformation of entitlements which took often years to acquire?

This cannot be settled immediately either. As a practical political matter (see also Conclusions, below) a transitional strategy will require a reasonable level of incentive (compensation) for holders of entitlements at congested airports.

What can be said is that even these entitlements are under current conditions not as valuable as they might superficially appear. First, if the airport is oversold, the quality of the slot (as a time product) is undercut, and costs of its operation rise. Second, under the current restrictions on transferability, managing slot holdings as economic assets is at best complex.

As long as the market fails to generate needed supply, scarcity administration will need to adjust constantly to an ever changing situation, and the quality of any public entitlement may be fragile; whereas, if the market becomes robust, a narrowly-formulated entitlement to use an expanding supply of infrastructure will lose relevance.

Substitution of private leases would give both parties, airlines as well as airports, recourse to a wide and established body of commercial laws under which respective rights and benefits of lessors and lessees could be negotiated under civil protections.

Possessing transferability or sub-lease rights would strengthen the capital and asset base of airlines as well as airports. For example, a carrier with a portfolio of attractively priced, long term, transferable slot leases at strategic airports would improve its credit standing and lower its borrowing costs. Moreover, in a liberalized market environment one could imagine innovative and interesting commercial arrangements.[4]

Conclusions and the Issue of Transition

Conclusions of this chapter really have to come in the form of questions, such as:

- Is reform of ownership and leasing a potential catalyst for stimulating elasticity of supply?
- Are there potential show-stoppers in the political or regulatory landscape, or simply fears of the unknown, which cannot be overcome?
- Is the idea of slot ownership reform sufficiently interesting to merit needed legal and economic policy studies of the potential cost/benefits as well as the feasibility of various implementation strategies?
- Finally, is ownership reform an all-or-nothing proposition, or could it be tested on a voluntary basis as part of a necessary future transition strategy?

---

4   Illustratively, airlines might seek to lock-in favourable slot charges at a fixed price across time, while airports might prefer agreeing on discounts against the reference fee (subject to annual adjustment). An airport with unused capacity might also price landing fee costs into slot fees (instead of vice versa as at present) meaning that it would waive or reduce the landing fee when the slot is actually operated.

Issues of Transition: Elements of a voluntary transition model:

The following elements could form part of a voluntary transition model:

1) Grandfather rights may continue, but holders pay the *full* reference slot fees.[5]
2) Alternatively, they may surrender grandfathered slots for time-limited leases.
3) Agreements unconditionally converting current grandfather rights to leases *cannot* be challenged by third parties. Maximum term is [say 12] years.
4) As an additional one-off procedure, operators who release grandfathered slots at congested airports that are placed at auction may claim resulting premiums; that is, any difference beyond the airport's reference slot prices that new operators are prepared to pay for full term leases.
5) EU restrictions against 'transfers' of traditional entitlements remain in force.
6) Secondary disposition of leased slot rights is, however, *fully* deregulated.
7) New slot series becoming available in slot pools must be leased subject to auction/other counter-bidding procedures except that:
8) 'New entrants' preferences as established under current regulation remain effective (perhaps in the form of closed auction rights), as long as more than [say, 50] per cent of that airport's slots still operate as grandfathered entitlements.
9) Once grandfathered entitlements at an airport fall below a set threshold [say 50 per cent] of total slot capacity, new entrant preferences are phased-out or abolished.
10) Nothing in the foregoing arrangements restricts the ability of the Commission under competition law to remedy defined problems of dominant position and contestability of markets.

Finally, a *conditional, effectiveness period* (a form of 'sunsetting') could be an aspect of the proposed model. That is, legislation might provide that if, after (say, 10) years, a solid majority of airport slots across the EU (say, 60) per cent had not self-deregulated – that is, if users had continued to hoard their grandfathered slots, then changes to the *status quo ante* would lapse after (say, four) years, unless adopted by new legislation. On the other hand, should a 'takeover' by the market have been experienced (to be narrowly defined as the voluntary liquidation of grandfathered entitlements below a level of 'x' per cent), then the new property relationships would stay.

---

5   See section on 'A Proposed Solution', above.

PART F
Reforming the Slots System

Chapter 18

Extraction of Economic Rent Under Various Airport Slot Allocation Approaches

Kenneth Button

Background

There are increasing pressures to make better use of transportation infrastructure and, in particular, encouraging suppliers to act in a manner more akin to a 'commercial' undertaking. Privitization, market liberalization, and new forms of economic regulation have thus emerged across virtually all modes of transport. In the context of airlines, the carriers in most major markets supply their services largely free of economic constraints, many airports have been privatized, and some air service navigation suppliers have been commercialized (often as not-for-profit enterprises). The motivating force for efficiency under commercialized conditions is mainly seen as economic 'rent seeking'; profit seeking to the non-economist. The retention of economic rent, however, can cause socially undesirable distortions, especially if it persists over a long period, and in some market conditions it can lead to inefficiency.

The analysis here focuses on how airport take-off and landing slot allocation procedures have implications that affect the levels and the distributions of these economic rents. The actors in the game are the airlines, airports and politicians (including the executive as well as legislators). The ultimate distribution is essentially a political decision, as is any form of allocation, but the form of specific procedures influences the nature of the benefits created and who enjoys them.

The institutional structures under which airports provide their services vary and, in many cases, has changed over time. There has, as a generalization, been a move to insulate their operations from government interference. The European Union, for example, in pushing for a greater decoupling of airports from public control, and in the context of slots there have also been measures to release capacity from incumbent carriers (European Commission, 2001). In a number of countries, such as the UK, there have been programs of airport privatized or, as in Canada, 'corporatization' as not-for-profit undertakings. Where airports remain in public hands, there have been efforts in a number of states to make them more efficient and market sensitive, often by initiating new regulatory regimes and initiating institutional reforms in their governance.

Given the importance of motivation in the success of any policy initiative, the issue of how the rent seeking process offers stimuli to the various major actors involved in airport activities is a central concern in evaluation commercialization strategies. The interest here, thus, lies much less with the detail of the institutional changes that are taking place at any single airport, although removing to a world of total abstraction is not intended, than with the potential reactions and attitudes of the airlines, airport authorities, and politicians to the levels of rent and their distribution more generally. The focus is on take-off and landing slot reforms and on looking at changes from the long-standing, essentially incumbent ('grandfather') dominated, structures, to procedures that make at least a minimum effort to mimic a market system. These reforms may involve more rational pricing within relatively unreformed allocation procedures, but they may also extend to such things as slot auctions with secondary markets.

Even within this framework, the focus is specific. Most of the attention on slot allocation has traditionally been on congestion and the peaks in demands that are placed on facilities. Early contributions to the peak slot fees literature were concerned with estimation procedures (e.g., Morrison and Winston, 1989), but more recently there has been an interest in the incidence of congestion charges on major hub dominating carriers versus minority slot holders (e.g., Brueckner, 2002; Daniel, 1995; Mayer and Sinai, 2003). There is also an extensive literature on the technicalities of slot allocation using auctions; Grether et al. (1981, 1989) are examples. This body of work has been more concerned with the way slot congestion is created and the ways in which it is, or may be internalized, than with rent seeking per se.

These are clearly important matters in their own right, and are related to the substance of what is discussed here, but is not central too it. The interest here is with the implications of various broad structures of slot allocation on the economic rent seeking activities of the various interested parties – 'stakeholders' to use the modern vernacular (e.g., policymakers, airports, and airlines) – that are directly involved. Those ultimately affected – namely the traveling public and freight consignors – are largely, and in the tradition of much work on air transportation, set outside of the scope of the chapter. The ways that airport slots are allocated, of course, is not neutral in its impacts on final customers of air services, but it is one step further along the production chain than we are concerned with here. Once slots have been allocated to airlines, how these are then used, and their costs passed on to airline customers, is outside of the purview of our discussions. Where the issue of congestion is important is in the distinction between location rent and economic rent. The former is a genuine resource cost. In the context of this chapter it is a reflection of the opportunity cost of using land as an input to air transportation supply rather than some other use. This is of limited interest here. Economic rent is associated with the exploitation, in a neutral sense, of market power.

The chapter also largely looks in detail at the economic rent extraction from a given capital base. It follows the normal approach of assuming that the physical capacity of the airport (or airport system) is fixed. There is only limited discussion of how either the total amount of economic rent generated by the airport, or the ways in which it may be distributed, can affect investment decisions. The excuse is simply one of space. Quite clearly their acquisition of economic rent affects how various actors view investment, its scale, its form, and its method of finance. But in practical terms, airport capacity decisions in industrial countries are seldom constrained by a shortage of money. Planning procedures and environmental considerations mainly drive them. Where there is some relevance to the long-run, or at least the medium-run, is in terms of the role of 'secondary airports'. These can provide alternative, latent capacity that may be taken up in some cases if the rent seeking activities at the primary facility makes them viable commercial options. The up-surge in low cost airlines in Europe, and the nature of their non-frills services, means that there may be significant game playing between interested parties involving trade-offs between short- and long-term rent seeking.

The emphasis is also on slots and the economic rents associated with their various allocations. These are not the only assets associated with an airports, however, and they are not the only ones that raise issues of rent seeking, efficient allocation, and distribution – there are hotels, car parks, retail outlets, lounge facilities, ground handling, etc. (Forsyth, 2004; Golasazewski, 2004). Treating slots in a largely independent manner is perhaps somewhat unrealistic. There is inevitably a degree of joint supply of many outputs at an airport. But again, a focus just on slots does mean discussion of the slot issue is much easier.

Finally, the discussion says little about network effects. Airline services provide numerous links between the airport hubs and, therefore, ideally the topic of economic rent attainment should be couched within a network economics context. (For a discussion of the economics of networks see Shy, 2001). This can be particularly so if there is a single entity (public or private) with control over a set of complementary or competing airports. Networks are complex and, although adding important nuances to some arguments, can be largely avoided if one thinks in terms of airports being providers of services and airlines being their customers that are also interested in rent seeking.

Rent Seeking and Institutions in the Air Transport Sector

Definitions

It is useful at the outset to define what economists mean by rent seeking. Perhaps the clearest definition of what traditional neo-classical economists understand by economic rent is to be found in Alfred Marshall's (1920) *Principles*. Essentially, in its purest form it is the return to a factor that is completely inelastic supply. More generally it is said earned whenever a factor of production receives a reward that exceeds the minimum amount necessary to keep the factor in its present employment. The air transport literature has recently seen a resurrection of discussions about exact definitions of rent, and in particular where actions such as airlines' yield management strategies fit into earning a normal profit (Levine, 2002). This type of debate is, however, outside of the remit of this chapter.

Measures of Economic Rent Earned

Isolating what in the real world is rent and what is not is no easy matter. In particular, there is the problem of defining normal profits that fit within the cost function rather than being the excess above it. Nevertheless, as a starting point it is useful to look at the financial performances of airlines and airports in recent years. This does not give a full indication of the economic rents that are being earned because the cost benchmark requires a normal profits component. This is not readily measurable in industries where there are very different levels of risk involved. There are also difficulties in most of the standard measures used to look at financial performance. Some require an initial input of the opportunity cost of capital that is largely a subjective matter – e.g., the Economic Value Added or Cash Value Added approaches.

Operating margins are used here as an indicator of rent, in part because they are a relatively standard measure, but also because they are available for airports and airlines in a fairly consistent manner over a period of time. They suffer, however, from the varying costs of equity capital incurred in different sectors, and from the differing capital/equity mix that the airlines and airports tend to have (Turner and Morrell, 2003). Considerations of whether an airport or airline is privately owned or publicly owned (or a mix) also influence the importance of the operating margin as an indicator of rent.

Figure 18.1 provides details of the financial performance in terms of their operating margins of major US, European, and global scheduled airlines over the recent past. It is clear that over the past 15 years or so the returns of the airlines have fluctuated considerably due mainly to business-cycle effects. They average out at round about a zero operating margin, or even slightly less. What a viable long-term return should be for scheduled airlines is debatable, but for an industry of this kind it is likely to exceed 4 per cent and is probably well over 6 per cent. The indications are, therefore, that very limited amounts of rent are being enjoyed by airlines as an entity, although individual carriers have either consistently (e.g., Southwest in the US or Ryanair in Europe) or periodically defied this pattern.

Making comparisons of the returns enjoyed by airlines with the potential economic rents being enjoyed by airports is far from easy, and the findings reported in Figure 18.2, although broadly in line with a number of others studies,[1] have to been treated with care. The figure looks at the returns that airlines are enjoying vis-à-vis other actors in the air service supply chain – or 'value chain' to use Porter's (1985) terminology. The data offers cross sectional information for the major European providers of inputs to the value chain. While in financial terms the airlines have performed relatively poorly, those suppliers of other elements in the chain did relatively well. Airports in particular would seem to have been earning high returns, and unlike the airlines, none have gone bankrupt. Table 18.2 provides some nuaning of this by offering details of the operating margins at many of Europe's larger airports.

---

1   For example, by McKinsey and *Airline Business* – see Button, 2004.

Some caution must, however, be exercised in interpreting these results. Airports are highly capital intensive, require significant periodic capital injections, and have extensive amounts of long-lived assets, whereas airlines have a smaller capital base and can lease or outsource many more of their activities. The latter practices having become more common since markets have been liberalized and airlines have sought greater flexibility in their operations. There are also institutional differences that make comparisons of the intrinsic underlying market structures difficult. The airlines tend to compete in their markets, either along links or across networks, whereas many of the airports in the studies were state owned or regulated. Nevertheless, the findings do raise important issues about where rent is being extracted in the air transportation services supply chain under the current regimes of slot allocation.

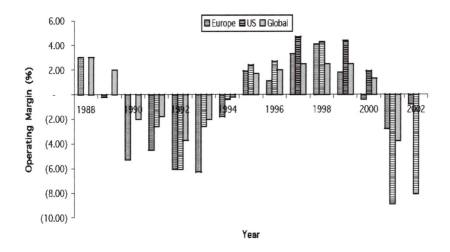

Figure 18.1 Operating margins of airlines 1988–2002

Note: An omitted observation indicates missing data and not a zero operating margin.

Source: Data from Boeing, Association of European Airlines and International Air Transport Association.

The Economic Environment

Another way of looking at the ability of airlines and airports to engage in rent seeking is to consider the market powers that each enjoy. Prevailing institutional and market structures are important determinants of both the overall levels of rent extracted and the distribution of the resulting rent.

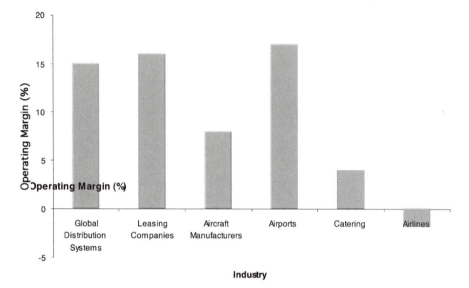

Figure 18.2 The returns for various elements in the European air transport supply chain (1999–2001)

Source: Button, 2004.

In Table 18.1 and Figure 18.2 the margins earned by airports have already been seen and, even allowing for the need for capital expenditures, there appear high. One reason for this is the degree of local monopoly power that many airports enjoy. Table 18.2, for example, provides an indication of the dominance of Frankfurt Main (Flughafen Frankfurt) airport as the major node for that area and catering for the vast majority of traffic within a 2 hour zone of the city. Similar pictures emerge elsewhere; Barcelona, for example carries 89 per cent of passengers within a 2 hour zone of the city, Berlin Tegel carriers 66.0 per cent, München, 64.6 per cent, and Paris Charles de Gaulle 66.4 per cent (McCarthy and McDougall, 2005). The picture is, however, more complex than these figures suggest. London Heathrow International Airport, for example, on carriers 49.1 per cent of the passengers within a two-hour zone of Central London (other major carriers are Gatwick, 21.5 per cent and Stanstead, 13.4 per cent) but Heathrow has a far larger share of intercontinental and scheduled traffic. Similarly, Paris Charles de Gaulle may have 66.3 per cent of Paris traffic and Paris Orly 31.0 per cent but Charles de Gaulle dominates scheduled and long haul traffic be a much larger margin. The implications are that the potential monopoly power exercised by airports could be large.

But what of the airline's market, or countervailing, power? The airline market, with the exception of a small number of international routes has largely been deregulated in terms of market entry, capacity, and fares in both the US and the

Table 18.1 Operating margins at European airports

Airport group	Operating margin (2001)	Operating margin (2002)
BAA plc (UK)	29.8%	30.6%
Fraport (Germany)	18.0%	15.8%
Aèroport de Paris (France)	6.0%	9.2%
Schiphol Group (Netherlands)	31.7%	32.0%
Luftartsverket (Sweden)	3.7%	9.1%
Flughafen München GmbH (Germany)	11.8%	3.7%
Avinor (Norway)	22.9%	17.1%
Aeroporti di Roma Spa (Italy)	16.8%	21.2%
SEA Aeroporti di Milano (Italy)	11.5%	10.4%
Manchester Airport Group (UK)	19.2%	19.3%

EU.[2] But the market, and particularly long-haul, is still somewhat distorted by the existence of large carriers that result from the combination of legacy effects of government protection and subsidies, and the natural economies of scale, scope and density that larger operations often enjoy. In some contexts this gives larger carriers the ability to exercise a degree of power over airports that they serve and influence slot allocation procedures.

While a rather primitive measure of market power, the proportion of throughput of traffic at an airport associated with each carrier offers a proxy of his or her monopsony power in Europe (Table 18.3). Comparable data for the US shows that, in many cases, airlines have an even larger share of passengers carried through their hubs. The situation in Table 18.3 is also distorted and underplays the concentration of airline power because it does not take into account the various global strategic and other alliances (Table 18.4) that allow for coordinated actions by groups of airlines (e.g., Air Dolomiti's link with Lufthansa increase their combined passenger share at München significantly, as does Lufthansa and Austrian Airways at Frankfurt).

The power of airlines at smaller airports is also changing. The recent expansion of low cost carriers, and their innovative business models, suggests that their power for rent extraction may be growing, especially at smaller airport (Mason et al., 2000). The low cost carriers pay less for landing fees than do the full cost airlines and also make less use of other airport revenue earning facilities (e.g., buses and skyways) than legacy carriers. This may in part be explained by the fact that many of the low cost carriers, such as Ryanair, do not use major termini, but even where they do they have often changes the basis of negotiations with the airport management.

---

2   The transatlantic market is largely deregulated as the US has pushed through its Open Skies policy. The only major restrictions involve some UK-US origin-destination pairs and reform of this has been under negotiation as part of a European Union initiative.

Table 18.2   Passengers using airports within 2 hours of Frankfurt (2003)

	Total distance	Travel time	Terminal passengers	%
Flughafen Frankfurt	13.0 km	15 minutes	48,107,669	56.
Flugplatz Egelsbach	16.4 km	15 minutes	71,689	0.1
Flughafen Hahn	114.7 km	1 hour 18 minutes	2,410,952	2.8
Flughafen Karlsruhe/ Baden-Baden	162.5 km	1 hour 29 minutes	298,825	0.4
Flughafen Koln/Bonn	167.4 km	1 hour 32 minutes	7,697,725	9.1
Flughafen Saarbrücken	162.9 km	1 hour 33 minutes	409,079	0.5
Flughafen Stuttgart	197.7 km	1 hour 49 minutes	7,464,903	8.8
Flughafen Dortmund	211.0 km	1 hour 53 minutes	1,023,339	1.2
Flughafen Nürnberg	213.5 km	1 hour 58 minutes	3,240,787	3.8
Flughafen Dusseldorf	227.3 km	2 hours 3 minutes	14,172,922	16.7

The opportunity costs, even with the existing structure of slot fees, is thus less at the regional airports being served. In addition to this the low cost carriers are often short-term monopolists or quasi-monopsonists at regional airports, and as such can exercise local market power in circumstances where there are numerous alternative facilities that they can fly to.

How much power an airline can exercise at an airport to extract as much rent as possible from its operations depends in part on its ability to use the threat of moving elsewhere. The major carriers often have significant potential stranded costs to consider if they move their basis of operation, and congestion at other European airports limits potential alternatives. This generally shifts the balance of the game in favour of the airport. The situation in the US is somewhat different because of the geography of the markets and the continuing excess capacity at some major airports. The effective move of Delta and US Airways out of Pittsburgh is illustrative of this. Low cost carriers, however, are not completely footloose. The 'radial' pattern of services favoured by Ryanair, for example, where all crew are based at a particular airport with non-interlining flights radiating out from it, to allow maximum flight hours to be exploited for plane and labour, does imply some fixed costs at that node, if not at the end of each spoke.

Institutional Structures

Rent seeking is a motivational force and is influenced by the institutional context in which people make decisions. The potential for rent seeking stems form both the formal and informal structure in which transactions take place. Figure 18.1, borrowed unashamedly from Williamson (2000), illustrates the issue.

Table 18.3   Market share of passengers by airline at European's 10 largest airports (2002)

Airport	Carrier 1	Carrier 2	Carrier 3
London Heathrow	British Airways 41.6%	bmi 12.1%	Lufthansa 4.8%
Frankfurt	Lufthansa 59.4%	British Airways 3.6%	Austrian 2.9%
Paris Charles de Gaulle	Air France 56.6%	British Airways 5.15	Lufthansa 4.9%
Amsterdam	KLM 52.2%	Transavia 5.5%	easyJet 4.3%
Madrid	Iberia 57.0%	Spanair 12.7%	Air Europa 7.1%
London Gatwick	British Airways 55.1%	eastJet 12.8%	flybe British European 5.6%
Rome	Alitalia 46.2%	Air One 10.0%	Meridiana 3.9%
Munich	Lufthansa 56.8%	Deutsche BA 6.6%	Air Dolomiti 6.5%
Paris Orly	Air France 64.2%	Iberia 8.2%	Air Littoral 3.6%
Barcelona	Iberia 48.5%	Spanair 9.4%	Air Europa 5.5%

Table 18.4   The main US and European airlines that are part of the three global strategic alliances

Star Alliance	oneWorld	SkyTeam
United Airlines	American Airlines	Delta Airlines
Lufthansa	British Airways	Continental
BMI	Aer Lingus	Northwest
TAP Portugal	Iberia	Alitalia
Finnair		Air France
Lauda Air		CAS Czech Airlines
LOT Polish Airlines		KLM Royal Dutch Airlines
Spanair		
SAS Scandinavian Airlines		
Austrian Airways		
Tyrolean Airlines		

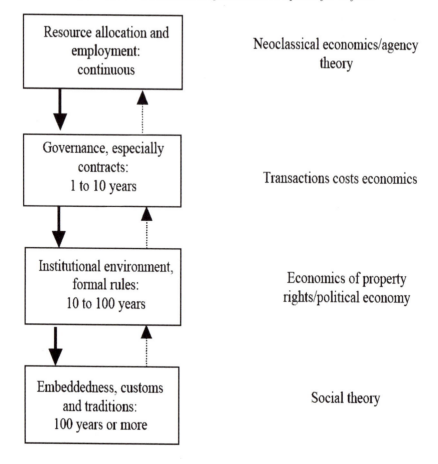

Figure 18.2 Williamson's characterization of where institutions become important

The figure is largely self-explanatory. Each box relates to a particular time frame within which a particular institutional structure is relevant. The dark arrows show constraints that come down from higher levels of analysis and the lighter arrows the direction of feedback. There is also information in the boxes on the general frequency in years over which change takes place; but this is only meant to offer a broad indication. The traditional situation of slot allocation results in market clearing at any point of time and can be examined using the tools of neoclassical economics. Allocation of slots seldom uses market mechanisms and it is not difficult to illustrate the sub-optimality of the situation in terms of resource allocation. The scheduling committees that allocate slots are structures that use established legal and informal procedures (e.g., 'grandfathering) to go about their business. Various types of formal and informal contracts emerge within this governance structure. In some cases there are international constraints on what can be done; for example as imposed under International Civil Aviation Organization agreements. Rent seeking takes place within this institutionalconstruction, and forms a large part of the discussion that follows.

Over time there are legal changes that affect the way contracts are developed; in the airport case this may embrace a range of actions from privatization and commercialization that can affect motivations of management to the changes in the way airports may be charged for run-way use that can introduce different constraints. Developments within the EU can be seen as one such process. Overall, the ways in which the system is viewed is influenced by larger societal considerations ('embeddedness') that takes generations to change; in concrete economic terms one may think of the divergence views of Christian and Muslims on interest rates. These societal shifts have clear implications for rent seeking motives but change so infrequently as to be well outside of our domain of interest.

The issue in the 1970s was that the economic institutional structures *in situ*, including regulations of network industries, were not bringing about desired resource allocations. *De facto* reforms were possible that would change the governance structure, essentially tinkering with the existing regulstions, but seemed unlikely to resolve the problem; such actions in the past had provided no enduring solutions to the underlying problems Hence fundamental institutional reform of systems largely intact for thirty or more years was deemed necessary in many countries. In addition to the economic forces that we look at below, there were wider social pressures in play that were shifting the 'embedded' ideas and culture, most notably disenchantment with socialism and communism as a mechanism for resource allocation. The outcome was a shift in the underlying paradigm away from the Continental philosophy of governmental control to the Anglo-Saxon philosophy where interventions in markets are only seen desirable if they bring about demonstrable benefits. This shift involved formal institutional changes and with them changes in property right allocations and motivations for rent seeking.

Distribution of Economic Rent

It is perhaps useful to look at some of the very basic implications of rent seeking, and in particular to look at the potential level and distributional implications of alternative allocative mechanisms in a very general way.[3] A simple diagrammatic approach is adopted for this purpose and thus Alfred Marshall's well-known words of caution certainly should be borne in mind – 'It is to be remembered that graphical illustrations are not proofs. They are merely pictures corresponding very roughly to the main conditions of certain real problems'.

---

3   Button (2005a) goes into these issues in more detail.

Airports and Airlines

Here we focus on the interactions between airport and airline markets to explore both how economic rents are generated and distributed in various context. The interest is in how actors seek to acquire rent when there is no genuine capacity problem. Figure 18.4 represents, for simplicity, the short-run costs (MC) and demand for slots (D) associated with a single runway airport. For ease of presentation the demand curve is drawn as linear, and the cost curve is horizontal. In most natural monopolies, because of various scale and scope economies, the cost curves would almost certainly be downward sloping. Introducing this, however, whilst more realistic, adds complexity but not substance to the argument.

There is assumed to be adequate capacity to meet all effective demand currently in the market. If there were a binding physical capacity constraint at a level before Q_2, as strict classic theory would require, then the MC curve would become vertical at the point that capacity is reached. This does not affect the fundamental tenure of the rent seeking argument as presented here, and is thus ignored. But in these circumstances charges become a rationing device for the limited capacity as well as a reflection of the costs incurred in providing the existing capacity and as a means of extracting economic rent.[4] The issue is the very simple one of the overall level and distribution of benefits from alternative allocating mechanism assuming that only airlines and an airport is involved. It is assumed, unless stated otherwise, that both the airlines and the airport are seeking to be economic rent maximizers. This may be a strong assumption for those in public ownership where there may be valid arguments for considering other possibilities such as 'satisficing', but it does provide the ability to keep the analysis manageable.

Competitive airlines/competitive airport This yields the standard Pareto first best outcome of a price of P_2 and an output of Q_2. The combined rent of buyers and sellers is maximized as the area afP_2. The airlines enjoy economic the surplus from the use of the airport in this case because of the horizontal nature of the cost curve; with a positive slope the airport would also enjoy some rent but with a negative slope it would be losing money[5]. The same outcome in terms of overall slot use would emerge if the airport were a monopoly that is regulated requiring it to marginal cost price its slots or allocate them using a Vickrey style of auction, although here the airport would extract the rents from the airlines (see below).

---

4   The figures, for simplicity, ignore the usual 'club good' issues – notably the divergence on the marginal private cost and marginal social cost curves – that accompany consideration of congested facilities. In that sense is can be seen as close to Knight's (1924) views on congestion.

5   This is rent going to the airlines in the sense that their willingness-to-pay as mapped by their demand for slots is a reflection of what they can earn from the final consumers of air services – passengers and freight consigners.

Monopoly airline (or strong strategic alliance)/competitive airport Since the airport has no market power, the airline, or cartel of airlines, will be able to force it to price its slots down to the airport's marginal cost of providing the slots, P_2. The airlines will again gain all of the economic rents, again afP_2.[6]

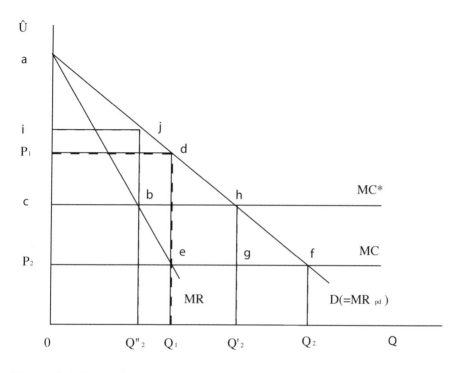

Figure 18.4 Illustration of cost and demand for airport slots

Competitive airlines/single pricing natural monopoly airport[7]. This is the neo-classic case of a monopoly with slot prices reaching P_1 and the number of slots limited by the airport to Q_1. The gains from the airport's activities are shared between the airlines (aP_1d) and the airport (P_1deP_2). But there is a welfare loss equal to the area dfe – the Harberger (1959) 'deadweight loss triangle'. This is the sort of institutional structure that one could imagine under a regime where the airport sets its slot prices at P_2 and then allows any airline to take up slots at

----

6   In the case of a declining cost situation, the airport would not be able to recover full long-run marginal costs and in the consequential absence of a core market instability would ensure (Telser, 1978).

7   For simplicity it is assumed that the airport is confronted with the same demand for its slots irrespective of whether the market is competitive or not and the only effect is on whether it can equate MC with MR or not. The single monopoly airport case may also be seen as representative of a number of airports that are commonly owned by one enterprise (public or private).

this rate with subsequent slot swapping allowed. If once the slot fees have been paid the airlines can buy and sell slots, then there would be transfers of benefits between them to the extent of aP_1d with no effect on either the airport's rent or aggregate welfare.

Competitive airlines/single pricing institutionalized monopoly airport Here the price and output combination is the same as the above but the airport will potentially enjoy a smaller rent and overall social welfare will be less. Tullock (1977) argues that there is likely rent dissipation in this type of situation. The airport, to retain its institutional monopoly, will be willing to expend up to P_1deP_2 in potential economic rent to defend this position – e.g., through lobbying, political support, advertising or legal actions. Even a small amount of rent is preferred to none. Since such actions are essentially non-productive, at least in a non-Keynesian world, not only does it cost the airport money but it also represent a loss of social welfare.

Competitive airlines/perfect price discriminating natural monopoly airport In this case, through such actions as the auctioning of individual slots in a manner that allows first-degree price discrimination[8], the airport can extract slot prices down the demand curve. This action produces for the airport economic rent of afP_2 with the airlines having none. The outcome is Pareto efficient and maximizes welfare. De facto this means the AR curve represents the airport's marginal revenue curve with price differentiation. The outcome is this Pareto optimal.

Competitive airlines/single natural monopoly airport with X-inefficiency Leibenstein (1966; 1979) argued that monopolies have limited incentive to minimize their costs but rather they often operate in an 'inert area' where there is little incentive to seek maximum efficiency. This inertia may be brought on by a variety of factors such as the considerable managerial effort required to negotiate labour contracts when confronted by labour union demands. If this is the case then the cost curve will rise to MC* and the single profit maximizing monopolist will limit the supply of slots to Q''_2 and extract rent amounting to ijbc. The airlines will be left with aij, but perhaps most important, the social welfare associated with slot use will be $cbjfP_2$ less than when there is competition in both the airline and airports markets.

Competitive airlines/perfect price discriminating natural monopoly airport with X-inefficiency The situation here is that the airport will provide Q''_1 and enjoy economic rents of ahc, with the airlines taking none. Compared with the situation where there is no X-inefficiency, the total social welfare is, though, reduced by $chfP_2$.

This categorization does not consider all possibilities – e.g., there is no consideration of X-inefficiency amongst the airlines or of collusion between

---

8   The literature on slot auctions has a long pedigree (e.g., Rassenti et al., 1982; Brander et al., 1989) and is expanding.

the parties – but it does indicate the importance of the institutional, and well as technical, characteristics of the situation in determining the levels of economic rent generated and the recipients of the rent.

Taxation

The simple diagrammatical analysis is devoid of any consideration of taxation; but the government may be rent seeking as well. It assumes that the undertaking (airline or airport) that extracts economic rent will enjoy this rent. But there may be institutional constraints imposed. In some cases, as with many US commercial airports, the authorities circumscribe how these rents may be used and hypothecation to airport enhancement is a common regulation. There is also the issue of the government extracting money for general revenue or for sumptuary taxation purposes. In a slightly different context, for example, the New Zealand Government extracts a return from its air navigation services provider once the provider's costs have been recovered.

Prudent fiscal policy indicates that taxes result in the least distortion to the economic system when applied to goods or services that have low demand elasticities, and where collection costs are low. Airports and airlines are limited in number and their accounts are relatively simple and thus meet the latter criteria.[9] It is less certain whether they meet the first, and there are clear market segmentations that make generalizations problematic. The work on airline fare demand elasticities in general is extensive, but rather inconclusive, and there is very little work on airport elasticities.

The evidence from a survey of a large number of previous studies conducted by Gillen et al. (2002), however, suggests elasticities slightly in excess of unity for most types of air travel, but with quite a lot of variation. From a distributional perspective, air transportation as a whole also emerges as highly income elastic and thus the imposition of indirect taxes is unlikely to be excessively regressive. If shadow prices are attached to reflect social welfare preference functions, then this provides a normative justification to tax rents irrespective of whether the airlines or airport is acting as a 'till'; the choice is often more a pragmatic one related to transactions costs and public perceptions.

The tax regime can thus limit the rents that airlines and airports receive, but they may also impact on the distribution of these rents through their incidence. Since, by definition, the tax authority is a monopolist, if its aim, following Ricardian principles, is to maximize its revenue then it will devise a tax structure that extracts all the rent from the airlines and airports irrespective of which is the original recipient. It would, in conjunction, foster a regime that would allow the aggregate rents to be maximized by allowing prices to be determined by demand – for example giving a monopoly license to an airport and allowing it to auction slots or engage in some other form of price discrimination. If, as an alternative, the government seeks a particular target level of revenue, then this raises problems

---

9   Strictly, a tax on pure economic rent will not have any allocative effects in a neo-classical framework because of the fixity of supply.

of defining a tax structure that also meets secondary goals regarding the incidence of the taxation and also entails the nature of any supplementary regulations that may be imposed.

In general, the issue of the role of air transport taxes has been neglected other than in cases where they currently exist and there are debates about their use in financing infrastructure – e.g., to finance the Airport and Airway Trust Fund in the US (Button, 2005b). Taxes on airlines, are, however, generally significant and may influence the incentive of airlines to rent seek as well as the ways in which rents are distributed.

The Simplifying Assumptions

The simple economic diagrammatic representations are insightful in highlighting the role of various market structures in determining the aggregate rent available and how it may be distributed. But they have their limitations. Removing some of the simplifying assumptions can offer some additional information.

Decreasing Cost Problem

Airports, and in particular their runways, are not normally seen as constant cost entities. Runways are long-lived capital-intensive investments with low marginal maintenance and operating costs. The resultant downward sloping cost function of providing the hardware to make take-off and landing slots possible makes full cost recovery difficult without some degree of monopoly power.[10] The airlines may in these circumstances, if they are not themselves under excessive competitive pressure, enjoy economic rents by pressuring the airport to price down to short-run marginal cost.

This problem, however, is far from unique to airports and many industries have been forced to find innovative ways of tackling the problem. Indeed, seeking mechanisms for full cost recovery have provided a fruitful area of study for economists from the inception of the railways. There are in fact numerous *ad hoc* ways in which full costs may be recovered. Briefly these option encompass in the airport context:

- *Direct subsidies.* On the premise that airports meet a public need and sometimes, although erroneously, their services are seen as public goods, governments have provided direct subsidies for airport development and operations. This approach has tended to become less popular as the commercial potential of airports has been realized and issues of X-inefficiency have raised

---

10   A straightforward analysis of how an undertaking with such a cost structure should prices its facilities, albeit couched in terms of public transit, is found in Turvey (1975). Airlines offering scheduled services may also require a degree of monopoly power if they are to recover their full costs (Button, 1996). This latter issue is not discussed here.

questions about the effectiveness of direct subsidies in helping with efficient management.

- *Natural market power.* This is the classic case of having market power. Provided the demand curve is at some point outside of the airport's cost function there will be a single slot price that will ensure costs can be recovered and possibly earn rent. Perfect price discrimination again extracts all economic rent for the airport.

- *Institutional market power.* This gives the same results as a natural monopoly airport, although the legal structure of the monopoly arrangement may limit the power of an airport to exercise extensive monopoly power. One would include as examples here international gateway airports designated under air service agreements, although there are also other institutional structures that confer considerable monopoly power – e.g., the Wright Amendment gives Dulles/Fort Worth considerable monopoly power over long haul services into and out of the region. This type of situation may also involve cases where a group of airports in a region is operated by a single entity with institutionalized monopoly power across the group. The BAA structure in London may, in very broad terms, be seen to operate in this way.

- *Internal coalitions.* Airports in a competitive situation could theoretically combine to seek the power of a cartel to extract sufficient revenues to cover their full costs. Such coalitions between airports are rare and when they do exist they are often aimed at reducing costs rather than generating cost recovery revenues.

- *Internal cross-subsidization.* If some activities are highly profitable but their viability is dependent on other activities then the full costs of the latter may be partly covered by transferring revenues from the profitable activities. This is possible, for example, when airports gain significant revenue from commercial concessionary activities. It is the standard joint product case. The interesting issue then becomes that of optimization of prices and investments between the two activities if they are not strictly in joint supply – which can lead to policy debates such as those involving the 'Two Till' versus 'Single Till' argument (UK Civil Aviation Authority, 2002). In some cases, such as Spain, there is extensive cross-subsidization across airports. The short-run objective of the 'airport system' in this case can be seen in terms of maximizing the rent across the system so that full costs are recovered and a surplus earned. From an efficiency perspective this will penalize carriers using airport where demand is strong and advantage those carriers using airports where costs are not being fully recovered.

- *Long term contracts between supplier and customer.* The normal aim of this strategy is to obtain a guaranteed revenue flow that allows the supplier to adjust capacity to meet this predetermined demand. There is limited scope for doing this at airports because of the physical nature of the infrastructure. Whereas it application to slots is limited, long-term contracting and leasing is common for some other aspects of airport activities, such as terminal buildings.

- *Unbundling of services.* Airports provide a range of services, not all of which are directly related to airline operations or slot use. Airports have increasingly clarified their costs by outsourcing or privatizing some of their activities. As noted above, many major US airports, for example, have airline owned/leased terminal facilities. This removes some items from their cost function and makes clear where core costs lie, including those associated with runways.

- *Two-part tariffs.* Two part tariffs involver an explicit elements to cover fixed costs. This has been a long-used mechanism for telecommunications companies but is not in widespread use at airports. This, however, may be viewed more as a tradition than a technical issue. Since runway use is essentially a 'club good' problem there seems no theoretical reason why two part tariffs are not used; the vested interests of existing club members is a more pragmatic matter.

- *Vertical integration.* Vertical integration allows control over the 'value chain' that feeds costs into the activity under review. Airports seldom have ties with links further up the chain (e.g., air traffic control and global distribution systems) or further down the chain (airlines) but this does deviate from practices in similar sectors. There are seaports around the world where shipping companies have large financial interests (e.g., P&O have interests in the Port of Felixstowe in the UK) and in many countries there was a long standing link between railway operations and terminals.

- *Discriminate pricing.* Airline through their yield management systems have engaged in price discrimination but airports have seldom been able to do this, or have been reluctant to do so where publicly owned. In some instance, such as the US there are legal constraints that prevent such actions. The accountancy-based-rules of the International Civil Aviation Organization also make any form of discriminatory pricing for international services difficult.

The difficulty for any capital intensive industry, including airports, is to find a cost recovery mechanism that is practical, ensures sufficient income is generated, and has low transactions costs associated with it. In many cases a combination of approaches are adopted – in part these may be seen as strict cost recovery measures, but also, in part, there can be rent seeking motives involved with much depending on the incentives provided by the institutional structure in which they operate. In many cases an essentially 'not-for-profit' remit tends to dampen the rent-seeking element for airports, but it may also have adverse effects on carriers if results in 'gold plating'.

Objective Functions

The diagrammatic analysis in Figure 18.4 explicitly assumes that both airlines and airports seek to maximize their profits. This may not be the case. Airlines have recently been the subject of numerous measures to inject much more competition into the sector, but in the past they have generally been viewed as providing public services. The US domestic air cargo industry was deregulated in 1977 and the Airline Deregulation Act of the following year removed economic controls from the passenger market. The three 'Packages' of reforms from the late 1980s have resulted in a liberalized intra-European market. These changes, combined with the privatization of airlines in Europe and the imposition of anti-trust and mergers policies in many major markets have resulted in significant market entry and reductions in air fares – Button (2004) provides details of the EU situation. One may safely conclude that airlines are, in general, in both the North American and EU markets seeking to maximize their economic rents within a competitive environment. For sure, in both the US and the EU there are programs of financial support to ensure that a limited number of specific services meeting social needs are provided but these are put out to tended in a manner that reflects competition for these markets.

Airports, however, often remain in public hands (albeit frequently at the local or state rather than national level) and are often managed on a system-wide basis with considerable amounts of cross subsidization (e.g., as in the Spanish system.). The motivation driving the management of airports is thus not always that of rent maximization even when there is a high degree of monopoly power involved. In many cases the airports are seen as part of regional policy and their strategic managerial goal is to assist in a larger planned development of a region. To attract industry to an area, a local airport often set landing and take-off slot rates at below long-run cost. In some cases airlines are subsidized to fly to the airport[11]. In public policy terms this is often referred to as the 'Continental Approach'. It treats airports as a part of a larger entity and strict internal economic efficiency of airport operations is not a necessary criterion in this context.

How an airport approaches rent seeking in these circumstances is not clear, and almost certainly varies on a case-by-case basis. Many airports are given the explicit objective of meeting demand at the prevailing rate structures and to recover costs in doing so. This offers little incentive for X-efficiency let alone rent seeking. In simple terms, provided that sufficient revenue is forthcoming, the objective is to push out utilization of the airport to the point in Figure 18.4 where average costs are recovered but leaving the rent to the airlines with the aim of stimulating lower fares. There is no incentive when there is limited capacity to ensure that airlines that could enjoy the greatest benefit would gain use of this slot capacity; it results in such conventions as 'grandfather rights' that favour incumbents irrespective of their efficiency.

In other circumstances, even when there is private ownership, fear of monopoly exploitation can call forth regulatory controls that act as *de facto* restrictions on the ways in which rent is allocated between the airport and customer airlines. In general, the aim of such regulation is to ensure a distribution of rent that maximizes aggregate welfare and offers an incentive for acceptable future levels of capacity expansion. The price-capping mechanism deployed by the UK government over the privatized BAA is an example of such a measure, although its details have been criticized (Starkie, 2001). Price capping was originally devised for industries such as telecommunications that were enjoying significant technical progress, capacity expansion, and that would ultimately operate in a competitive environment. It was

---

11   There is a fundamental legal difference here between whether the airline is subsidized from state funds (which is a subsidy as the EU Court of Justice found in the case of Strasbourg Airport) or from private companies that collectively guarantee the sales of a minimum number of tickets (as happened at Wichita in the US).

seen as requiring minimum informational inputs. Airports, however, cannot be easily and rapidly expand their capacity – environmental and other constraints being a major concern – and competition, although possibly greater than is some times thought, is not as intensive as in ideal conditions. As the X-inefficiencies are driven from the system by price capping, so more information is needed to adjust the price-cap; it *de facto* becomes a rate-of-return regulation. And these problems are in addition to the one-till/two till issues alluded to earlier.

The Political Perspective

Even this very basic discussion of the magnitude and distributions of the rents associated with airport slots highlights the importance of the underlying market structure. But market structures are not neutral entities; governments can influence them. Much has been written on the various mechanisms of slot allocation – the 'grandfathering approach', slot charges, slot auctions, secondary markets, slot-trading etc. But the implications of each of these must be set within the overall underlying legal basis for the market structure and the governance associated with this structure. This in turn leads to the political motivations that underlie the *de jure* structure that is in place and its de facto operation. Put simply, institutions are important.

As the Public Choice School of economics emphasizes, rent seeking is not just the prerogative of industrial players. The policy-makers themselves have a vested interest in the ultimate market structure and the way it performs in terms of the rents they enjoy (Buchanan and Tullock, 1962). The form this rent takes depends upon on how the political decision-making process is viewed, and it may well not be in terms of a financial return to all actors.[12] There is, for example, a well-established body of thought within the Philadelphia School of economists about the objectives of corporate leaders that transcends simple monetary rewards, and this approach can be extended to political decision-makers with little modification.[13]

If there is the belief that politicians seek to serve the public interest then a definition of public interest is needed. If this is the simple one of maximizing social welfare then the Hicks-Kaldor criteria would indicate that any move towards a

---

12   In the case of dictatorships the aim may well be the basic one of the dictator extracting the maximum financial rent. But Stigler (1971) views regulation and control by politicians within a democratic system as 'commodities' upon which the forces of supply and demand play. Interest groups, wishing to benefit from particular policies demand them. Politicians seeking election or re-election supply them. Within this framework there is the scope for coalitions of interests to emerge, in our context either within airlines or airports, or embracing participants from both groups to define a slot allocation procedure that maximizes their joint objectives.

13   Isolating 'political rent', however, is even more difficult than financial rent but the recent cases before the European Court of Justice do offer tentative evidence that local politicians see economic rent in supporting their regional airports.

regime that increases the combination of rents enjoyed by airlines and airports is desirable – who ultimately gains is then a separate political decision. (This is in effect simply a refection of Robbins' (1932) view of the distinction between normative and positive sciences.) From a purely efficiency perspective it does not matter whether the airport allocates slots using marginal cost principles or whether there is a monopolist supplier employing perfect price discrimination. The political consideration in a democracy, however, is which of these alternatives is more attractive to the median voter, or more generally if the electorate is sensitive to various ways of taxing that can bring about an alternative allocation of the rents (Pattanaik, 1971). For example, from the political perspective, the minimization of delays, truncation of the delay distribution function or some similar criterion may be for relevant than any larger welfare maximization concept.

There is a significant amount of evidence that maximizing potential economic rents may not be a first-best political objective to attract the median voter. In particular, work by Bruno Frey (2001) and others over the years have highlighted the prime importance politicians and administrators put on the immediate distributional implications of policies – subsequent redistribution having a second order effect. Matters of equity and fairness can be a major driver. In many cases monetary compensation may also not be valued in the same way as actual physical implications. There is also mounting evidence that individuals are often more interested in the relative impacts of policies than in the absolute effect. In other words, the neoclassical assumption of independent utility functions may not be valid.

In these circumstances politicians may not seek to maximize overall social welfare, but rather they may seek to ensure that the expectations, or desires of specific target groups of the electorate are met. The slot allocation procedure that is politically supported may thus see specific groups benefit not only at the expense of other groups but also at the expense of overall economic efficiency. The motivation for doing this and interfering with the allocation mechanism may, for example, involve protection of a national carrier for strategic purposes and defense reasons, or of an airport in an electorally marginal region.. This is not just a matter of economic rent distribution between airlines and airports, but rather it favours specific carriers or airports. There may be certain groups in society that politicians wish to attract, for example regular business users around the airport. In this case the allocation process may be designed to assist the retention of carriers meeting this demand at the expense of other airlines, and of the airport.

Policies are also influenced by the actions of administrations. Although administrations in democratic systems are meant to implement policies passed by legislatures, in practice control over information flows, coupled with an inevitable flexibility in interpretation, provides administrators with the ability to pursue their own agendas to some extent. Thus, just as politicians may not have agendas that maximize the rents from slots, so do administrators of any legal system (Posner, 1975; Niskanan, 1971). A principal-agent issue almost invariably arises.

In particular, administrators, by capturing parts of the allocation process, have the opportunity to enhance their own utility. This may take a variety of forms, for example in terms of developing large bureaucracies and complex administrative structures offering career ladders or 'gold plating', with its associated prestige effects (Williamson, 1963), of the facilities that they are responsible for. In the case of slots, the gold plating can take the form of proving excess capacity, runways that are over engineered, or terminals that resemble art galleries more than waiting areas. It may be argued that in doing this, the administration is seeking to overcome potential market failings, but in itself it has immense potential for creating even larger market intervention failures. There would also seem to be the tendency to support the constituency that has most affinity with administrators, namely the management of monopoly undertakings.

Conclusions

This chapter has sought to offer a broad view of the complexities involved in analyzing the economic rents that can be associated with airport slot provision and allocation. The emergent climate under which air transportation services are being provided moves airlines and airports away from entities that see their rationale in terms of meeting a public service obligation to one of commercial viability and profit making. Some public sector involvement and regulation of more generic kinds do still limit the actions of airlines, and more markedly airports. In a commercial environment economic rent extraction can act as an incentive to allocate resources more efficiently, but much depends on the institutional umbrella under which the competition takes place and on the abilities of the players to themselves act and price efficiently.

The challenge for policymakers confronting this situation is that while there is an intellectual tool kit, although neither complete nor perfect, to help understand the issues and to put forward general policy approaches regarding the use of airport capacity, there is a lack of quantification regarding the scale of the distortions and benefits that rent seeking produces. The obvious problem is the lack of a genuine counterfactual against which current practices and policies may be judged. As increasing numbers of airports are freed from tight regulations, and as privatization and commercialization occurs in various guises, an inevitable process of trial and error is likely to provide greater insights and, in the longer term, the more efficient delivery of airport services.

References

Brander, J.R.G., Cook, B.A. and Rowcroft, J.E. (1989), 'Entry, Exclusion, and Expulsion in a Single Hub Airport System', *Transportation Research Record*, 1214, pp. 27–36.

Bruckner, J.K. (2002), 'Airport Congestion When Carriers Have Market Power', *American Economic Review*, 92, pp. 1357–75.

Buchanan, J. and Tullock, G. (1962), *The Calculus of Consent, Logical Foundations of Constitutional Democracy*, Ann Arbor, MI: University of Michigan Press.

Button, K.J. (1996), 'Liberalising European Aviation: Is There an Empty Core Problem?', *Journal of Transport Economics and Policy*, 30, pp. 275–91.

Button, K.J. (2004), *Wings Across Europe: Towards an Efficient European Air Transport System*, Aldershot: Ashgate.

Button, K.J. (2005a), 'A Simple Analysis of the Rent Seeking of Airlines, Airports and Politicians', *Transport Policy*, 12, pp. 47–56.

Button, K.J. (2005b), 'Taxing the US Airline Industry – A Time for Change?', *Aerlines*, 13, pp. 1–4.

Daniel, J.L. (1995), 'Congestion Pricing and Capacity of Large Hub Airports: A Bottleneck Model with Stochastic Queues', *Econometrica*, 63, pp. 327–70.

European Commission (2001), *European Transport Policy for 2010: Time to Decide*, Luxembourg: Office for Official Publications of the European Communities.

Forsyth, P. (2004), 'Locational and Monopoly Rents at Airports: Creating Them and Shifting Them', *Journal of Air Transport Management*, 10 (1), pp. 51–60.

Frey, B. (2001), *Inspiring Economics: Human Motivation in Political Economy*, Cheltenham: Edward Elgar.

Gillen, D.W., Morrison, W.G. and Stewart, C. (2002), *Air Travel Demand Elasticities: Concepts, Issues and Measurement*, Ottawa: Department of Finance.

Golaszewski, R. (2004), 'Location Rents and the Experience of US Airports – Lessons Learned from Off-Airport Entities', *Journal of Air Transport Management*, 10 (1), pp. 61–9.

Grether, D.M., Isaac, R.M. and Plott, C.R. (1981), 'The Allocation of Landing Rights by Unanimity Among Competitors', *American Economic Review, Papers and Proceedings*, 71, pp. 166–71.

Grether, D.M., Isaac, R.M. and Plott, C.R. (1989), 'The Allocation of Scarce Resources: Experimental Economics and the Problem of Allocating Airport Slots', in Arrow, K., Heckman, J., Pechman, J., Sargent, T. and Solow, R. (eds), *Underground Classics in Economics*, Boulder, CO: Westview Press.

Harberger, A.C. (1959), 'Using the Resources at Hand More Effectively', *American Economic Review Papers and Proceedings*, 59, pp. 134–47.

Knight, F.H. (1924), 'Some Fallacies in the Interpretation of Social Costs', *Quarterly Journal of Economics*, 38, pp. 582–606.

Leibenstein, H. (1966), 'Allocative Efficiency vs 'X-efficiency', *American Economic Review*, 56, pp. 392–415.

Leibenstein, H. (1979), 'A Branch of Economics is Missing: Micro-Micro Theory', *Journal of Economic Literature*, 17, pp. 477–502.

Levine, M.E. (2002), 'Price Discrimination without Market Power', *Yale Journal on Regulation*, 19, pp. 1–36.

Marshall, A. (1920), *Principles of Economics* (8th edn, 1949 reset)London: Macmillan.

Mason, K., Whelen, C. and Williams, G. (2000), *Europe's Low Cost airlines: An Analysis of the Economics and Operating Characteristics of Europe's Charter and Low Cost Scheduled Airlines*, Air Transport Group Research Report 7, Cranfield University.

Mayer, C. and Sinai, T. (2003), 'Network Effects, Congestion Externalities, and Air Traffic Delays: Or Why Not All Delays Are Evil', *American Economic Review*, 93, pp. 1194–215.

McCarthy C. and McDonnell, J. (2005), 'Competition or Regulation for European Airports?', in Pickhardt, M. and Sarda Pons, J. (eds), *Perspectives on Competition in Transportation*, Berlin: VWF.

Morrison, C.A. and Winston, C. (1989), 'Enhancing the Performance of the Deregulated Air Transportation System', in Bailey, M.N. and Winston, C.A. (eds), *Brooking Papers on Economic Activity: Microeconomics 1989*, Washington DC: Brooking Institution.

Niskanan, W.A. (1971), *Bureaucracy and Representative Government*n, New York: Aldine-Atherto.

Pattanaik, P.K. (1971), *Voting and Collective Choice*, Cambridge: Cambridge University Press.

Porter, M.E. (1985), *Competitive Advantage; Creating and Sustaining Superior Performance*, New York: Free Press.

Posner, R.A. (1975), 'The Social Costs of Monopoly and Regulation', *Journal of Political Economy*, 83, pp. 807–27.

Rassenti, S., Smith, V. and Bulfin, R. (1982), 'A Combinatorial Auction Mechanism for Airport Time Slot Allocation', *Bell Journal of Economics*, 13, pp. 402–17.

Robbins, L. (1932), *An Essay in the Nature and Significance of Economic Science*, London: Macmillan.

Shy, O. (2001), *The Economics of Network Industries*, Cambridge: Cambridge University Press.

Starkie, D. (2001), 'Reforming UK Airport Regulation', *Journal of Transport Economics and Policy*, 35, pp. 119–35.

Stigler, G.J. (1971), 'The Theory of Economic Regulation', *Bell Journal of Economics and Management Science*, 2, pp. 3–19.

Telser, L.G. (1978), *Economic Theory and the Core*, Chicago, IL: University of Chicago Press.

Tullock, G. (1967), 'The Welfare Costs of tariffs, Monopolies and Theft', *Western Economic Journal*, 5, pp. 224–32.

Turner, S. and Morrell, P. (2003), 'An Evaluation of Airline Beta Values and their Application in Calculating the Cost of Equity Capital', *Journal of Air Transport Management*, 9, pp. 2001–210.

Turvey, R. (1975), 'A Simple Analysis of Optimal Fares on Scheduled Transport Services', *Economic Journal*, 85, pp. 1–9.

UK Civil Aviation Authority (2002), 'The "Single Till" and the "Dual Till" approach to the price regulation of airports', consultation paper, CAA, London.

Williamson, O.E. (1963), 'A Model of Rational Managerial Behavior', in Cyert, R.M. and March, J.G. (eds), A *Behavioral theory of the Firm*, Englewood Cliff, NJ: Prentice-Hall.

Williamson, O.E. (2000), 'The New Institutional Economics: Taking Stock, Looking Ahead', *Journal of Economics Literature*, 38, pp. 595–613.

Chapter 19

The Slot Allocation Philosophy at the EU

Frederik Sørensen

Introduction

Air Transport and the EU Treaty

Ever since the EC was created the European Commission[1] was of the opinion that the Treaty applied to air transport. However, in spite of this nothing happened in respect of air transport for many years. It was only when the Commission won a case[2] against France in 1974 that the development of an air transport policy began to take shape. The ruling stated in clear terms that the general rules of the Treaty did apply to both air and maritime transport. This produced a protracted series of consultations on the need to create an air transport policy. Finally in 1978 the Council of Ministers[3] found it necessary to adopt a priority list for such a policy. In 1979 the Commission followed this up with a Memorandum on Air Transport, 'A Community Approach'.

Evolution of Air Transport Legislation

Although everybody agreed that air transport is vital for economic development or perhaps because of this it took a long time before any significant legislation was adopted. The member states clearly did not like to involve the EC in these matters but on the other hand could not in the long run maintain a totally negative attitude. The first legislative instrument addressing the EC air transport market was adopted in 1983 on Interregional Air Services and this was followed in 1987 by the first package.[4]

At that point in time nobody had thought of legislating in the area of slot allocation. On the other hand airport capacity had been identified by the Commission as a potential problem. In 1982 a report was produced by Metra

---

1   Hereinafter called the Commission.
2   Case 167/73 at the European Court of Justice, the so-called French Seamen's case.
3   Hereinafter called the Council.
4   Consisting of Council Directive 87/601 and Council Decision 87/602 of 14 December 1987.

Consulting on 'Capacity Constraints for Air Transport in the Community'.[5] This report recognized the capacity constraints that growing traffic volumes would generate and urged the EU to create an air transport policy. The first legal measure on market access followed shortly afterwards, namely the Directive on Interregional Air Services.

This changed in 1990 when the Council adopted the second package[6] because it was clear that a total liberalization would follow in 1992. This focused the attention on the growing capacity problem and the need to ensure a level playing field. The Council therefore in Regulation 2343/90 included a safety clause that allowed a member state to refuse traffic rights to air carriers from another member state if their own air carrier(s) were prevented from competing effectively because of capacity problems in that other state. This safety clause was valid:

> pending the adoption by the Council and the coming into force of a Regulation on a code of conduct on slot allocation based on the general principle of non-discrimination on the grounds of nationality.

The third package[7] was adopted in 1992 with effect from 1 January 1993. Soon thereafter the first EU Regulation on Slot Allocation was adopted.

I have been involved in the legislative process on slot allocation and on air transport in general since 1977 and until five years ago. Although it is difficult to stand aside and look at this process in a dispassionate manner I shall try to describe the development of legislation in this area and define the key issues that influenced the way the legislation was developed and adopted and which even today are still at play. I also intend to discuss whether and in which way a reform of the existing system might be possible in a realistic way.

Development of Legislation on Slot Allocation

The key year in the development of the legislation on slot legislation was 1990, when the Commission began contemplating a possible modification of the traditional system. After a period of very thorough consultation this led to a Council Regulation in 1993. This was followed by consultations in 1996 and a further set of consultations in 2000 with an official Commission proposal in 2001 and a new Council Regulation in 2004. This development will be described and assessed in the following. In this assessment the position and reactions of the main interested parties will be included as far as possible. The key players in the discussion on slot allocation were:

* air carriers;

---

    5   Metra Consulting Group Limited, Capacity Constraints for Air Transport in the Community, Final Report, November 1982.

    6   Council Regulations (EEC) No 2342/90 and 2343/90 of 24 July 1990.

    7   Council Regulations (EEC) No 2407/92, 2408.92 and 2409/92 of 23 July 1992.

- airports;
- slot coordinators;
- member states/Council;
- European Parliament;
- European Commission.

The principle underlying the efforts from 1990 was basically to ensure the most efficient use of scarce capacity at airports. This was gradually joined by a second principle namely that supply and demand should be balanced by economic mechanisms. In this context it is worthwhile to note that for many this meant that a certain movement of slots should take place and 5 per cent annually was in many instances put forward as a benchmark. From an economic point of view it is, however, more reasonable to state that it is availability of slots that is important rather than actual movement. If an air carrier, willing to pay the market price, is able to obtain slots then the market can be said to be balanced.

Commission Proposal 1991[8]

Existing set-up In addition to the aim that a level playing field should be created the Commission had additional concerns with respect to the effects on competition of a capacity straight jacket at major EU airports. At a time when the market was liberalized it could not be accepted that competition was more or less eliminated at these large airports. This translated into the aim that room had to be found for new entrants. The competition services at the Commission wanted the EC rules on slot allocation to counteract this logjam by ensuring a certain turnover which started the whole discussion on withdrawal.

Consultations The Commission sent out a Consultation paper in August 1990 to all interested parties. A number of consultation meetings followed and a large number of position papers were received.

Many submissions in particular from air carriers defended the existing system and warned against touching it since it had been developed over many years and were an expression of a carefully drafted equilibrium. It has been suggested that the situation represented a state of scarcity rent at the airport[9]. This may well be so but it was not brought up in the discussion and in my opinion the air carriers did not look at this in any concrete way in Europe. However, they were keenly aware of the fact that without slots they could not operate at an airport especially if the international rules were given legal power in the EU. From their point of view the existing international system introduced a form of mutual deterrence in the sense that an air carrier being responsible at its own home base for the slot allocation would not dare to bias the allocation process (too much) since in such a situation it would have to fear retaliation at other airports where its competitors

---

8   Official Journal No C 43 of 19.2.1991.
9   The issue of scarcity rents was taken up in the context of creating rules for airport charges.

would be responsible. AEA[10] and IATA[11] claimed that the existing system based on grandfather rights also provided some stability without creating a blockage for new air carriers. However, the 'do-nothing' option was not acceptable to the Commission since the second package required a proposal on slot allocation.

The airports were rather cautious at this time since traditionally they had been under the thumb of the air carriers. This position in my opinion can be explained by the fact that airports were usually owned by the states and that the states were masters of market access through bilateral agreements. The airports therefore were in a situation where they were totally dependent on state action. This situation was changing for some countries where liberal bilateral agreements were being pursued and airports in these countries were developing more active attitudes. However, in general they did recognize that the existing system made it difficult for new entrants and that it was not optimal in ensuring an efficient utilization of capacity. Therefore, the airports did ask for direct involvement in the slot allocation process.

A specific study was carried out for the Commission by the Stanford Research Institute (SRI) on 'Airport Capacity'.[12] This study was intended to further look into the capacity problems as already observed by the Metra study. In addition SRI looked at the draft Commission proposal and recommended that the Commission should look further into the problem and that regulatory intervention was not really necessary. This advice was somewhat at odds with the existing atmosphere in the Member States and the Council and was not taken seriously. The CAA of the UK however offered substantial constructive advice.

In fact, at the time no major problem was identified which would block progress in the Council. It should be recalled that at that time the European Parliament (EP) only had a consultative role. It was also felt as reassuring that the Commission took the position that it would adhere to the existing international guidelines as much as possible.

Structure of the proposal The Commission did not have any great difficulties in developing the proposal since the consultation had been constructive and to all intents and purposes rather comprehensive. Furthermore, the Commission had to work towards a schedule created by the Council i.e. the Member States. Certainly there were Member States which were less than enthusiastic in this respect but they could not very well argue against a decision taken by themselves as mentioned earlier.

Scope When submitting its proposal the Commission based it on the following elements.

Firstly the Regulation could not be limited to Community Air Carriers because international air transport accounted for very high percentages at the major airports which would become slot coordinated as a consequence of the

---

10   Association of European Airlines.
11   International Air Transport Asociation.
12   SRI International, Study on Airport Capacity, Final Report, September 1991.

Regulation.[13] This was contrary to the approach in the US where the slot rules applied to US air carriers while slots for international air carriers would be provided when traffic rights were provided. When open skies was agreed with Canada the approach was changed in the direction of the European approach since Canadian air carriers after an introductory period would have to obtain slots in the same manner as US air carriers.

Secondly, the international guidelines i.e. the guidelines established by the Scheduling Conferences[14] would be used as much as possible. In fact these guidelines were included to a large extent in the annex to the proposal.

• Independent allocation
In contrast to the existing situation the Commission introduced the requirement that slot coordinators should be able to conduct their business in an independent manner. The independence of the slot coordinators would be safeguarded both functionally and financially from air carriers, airports and governments.

• Avoid state patronage
This independence needed to be established in respect of not only air carriers and airports but also states. It was all too evident that state ownership of an air carrier created a strong temptation for governments to interfere and protect their air carrier.

• Capacity utilization
Capacity problems were at the heart of the matter. The Commission was sceptical about the possibility of increasing capacity so that the capacity problems would disappear. The liberalization of the EU air transport market could therefore be neutralized unless action was taken to ensure that new entrants and air transport market developments could be accommodated by the slot allocation mechanism. It was also necessary to ensure that the existing capacity would be used as efficiently as possible.

• Liberalize market access
The Commission had looked at peak hour pricing and thought that this might be a valid approach to balance demand and supply but refrained from using it since it was felt that it would probably be costly for consumers. This was not seen as an acceptable message since the Commission argued that the liberalization of air transport would lead to lower prices. However, the Commission left the door open for later application of such a mechanism. This is significant since it demonstrates that the idea of market mechanisms for slot allocation or capacity

---

13   Later on when the European Court of Justice looked at Community Competence in air transport this total coverage had as a consequence that slot allocation clearly became an area with Community Exclusive Competence.

14   This framework is often called an IATA system. This is not correct. The Scheduling Conferences are arranged by IATA but any air carrier can participate whether or not it is a member of IATA or not.

utilization is not a newfangled idea. The principal idea of market forces was introduced by the third (and second) package in a regulatory sense in the context of air transport services in situations where the competitive market would not provide satisfactory service to peripheral regions where air transport was vital for the local economy. The concept of public service obligations (PSO) was introduced with a selection of air carriers being based on public tender. This concept was similar to the essential air services concept in the US. The proposal introduced the concept that slots could be reserved for these PSO.

- Level playing field

The international rules would be placed in a legal framework which would ensure a level playing field.

- Maintain competition

Grandfather rights were accepted but with some reservations as mentioned below. The Commission specifically stated that they have become more and more stifling with respect to the possibility for mobility. In other words an important issue in the context of loosening the capacity straightjacket.

Competition must be preserved. This was an important element for two reasons. First of all because the competition rules do not permit enterprises to reach an agreement which in fact will eliminate competition nor do they permit member states to accept or introduce such rules. The same is naturally the case for EC secondary legislation. Under the competition rules there was therefore a need to grant an exemption for the scheduling conferences since they were seen as akin to a cartel like activity. However, an exemption could not be granted if this activity would eliminate competition. This concern led directly to a reconsideration of the grandfather rights.

In order to avoid that the grandfather rights would eliminate competition a provision was included in the proposal forcing member states to produce a certain number of slots for redistribution every season if necessary by withdrawal of slots. This idea clearly had been taken from the US where such withdrawals (by lottery) had been taking place with the purpose of ensuring competition (mobility). New entrants should be given a large share of these 'new' slots. This mechanism had been introduced by the competition services in order to facilitate the granting of a block exemption.

Council Regulation 1993

Political process The member states in the Council may have wanted to forget about the proposal but could not endanger the existing slot allocation mechanism which was in need of a block exemption and the Council had acknowledged the need for a slot allocation Regulation. The negotiation in the Council between the member states and the Commission therefore became a question of subsidiarity[15]

---

15   The principle of Subsidiarity means that the EC should not take up issues which could better be dealt with by the member states individually.

and detail. The Subsidiarity principle could not really be argued since slot allocation automatically concerns not only one member state but both the departure and the arrival airport. It was therefore clear from the outset that this negotiation would end up in an agreed text for a slot allocation Regulation. It is, however, illuminating to see where the Commission's proposal was modified.

The negotiations took primarily place in the Council but although the European Parliament had only a consultative role it did nevertheless influence the result because discussions in the EP took place at the same time as the negotiations in the Council. The Commission acted to ensure that the ideas[16] in the EP were introduced in the Council so as not to delay that negotiation by having to consider the EP formal amendments at a later stage.

The new Regulation was adopted early in 1993 shortly after the entry into force of the third package. From an EU point of view the negotiations were fairly easy and consequently they took only about one year and a half. The Commission had no difficulties in accepting the result but it was less than happy and so was ready to accept an invitation to present a report and a proposal for amendments by 1997. It soon started developing new ideas or rather new ways of reintroducing concepts which had not survived the political process of the first proposal.

Changes from proposal The general structure of the Commission's proposal survived in the Council but a number of modifications were introduced[17]. In the following only the changes from the proposal are discussed since all the points where agreement existed cannot be seen as blockages to reform.

• Definitions
The definition of a new entrant was modified but not drastically in order to define a more operational concept.

The definition of a congested airport disappeared but was to some extent incorporated in the main text on capacity in articles 3 and 6.

The definition of a coordinated airport was given more detail and distinguished between coordinated and fully coordinated airports. It was also made very clear that it is only at fully coordinated airports that a slot is required for take-offs and landings.

The grandfather concept disappeared from the definitions but was incorporated in the main text of articles 8 and 10.

All issues could be negotiated in a normal way.

• Article 3
One major change was introduced for fully coordinated airports. If no significant capacity problems exist for example if new capacity has been created then the status as a fully coordinated airport will have to be suspended. This makes it clear

---

16   These ideas in a number of instances were introduced in the EP Committee by the Commission but it was more efficient to be able to say in the Council that this idea came from the EP rather than building up a long list of Commission requests.

17   The reference to articles are from the text adopted by the Council.

that the liberalization of market access is serious and that the restrictions, which a slot allocation mechanism introduces, are not acceptable in a normal situation without congestion problems.

- Article 4

It was made clear that the coordinator is exclusively competent for slot allocation.

- Article 5

Instead of being optional the coordination committee was made obligatory. The scope of work for the committee was extended to also cover additional guidelines for allocation, serious problems for new entrants and complaints.

- Article 6

More precision was introduced to the text which stipulates that before each season the state responsible must declare the capacity available for allocation. The position of airports was weakened in that it was clarified that airports were not always the competent authority to formally assess the capacity of the airport. The Council, i.e. the Member States, did not want to weaken their key role in this respect.

- Article 8

Grandfather rights were spelled out directly in the manner developed in international practice. The 'use-it-or-lose-it' rule was also incorporated. The direct reference to the competition rules in this context were replaced by a general reference to the competition rules in the 'whereas' clauses.

The incorporation of the international rules in the annex was deleted and replaced by a more general reference to international and local rules.

- Article 9

The text on regional services was new. In many ways it was superfluous since grandfather rights were incorporated but the Council felt that these services and Public Service Obligations should be safeguarded at least for a certain time.

- Article 10

The proposal was given more precision in general. As a part of this the Commission's direct role disappeared. Although this was in line with the exclusive competence given to the coordinator it was also an example of the application of the Subsidiarity principle to which the member states throughout the negotiations paid considerable attention. This is not surprising since the Council in general in many other subject areas also was very keen to ensure that the Commission did not acquire too much power.

The 'use-it-or-lose-it' utilization limits were made more stringent than the proposal. Instead of the 65 per cent proposed they were increased to 80 per cent for scheduled services and 70 per cent for non-scheduled.

• Withdrawal

Withdrawal in order to ensure competition as proposed disappears. This was due to a strong pressure from the air carriers which already at this point talked about confiscation. The Commission was quite upset by this and made it clear that this might lead to drastic application of the competition rules.

The withdrawal discussion was the closest to a blockage that occurred during the negations of this the first Regulation on slot allocation. The Commission has come back to this element several times in the subsequent years and by now it must be characterized as a blockage to reform at least in the minds of the Commission.

• Article 11

Proposed text on safeguard mechanisms is modified so that the role of the Commission is minimized. Again this is an example of the application of the Subsidiarity principle.

• Article 12

Commission role in respect of problems with third countries disappears nearly completely. The Commission accepted this because it is better not to say anything than to risk to endanger existing jurisprudence.

Potential Blocking Issues for the Future

This first negotiation already identified one of the key issues for further reform namely the question of withdrawal. It is understandable that air carriers economically were loath to give up slots since they may have invested heavily in the services making use of those slots and in this respect they were able to influence the Member States especially those which owned the national air carriers.

Another potential blockage also surfaced namely the position of new entrants. Both of these issues would be pursued by the Commission in the preparatory work towards a new proposal

Commission Proposal 2001[18]

Consultations 1996–1999

• Technical improvements

The Regulation called for a follow-up proposal in 1997. In order to prepare this the Commission started consultations in 1996 by issuing a consultation paper on options for the further development. This consultation paper basically addresses a number of technical options to improve the first Regulation. It was to a large extent inspired by a report carried out by Coopers & Lybrand in 1995 which was made available to interested parties and discussed in detail before the consultation paper was finalized.

---

18   Official Journal C 270 E, 25.9.2001, p. 131.

• Withdrawal of slots

The paper is fairly inoffensive except that the Commission revives the idea of withdrawing slots if the necessary mobility cannot be obtained via the slot pool. This element was sharply refused by the air carriers which could surprise nobody. However, it should be noted that the UK CAA in CAP 644 (CAA, 1995) specifically supported a system of slot withdrawal if necessary to accommodate new entry. This slot withdrawal system was supposed to be based on a lottery system similar to the approach already used in the US.

• Ownership

However, as a consequence of these considerations a secondary discussion started namely on who owns the slots. In the beginning the air carriers took it for granted that they basically owned the slots since the grandfather rights had been confirmed by the EC Regulation. However, as soon as the air carriers began to make their claim the member states, the Commission and the airports began to develop their positions. It comes at no surprise that the air carriers made this claim and in fact continue to take that position because their whole business depends in many instances on having the necessary slots. If slots are taken away an air carrier may suffer substantial losses to its network. Furthermore, the air carriers in some instances paid for the slots either directly or by purchasing another air carrier or part of it. At Heathrow for example a prime slots could cost several million pounds. Air carriers which have spent that kind of money in order to obtain a slot will not lightly give it up (without compensation).

The airports were now beginning to make their own claim for ownership of the slots

The member states and the Commission took a different view. They considered slots a form of public service which belonged basically to the state. This argument was supported by the fact that states used to negotiate traffic rights bilaterally. On the other hand this argument has been weakened in the EU by the fact that free market access now exists within the EU.

• Slot Trading

Slot trading was not really an issue during this first round of consultations. However, the Commission had asked Putnam, Hayes and Bartlett (Starkie, 1992) to carry out a study of Slot Trading in the US. The report concluded that slot trading had been beneficial in the US but that it was not enough in itself to ensure market entry for new entrants. It needed to be combined with a vigorous pro-competitive policy.

The Commission concluded at the time that while slot trading should be analysed it was not yet time to include it in any specific proposal. The UK CAA in CAP 644 agreed with this position.

• Intermittent Period

Ad hoc consultations took place during the following years and gradually the Commission changed its position. It became convinced that the slot allocation itself should be improved in particular by introducing sanctions to counter

wasteful slot use such as non respect of slot timings (off slot operations), late
return of unused slots, operations without slots, etc. Furthermore, it became clear
that slot trading did take place in particular at London airports. Considering this
and taking into account the US experience the Commission gradually accepted
that a market mechanism had to be found. Slot ownership and withdrawal
continued to be a sensitive issue and gradually grew in importance.

A further study by IAURIF (1999) on airport capacity emphasized the
problems at the large airports and stressed that the slot allocation system did not
function in a satisfactory manner. The study report did not, however, advocate
slot trading.

Consultations July 2000 In 2000 the Commission had advanced its reflections
to a point where it was close to present a new proposal. A draft proposal was in
internal interservice[19] consultation when somehow it was leaked. The draft raised
such a hullabaloo among air carriers that the Commission decided to consult
anew with interested parties in July 2000.

• Concession system for slots
The leaked draft report included a number of technical modifications but
everybody reacted to one specific element of the draft namely to treat slots as
a time limited concession. This was the main idea for the market based system
of slot allocation that the Commission intended to propose. The idea of the
Commission was not just to introduce such a system but to gradually expand it
to all slots. This brought out in the open the discussion on ownership of slots and
withdrawal of slots. The air carriers once again were fundamentally opposed to
any system which included withdrawal and time limitation of slots. At this meeting
the airports joined the air carriers. Perhaps not because they were as opposed to
the new ideas but because they wanted to establish a constructive relationship with
the air carriers and they found the presentation of the new ideas to be too abrupt
and lacking the necessary study and preparation. Furthermore, the draft did not
allocate any share of revenue to the airports generated by the new system.

The member states reactions were more nuanced and some expressed an
interest in looking into the new ideas. On the other hand the member states with
a shareholding in the national air carrier were reticent.

The protests at the meting were so strenuous and vociferous that no constructive
discussion could take place.

Letter to Member States After this meeting the Commission asked the member
states in writing whether a) the current rules were satisfactory, b) the grandfather
rights should be reviewed and whether more market driven allocation ways
should be developed to promote a balance between supply and demand and c)
what alternatives could be proposed to ensure that competition would not suffer
from the lack of capacity.

---

19 The proposal was prepared by the air transport services of the Commission and
the draft was then sent to the other services of the Commission for consultation.

The result of this written consultation of the Member States did not block the possibility for the Commission to develop more market oriented allocation principles but a thorough analysis should be carried out first.

• Separation of technical and market based elements

On this background the Commission decided to split the efforts into a technical proposal to be submitted quickly, which would later be followed by a proposal on market based mechanisms. This later proposal was to be based on further studies.

Proposal[20] by the Commission 2001

The proposal was intended to be purely technical i.e. not to include elements of a market mechanism.

In order to fully comprehend the proposal it is necessary to understand that the terminology changed in the sense that coordinated and strictly coordinated airports became schedule facilitated and coordinated airports respectively. This means that it is only at a coordinated airport that a slot is required for landing and take-off.

Structure of the Proposal

• Ownership

While Regulation 95/93 did not specify who owns a slot the proposal attempts to clarify the situation. It characterizes a slot as an entitlement established by the Regulation for an air carrier to use the infrastructure of the airport. In other words the air carrier does not own the slot but is dependant on the Regulation. Since the Regulation includes the so-called grandfather right the difference between entitlement and ownership is not very large.

However, it is significant that the Commission attempts to modify the provisions on transfers and exchanges as seen below.

• New entrants

There have been a number of remarks on the inefficiency (non-use) of the new entrant provisions in Regulation 95/93. The Commission therefore attempted to expand these provisions by including a provision that allowed any air carrier irrespective of its holding of slots at the airport in question to be considered as a new entrant in respect of starting operations on a new route where no other air carrier was operating. This was accompanied by a proposal to increase the threshold for new entrants from a maximum holding of 3 per cent to 7 per cent of the total slots at the airport in question.

Furthermore, it was proposed that new entrants should receive the first 50 per cent of the slots in the pool.

---

20   COM(2001)335 final.

With these changes the Commission hoped to invigorate the new entrant provisions.

- The basis for designating an airport as coordinated

Regulation 95/93 allowed a member state to freely designate an airport as coordinated. This has had as a consequence that too many airports were designated as coordinated even when they did not have serious capacity problems. In many instances it became a matter of prestige for an airport to be designated as coordinated. This was unfortunate because operations at an uncoordinated or facilitated airport are much more flexible than at a coordinated airport.

The proposal would make it mandatory to carry out a thorough capacity analysis at first. The member state could designate the airport as coordinated only if this analysis showed that serious capacity problems existed which could not readily be remedied.

- Local rules

For some time there has been uncertainty as to the force of local rules since Regulation 95/93 only gave them the status of recommendations. The proposal states that local guidelines shall be taken into account by the coordinator when they have been proposed by the coordination committee and approved by the member state responsible for the airport.

- Clarification of transfers and exchanges

The text in Regulation 95/93 on transfers and exchanges of slots has always given rise to interpretation questions. The final text in that Regulation was drafted in the Council in perfect English or so we were told. As it happened translating it into other languages however proved difficult. The proposal therefore expanded the text and explained transfers and exchanges in some detail without any real change in the meaning except perhaps by allowing transfers between parent and subsidiary air carriers.

In addition to this the proposal also attempted to close down trade by including text that would prohibit monetary transfers in association with transfers as such; except between parent and subsidiary air carriers.

Furthermore, it was proposed that both parties to an exchange must commit to use the slots subsequently.

- Retiming

If a slot is given up it must be placed in the slot pool according to Regulation 95/93 which also sets rules for the distribution of slots in the pool for example to new entrants. The practice of retiming where air carriers give up a slot in order to get a better slot which normally should have gone to the pool was therefore illegal. With its proposal the Commission recognized that retiming could be useful and that it therefore should be accepted in a certain number of instances. The proposal, however, did not see retiming as acceptable in all situations.

• Use-it-or-lose-it rule

This rule was tightened in the sense that all air carriers, whether scheduled or non-scheduled, would have to respect the 80 per cent usage factor. In Regulation 95/93 non-scheduled was only subject to 70 per cent.

• Sanctions/withdrawal

In the context of sanctions the Commission proposed that the ATM services should reject the flight plan of an air carrier if it does not possess the required slots at the departure and/or destination airports. This is the first emanation of the principle that the Commission had started to promote namely that slot allocation and Air Traffic Management (ATM) should be coordinated.

The proposed sanctions come in two categories namely withdrawal and fiscal. Withdrawal is to be administered by the coordinator and the fiscal sanctions by the member states. Before a sanction can be imposed on the air carrier in question it must be given a possibility to present its case.

• Legal situation of the coordinator

The financial independence of the coordinator was to be further strengthened.

Council amd European Parliament Regulation 2004

This Regulation involved the European Parliament[21] in its new role with co-decision powers. The Parliament therefore had a much more direct influence on the final shape of the Regulation and not just a consultative role as in the past. It is therefore noteworthy to observe that air carriers, airports and slot coordinators via their European organizations carried out quite an extensive lobbying effort aimed at the Parliament.

Legislative process The proposal was in many ways strongly opposed by air carriers who saw it as going beyond technical modifications and including more fundamental proposals for change for example in the areas of ownership and new entrants. They took in general a rather negative approach and it was perhaps for that reason that they were not very successful.

The airports were strongly in support of the proposal. They naturally had areas where they could suggest improvements in particular with respect to new entrants, ownership and local rules. The airports would in fact have preferred to see the new entrant provisions deleted because they did not believe that they were in any way useful. On ownership they were adamant that air carriers did not own the slots that they were allocated. Finally, they wanted to see the specific characteristics of individual airports reflected in the Regulation by virtue of local rules. They went to considerable length to suggest direct wording to key participants and to the member states. On balance therefore the impression is that they had a reasonable success.

---

21   Hereinafter called the Parliament.

The slot coordinators also supported the proposal especially because of the introduction of sanctions. However, they wanted particularly to strengthen their legal position not only financially but also with respect to liability issues. As seen below they had considerable success.

The discussion was basically about whether the new or amended elements were technical in nature or whether they fundamentally changed the existing Regulation.

The proposal in 2001 resulted in a 'technical' amendment to the earlier Regulation. It was technical in the sense that the existing text was made more precise and generally sharpened up. There is no mention of withdrawal except as a consequence of abuse of slots.

• Ownership

The definition of a slot was modified in a way that weakened the claim for ownership by the air carriers:

> 'slot' shall mean the permission given by a coordinator in accordance with this Regulation to use the full range of airport infrastructure necessary to operate an air service at a coordinated airport on a specific date and time for the purpose of landing or take-off as allocated by a coordinator in accordance with this Regulation.

This is what the Commission wanted. It is, however, doubtful whether this change in definition will facilitate the introduction of general withdrawal rules. Therefore it should not alarm the air carriers, which after all retain the grandfather rights.

From an airport point of view some concern was voiced against the wording with respect to infrastructure. From their point of view it should have been added that the air carriers otherwise should respect payment of charges and live up to the rules of operation at the airport.

• New entrants

The original wording basically remained but the general threshold for remaining a new entrant at an airport or airport system was increased to 5% for individual airports and 4 per cent for airport systems.

• The basis for designating an airport as coordinated

The Commission's proposal was basically approved.

• Local rules

The concept stayed in the proposal and it was made mandatory for the coordinator to take local legislation into account. However, the wording is less exhaustive than desired by the airports.

• Clarification of transfers and exchanges

The proposal of the Commission was basically accepted. However, the attempts by the Commission to ban trading were not included.

• Legal situation of the coordinator

The Parliament and the Council followed the Commission in reinforcing the independence of the coordinator. However, they went further by granting, in general, a legal protection against claims for damages against coordinators except where the coordinator had behaved in an irresponsible way.

• Sanctions/withdrawal

The proposals from the Commission were basically accepted. However, the obligation on ATM services to reject a flight plan if the air carrier did not have the slots required became optional.

Sanctions have been introduced and are aimed at a more efficient use of capacity in three ways namely by sanctioning systematic operations of slot,[22] operations without slots and excessive non-operation of slots. In such circumstances the coordinator shall withdraw the slots in the following circumstances:

> 3. The coordinator shall withdraw and place in the pool the series of slots of an air carrier, which it has received following an exchange pursuant to Article 8a(1)(c) if they have not been used as intended.
>
> 4. Air carriers that repeatedly and intentionally operate air services at a time significantly different from the allocated slot as part of a series of slots or uses slots in a significantly different way from that indicated at the time of allocation and thereby cause prejudice to airport or air traffic operations shall lose their status as referred to in Article 8(2). The coordinator may decide to withdraw from that air carrier the series of slots in question for the remainder of the scheduling period and place them in the pool after having heard the air carrier concerned and after issuing a single warning.
>
> 6. (a)Without prejudice to Article 10(4), if the 80% usage rate as defined in Article 8(2) cannot be achieved by an air carrier, the coordinator may decide to withdraw from that air carrier the series of slots in question for the remainder of the scheduling period and place them in the pool after having heard the air carrier concerned.
>
> (b) Without prejudice to Article 10(4), if after an allotted time corresponding to 20% of the period of the series validity no slots of that series of slots have been used, the coordinator shall place the series of slots in question in the pool for the remainder of the scheduling period, after having heard the air carrier concerned.

These sanctions were to be supplemented by economic sanctions by the member states by July 2005.

Possibilities of action in addition to Regulation 793/2004 The new rules can be supplemented at individual airports by other schemes which together, if they are applied, might be considered as a mini-reform. They do not need any legal initiative but could be implemented at present.

I believe that these different elements should be used much more aggressively. Together with sanctions they could result in balancing supply and demand especially if secondary trading, where possible, would become much more

---

22   Operation off slot means that an air carrier is operating at another hour than defined by the slot received.

transparent. At present it is not evident that a potential buyer of slots will be able to identify which air carrier might be willing to sell some of its slots.

- Peak hour pricing

Presently, airport charges are generally too low to reduce peak time demand sufficiently at congested airports to match supply. However, at a number of airports a framework of peak hour pricing could be introduced. The extra revenue raised through peak hour pricing would allow airport charges to be reduced at times when the airport is not congested. The overall effect under present ICAO[23] rules should be revenue neutral since charges overall should be cost related. So, when airport charges are increased at peak times then this must be compensated for by lower charges at off-peak times. In fact such a system of peak hour pricing might allow airports to not only forfeit charges at off peak hours but to actually pay air carriers to operate.

It would be very useful to gain more extensive experience with peak hour pricing in order to assess to which extent a system of higher posted prices would influence the operations of air carriers.

- Infrastructure reservation charge

Whenever an allocated slot is not operated, it means that available capacity at the airport is not being used.[24] This has created a situation where the slot coordinators are overbooking the available capacity at coordinated airports in an effort to unused capacity. The airports are also taking this phenomenon into account and are in fact charging more per movement than justified by the costs related to the single operation.

The same problem exists for airlines. The fact that a certain number of full fare passengers do not show up at departure leads to a situation where airlines are overbooking in order to avoid empty seats on '*full*' flights. This has led to an unsatisfactory situation for both passengers, who are denied boarding, and airlines, which are forced to pay compensation for denied boarding. The airlines in AEA and other organizations have discussed whether it would be possible to apply a reservation charge instead of allowing passengers a full refund when they do not show up. Airlines have backed down from such a solution because of fears that such a system may lead competitors to enter into (tacit) agreements with important customers granting them the possibility of full refunds. Some countries, for example Singapore, have introduced a compulsory reservation charge but to my best knowledge it has not been a real success.

The reservation charge for slots is no different in principle than the reservation charge that has been discussed amongst airlines. However, the market situation is different for airports since it is unlikely that airlines will change airports because of the implementation of a reservation charge. However if they did it would probably be beneficial since it would free up capacity at the coordinated airport.

---

23   International Air Transport Organization.
24   This naturally only pertains to coordinated airports and not to facilitated airports. At facilitated airports nobody is turned away.

The implementation of this charge should also be in the interest of the airlines. Airlines which operate scheduled flights with few cancellations pay currently too much while air carriers that cancel late or simply do not operate get by without paying their fair share.

But do the air carriers receive a service from the airport when a slot is allocated? This question is important since airports in conformity with international rules can only charge for services which they provide. The slots are not allocated by the airport but by the slot coordinator who is independent of the airport. However, it is quite clear that an air carrier when asking for and receiving a slot is at the same time requesting a service. The airport should reserve infrastructure for that carrier so that it can operate.[25] If the air carrier cancels at a time when the slot can still be reallocated then no charge should be levied. However, if the air carrier returns the slot after that time the airport is entitled to charge a reservation fee since infrastructure and perhaps other resources were wasted. A reservation charge would provide an incentive to air carriers to use the capacity they have been allocated. In the case that the slot is operated, the reservation fee would be used towards paying for the full airport charge. If, however, the slot was not operated (either due to late hand-back or 'no show') the air carrier would not be reimbursed.

A slot reservation charge in fact could probably reduce the level of slot use abuses for which sanction possibilities have now been introduced. The late hand back of slots greatly jeopardizes the ability of the coordinator to re-allocate scarce airport capacity, other airlines' ability to improve their slot portfolios, and therefore the airport operator's ability to ensure efficient use of the infrastructure. Slot reservation charges would reduce the need to resort to sanctions and would lead to better utilization of airport capacity. However, the effect would not in itself be sufficient. It would need to be combined with other elements such as peak hour pricing and secondary trading.

• Aircraft size

Some airports (in particular Düsseldorf) have actively been considering a scheme for increasing the aircraft size at the airport. Legally such a scheme could be introduced at individual airports by adopting local rules under the slot allocation regulation, The reasoning behind the schemes is that more passengers could be moved per airplane operation and that the runway therefore would be better utilized. This seems appropriate for situations where the runway or environmental concerns constitutes the bottleneck but it would not be useful in situations where the bottleneck was found at other parts of the airport infrastructure for example the terminal. A scheme has emerged called OPUS[26]. This is basically a scheme where air carriers individually would have to use larger planes in step with growth in their traffic volume if they expand their service beyond a basic level where they could use any size aircraft with say up to four frequencies per day. If the air carrier would want to go beyond the four frequencies it would need to operate aircrafts

---

25 See the definition of a slot on p. 19.
26 Optimisation Programme for the Use of Slots.

of a specified minimum size. An alternative proposal to this scheme was based on the approach that aircraft size on a route would depend on the total collective traffic volume on the route. This last scheme, however would make it much more difficult for new competitors to enter that market.

A similar but much less acceptable scheme would be to cap frequencies on individual routes. Such an approach, however, is not appreciated by the air carriers since frequency rather than size of aircraft is an important competitive factor.

• Secondary trading

The Commission claims in its recent consultation paper in paragraph 2.3 that:

> *exchanges of slots for financial consideration is not permitted under Community law* and that no incentives exist to stimulate the efficient use of scarce resources by carriers using slots that could become available otherwise in the process of allocation from the pool e.g. through primary trading.

This is not correct for several reasons. As already stated peak hour pricing and slot reservation charges can be used to stimulate the efficient use of scarce airport resources.

Secondly, the current regulation specifically permits exchanges of slots and there is nothing in the old or the new regulation that would stand in the way of exchanging slots on a one-on-one basis with an accompanying financial compensation under the condition that it is a real one-on-one exchange. In fact as already mentioned above the efforts by the Commission to limit the scope for exchanges were defeated in the Council and the Parliament. In other words the secondary trade of exchanging a bad slot for a better one and paying the difference in value is not forbidden and the air carrier ending up with the bad slot is not obliged to operate it. Indeed, this practice is being undertaken at several airports either overtly or hidden and it has been conducive to creating an increased level of slot mobility at several airports. However, transparency is lacking in most instances.

Reform Options

The reform which is being discussed at this time concerns the need for and the potential shape of a market based system for slot allocation.

It should be underlined that everybody would like to see an increase in capacity in order to meet demand.

This however is unrealistic. Nobody expects the environmental pressure to ease up. According to existing investment plans for airports there will be some additional capacity in the next years but it will be far from enough to satisfy demand. Notwithstanding this, air carriers refuse the other schemes while saying that capacity must be increased. Their underlying argument is that they see the introduction of a market based system of slot allocation as simply resulting in increased costs for the air transport sector. This will probably be true for individual aircraft operations but not to the same extent per passenger since aircraft size will increase. Nevertheless in the absence of sufficient capacity growth it is clear that the Commission will further pursue the development of market based slot allocation mechanisms.

On this point a brief overview of the different concepts which are in discussion appears useful.

The basic idea is that demand should correspond to supply. Today demand far outstrips supply at a large number of airports.

However, it should also be kept in mind that the real test of a market is whether slots are available when needed. To simply look at the number of slot transactions is by itself not important. Any report on slot allocation in the US shows that an air carrier which is sufficiently interested has been able to find slots.

It should also be kept in mind that ample capacity exists at other airports than the coordinated ones. In fact by increasing the costs of operation of the coordinated airports, the non-coordinated airports would become more competitive without having to resort to state aid.

It should also be said that in order for a market based system to work transparency is necessary. If not the market would be imperfect in the sense that slot holders could not properly assess the opportunity costs of holding on to slots and buyers would not know which air carriers might be willing to sell.

What Happens to the Extra Money?

Some of the options described below will create extra revenue similar to a system of higher posted prices and auctions.

Before such a system can be accepted it is imperative to reach an agreement on what should happen with the extra revenue. Air transport is in crisis and it is very uncomfortable to discuss an increase in costs without being able to show any benefits.

The airports could claim this revenue. However, I have no doubt that air carriers in that case will try to develop the argument that this kind of revenue is tantamount to airport departure or landing fees. If that is so then the revenue would have to be matched to the underlying costs according to ICAO principles. This would turn the situation into a zero sum game since the introduction of revenue from auctions etc. would mean that existing departure and landing fees would have to be reduced. This would neutralize the whole idea behind establishing a market-based system for slot allocation. Therefore, it will be necessary for airports to develop an argumentation, which proves that this money is different from other airport fees if they want to control it otherwise the revenue would have to be allocated to a third party.

If the revenue has to be allocated to a third party, it is important to ensure that it does not disappear from the air transport sector. One possibility might be to establish a development fund that airports might use as a financing source for capacity enhancing projects, environmental programmes around the airport and possibly research into what can be done about the NOX[27] situation etc.

Market-based Systems of Slot Allocation

The reasoning behind a market based slot allocation system is that the slots are allocated to those air carriers which economically can make the best use of them. In general there would be a tendency for long haul traffic to be better able to pay for slots since it is clear that slot costs will constitute a smaller percentage out of the total costs of operation. This is a complex issue since it is claimed that long haul operations with large aircrafts in many instances are dependent on feeder traffic with smaller aircrafts. In these instances the long haul carrier might want to subsidise slots for the feeder operations (either their own or those of another air carrier). This has clearly been the case in the past but other modes of transport are able to carry feeder traffic although the full operational synchronization with air transport has still to be arranged. A long haul air carrier like Virgin Atlantic has not by itself any feeder traffic by air.

In this context it should be kept in mind that charter traffic is mainly long haul. On the other hand charter traffic is price sensitive so could only compete for slots to a certain extent.

The argument has been put forward that low cost carriers might benefit in a situation with increased airport costs. I tend not to believe this when considering the US and European experiences. Southwest and Ryanair are mainly operating to secondary airports. EasyJet left Zurich when it became too expensive. Virgin Express is merging with SN Brussels. However, some LCCs such as Jet Blue are operating to cat. 1 airports where they cannot avoid it.

The losers in such a system could easily be those air carriers for whom traffic would not be sufficient to allow them to purchase the necessary slots. This would in particular cover the services from regions where a public service obligation was introduced. In such instances it will have to be decided whether a certain amount of slots will have to be set aside or whether for example it should be possible for regions to purchase slots.

Certain regional services operate to markets where no other mode of transport offers a reasonable alternative. For these services the price elasticity would be low.

Secondary trading When talking about a market-based slot allocation system slot trading, what is nearly always meant is secondary slot trading.

This will probably be a fairly easy option to introduce. Several air carriers and the airports have expressed their interest and politically there appears to be only limited resistance. The cost effects would be gradual. It would have to be made very clear that the normal 'use-it- lose-it-rule' would apply and that the

---

27   Nitrogen oxide, a serious pollutant.

feasibility checks[28] would continue as well as sanctions. However, the system would function in a way which is already well understood although transparency would have to be ensured.

The system will reduce demand but maybe not right away to the extent that supply will balance demand. During economic upturns air carriers would want to expand and with this in mind they will hold onto their slots even when offered a high premium. The 'use-it-or-lose-it' rule will probably not result in freeing up many slots since the air carrier will try to sell the slots. The pool will therefore only reflect new capacity. In fact, based on the US and European experiences it will take considerable time before such a system modifies the existing operating pattern at the congested airports sufficiently to create an economic balance between supply and demand.

There should not be any significant blockages since secondary trading is currently already possible to a large extent. It has been claimed that allowing air carriers to trade slots immediately after allocation might give the air carriers involved a windfall profit. This is however only a small problem since the number of new slots to be distributed through the slot pools will continue to be small. Furthermore, incumbent air carriers can already today realize a certain windfall profit by using the existing possibilities of secondary trading.

The money will flow between air carriers if trading is introduced. It might be possible to levy a tax or fee on such transactions for example to pay the costs of the slot allocation. However, it would be highly unpopular amongst air carriers and it might turn out to be impractical in a slot allocation system, which is based on the scheduling conferences where it would be almost impossible to keep track of exchanges for the purposes of paying a fee.

The airports will want to obtain a share of the money streams but it is doubtful that they will succeed and it would be very difficult to monitor.

There will be no withdrawal so the ownership question will not be important except for the airports in the context of obtaining a share of the money.

New entrants may find life difficult since slot openings will be rather rare at the busiest airports. However when costs increase at major airports then the competitive position of the secondary airports will improve.

Higher posted prices If higher posted prices[29] would be introduced the impact would be immediate and a number of carriers would find that their services or some of them would become unprofitable. This system can be expected to fairly quickly establish a balance between supply and demand. However, the legal feasibility of this system needs to be further analyzed (see below) since it might be characterized as just another airport fee system. It is also clear that the higher such a fee would be the lower the slot value would be in the market.

---

28   Any slot transaction between air carriers must be approved by the coordinator with a view to whether the it would jeopardize the airport operations and in other ways would be compatible with the Regulation.

29   A kind of slot use fee.

As seen above and in the NERA study[30] this is a straightforward scheme which could incorporate normal airport charges, slot reservation charges and peak hour pricing. As stated above it will have an immediate effect although it will be difficult to set the prices

The effects of this system would be felt by all air carriers at the same time. This would therefore create a situation where a number of air carriers would be operating with a thin profit margin and could be tempted to sell. This might improve the situation for new entrants. Apart from the difficulty in setting the right price level the effect would be that slots lose to a large extent their market value. In fact the higher posted prices should probably aim for the level where the slots no longer have a market value which would indicate that a balance between supply and demand has been achieved.

Would revenue generated by such a market-based slot allocation system be considered as just another form of departure or landing fees? A further question might be asked namely to which extent airports would want that income or part of it since the suspicion might be raised that airports are hesitant to increase capacity in order to maintain that revenue.

Auctions

The next step up in intensity would happen if slot auctions were introduced in combination with trading. However, since there would be no immediate increase in costs for grandfathered slots the effects will not be very different from a general trading system except for slots allocated through the pool.

Although auctions have to my best knowledge never been implemented for slot allocation the amount of literature on this subject is large. However, the options developed in this respect in the NERA study are not convincing and do not appear realistic. Air transport is highly infused with information systems and slot allocation concerns a very large number of transactions so it is clear that an auction approach based on information technology must be developed otherwise a system covering auctions of all the slots or a certain percentage would be extremely difficult to handle. I am not aware that a practical system has been developed although a lot of research has been done. For this reason alone I do not believe that an auction system is currently a viable option.

Eliminate slot allocation In the US slot allocation is being phased out by 2007. Already this would seem to create difficulties at the Chicago airport. The cost of chaos will determine which air carriers will find life so difficult at that airport that they stop or reduce operations. In the US the situation is facilitated by the fact that the air carriers in many instances own the terminals. Therefore if an air carrier cannot get terminal space it cannot operate to the airport. This is not so in Europe where airports are being operated under common use rules. In principle there is access for all air carriers. However, the chaos that might be created is difficult to accept for air carriers and passengers alike. On the other hand the

---

30   See later.

example of Schiphol airport in Amsterdam shows that with proper management (facilitation) much is achievable.

There would be no problems of ownership or for new entrants since all in principle have access.

Fixed time allocation Any of the above systems could be strengthened if combined with a leasing or concession system where a slot is being allocated for a fixed time. It is clear that the Commission's idea of the length of such a concession was unrealistic. A more realistic time would be around the 20 years mark, which in itself would produce a minimum of 5 per cent annually in the slot pool. A transitional arrangement for phasing out the traditional grandfather rights would have to be developed since otherwise too many slots would come up for renewal or redistribution on certain years.

The payment for a concession could be established as a flat fee or by an auction. It is clear that the airlines would be against such a system. However, they probably look at it in the framework of the present regulation and in particular the provisions on new entrants. It is very probable that the application of a 50 per cent priority for new entrants would lead to a predominance of new entrants and constant change at the airports which would harm air carriers and airports alike. It might also damage air transport networks. This however is not a given thing and it may well be that the new entrant provisions would have to be stripped from a concession system (and several of the other options) which would open the door for air carriers to re-acquire slots from the pool.

Further remarks Several of the different concepts outlined above could be combined. For example a combined system of higher posted prices and secondary trading, such a system could be expected to create a balance between supply and demand. This would mean that the present concept of new entrants would not serve any useful purpose any longer. This effect of a market based slot allocation system has, however, not been studied thus far.

At present we have very little or no experience with how a fully fledged application of all the elements outlined above would affect the market and the utilization of capacity. This experience is needed in order to be able to discuss a further extension of secondary trading, higher posted prices, auctions or any other option that would increase costs.

Third Countries

In this context it must also be stated that it is necessary to have an idea of how third countries and/or their air carriers will react. It is unfortunate that the NERA report has not looked into this question and that it is not raised in the last consultation paper of the Commission. I believe that a system of higher posted prices could be accepted by third countries and probably also secondary trading although some retaliation might take place. However, withdrawal of slots whether related to auctions or not will no doubt create strong hostile reactions abroad apart from in the context sanctions connected to wilful abuse of slots.

Should the idea be that the market mechanisms would only apply to intra-EU routes questions of discrimination would be raised by both foreign air carriers and Community Air Carriers.

Commission Studies and Consultations from 2004 until Beginning 2006

NERA[31] study[32]

Immediately following the decision to divide the efforts of further developing Regulation 95/93 into a technical part and a market based part the Commission decided to ask for a study on market based slot allocation possibilities. The study contract was given to NERA who produced a report in early 2004. The executive summary of this study is included in the annex.

Commission consultation paper September 2004 After the NERA report had been accepted the Commission submitted a consultation paper. This paper sets out the ideas of the Commission for reform. In fact the Commission concluded that full secondary trading combined with auctions from the pool is its preferred system. Full secondary trading means that it could involve a clean transfer and the subterfuge of artificial exchanges would become unnecessary. This system would be supplemented by withdrawal of some slots if a certain level of mobility is not reached.

It would seem that the Commission's consultation paper also tacitly acknowledges that secondary trading is currently possible by virtue of arranging for exchanges since the proposed new system in chapter 4 of the paper only concerns secondary trading in the form of transfers against financial compensation. The Commission may still harbour hopes that it will be possible to succeed before the Courts with the argument that the exchange of slots where the bad slots will not be used, constitutes a de facto transfer in instances where transfers are forbidden.

The reaction to the consultation was immediate and unanimous from the industry. In a joint letter they pointed out that the new system based on Regulation 793/2004 had only just come into effect. It should be given the necessary time to show which effects would be created by the new rules and the effects would have to be properly analyzed. The reaction also challenged the Commission's conclusion that withdrawal would have to be a part of a market based slot allocation system.

Recent study 2006 As a follow-up to the NERA study the Commission has asked for a more direct study on the effects of secondary trading and the design of such a system. On the basis of the following published message:

---

31   Economic Consulting Firm.
32   The full study report is dated January 2004 and is available on the Commission's website for DG TREN.

The Commission is considering a further modification to Council Regulation (EEC) 95/93 on common rules for the allocation of slots at Community airports, as amended by Regulation (EC) 793/2004. This modification could entail the introduction of a commercial mechanism for slot allocation to the extent that air carriers would be allowed to engage in secondary trading of slots i.e. to exchange slots for financial consideration. The study should provide an overview of the possible effects of such a mechanism on the air transport market.

The firm of Mott McDonald was chosen to carry out a study with the above terms of reference (TOR) and a number of more explanatory guidelines (only the main points are shown).

In the above TOR it seems that the Commission only considers secondary trading in the context of exchanges of slots. However, the guidelines make it clear that in fact the intention is to study secondary trading in the context of transfers of slots.

The guidelines also conclude on the present situation:

> The measures laid down in Regulation (EC) 793/2004 are not sufficient to remedy situations where the level of saturation at an airport is such that i) new entry or ii) optimisation and maximation of the use of allocated slots by incumbent carriers is not feasible without a structural change to the system. New entry can not take place because there is not sufficient turn over of slots into the pool to be allocated to new entrants for them to economically launch air services. Incumbent air carriers are not compelled to make the most efficient use of their slots and this has a negative impact on the optimal and maximal use of airport capacity.

This conclusion cannot yet be backed up by data since Regulation 793/2004 has only come into effect with the summer season 2005 and the financial sanctions even later. This is unfortunate but hopefully Mott McDonald's report will include an analysis of the available data concerning the application of the new rules.

However, it is even more unfortunate that the Commission repeats its fascination with the withdrawal option:

> In a wider sense, it may be considered that, in general, at extremely congested EU airports, either throughout the day or at certain peak times, slot trading will not lead to a minimum[33] of slot mobility as few or no air carriers are prepared to sell slots. This mobility could be achieved by establishing a fixed percentage – for instance 1% or 2% per scheduling season[34] – that would apply to grandfathered slots. These reclaimed slots would be returned to the pool and become subject to re-allocation. In such case, the 1% or 2% of slots would be withdrawn by the slot coordinator.

In fact the TOR go further and asks the consultants to recommend how to arrange such a withdrawal system. There is no clear-cut question on the future of the new entrant provisions in the context of a market based slot allocation system. The study report is not yet available.

---

33 A minimum to be defined.
34 Or another percentage that would lead to a minimum of slot mobility.

Conclusion

The present stage of regulation for slot allocation are based on the worldwide guidelines but goes further and includes a number of efficient sanctions and specific rules for the EU. These rules appear well accepted by the industry, member states and the coordinators. The system allows for secondary trading in the context of slot exchanges but it is not known to which extent this option is being used. However, discussion concerning the development of a fully fledged market based system of slot allocation remains on-going.

Consultations by the Commission are also still being undertaken. It is uncertain whether they will result in a proposal later this year since the consultation paper has raised calls from all stakeholders to wait until the existing system has had a chance to function for a while. If the existing possibilities are being used properly then it may get closer to create a balance between supply and demand and more drastic changes may not be necessary. The change of Commissioners may also somewhat delay matters. However, sooner or later there will be a proposal from the Commission. In the following I will provide a guess as to the outcome of a reasonable reform.

In general everybody seems to agree that a buy and sell system:

1) will create possibilities for the economically strong (long haul) air carriers;
2) will not create confiscation (withdrawal) problems;
3) is not liked by most third country air carriers;
4) will allow market forces to play if transparent (which the US system is not).

The preceding discussion has revealed that one of the most difficult issues so far is the question around withdrawal. Withdrawal does exist in the present system as the ultimate sanction but not as a common tool. I believe this will turn out to be a strong blocking issue if it is taken up as a condition for further development. Unfortunately the Commission seems to be determined on this issue.

As stated above secondary trading is currently possible but limited to the context of exchanges. I believe that this will be increased to also cover direct trading. A consequence of this will be that the costs of operating to congested airports will become more expensive either because an air carrier will have to purchase slots or because of opportunity costs.

One of the difficult issues for such a solution will be to distribute new capacity. If the present system continues this will create windfall profits and I am sure that this will be viewed as unacceptable by for example the European Parliament. A proper application of peak hour pricing would reduce this problem and so would a system of higher posted prices.

In my opinion higher posted prices will only be accepted if a proper use is found for the extra money. It is clear that the state treasuries would like to see such a system if they would be the beneficiaries but this would be unacceptable for several reasons. It would appear similar to a tax system which would be incompatible with the Chicago Convention. Air carriers, airports and consumers would not like it since it would simply make air transport more expensive without any accompanying benefits. On the other hand it might also be difficult to create a fund which could be used to improve infrastructure at the airport. Earmarking revenue has never been very popular in state administrations and it might also be difficult to define the potential uses. However, the benefits of the introduction of higher posted prices might overcome some of these more ideological difficulties. There is no doubt that higher posted prices would have the potential to balance supply and demand at the congested airports. It would increase demand at the secondary airports currently an abundance of unused capacity exists. The combination with secondary trading would give it a certain flexibility which would facilitate changes in the market.

If higher posted prices are introduced I also believe that the concept of new entrants might disappear because supply and demand would balance out. If this results in vital air services disappearing then perhaps it would be useful to introduce the possibility for regions to acquire slots.

The *'use-it-or-lose-it'* rule will remain and in order to avoid hoarding or similar practices, sanctions may even be strengthened. The same will be true for the feasibility checks. In fact the scope of these checks might be expanded to not only cover infrastructure but also airport planning features.

I do not believe that it will be necessary to decide who owns the slots in order for such a configuration to work. However, in order to avoid complications the present clause that application of the competition rules will not entail compensation may be expanded to also cover other forms of EU law.

It is to be hoped that the forthcoming discussion on and development of a market based slot allocation system will be constructive and not distorted by the issue of withdrawal. In the meantime it will be interesting to see whether the present possibilities of secondary trading will regarded as useful in a significant way at coordinated airports and not only in the UK.

References

Civil Aviation Authority (1995), 'Slot Allocation: A Proposal for Europe's Airports', CAP 644, Civil Aviation Authority, London.

Institut d'aménagement et d'urbanisme de la région d'Ile–de-France (IAURIF) (1999), 'Étude sur les Capacités Aéroportuaires Alternatives', July.

Starkie, D. (1992), 'Slot trading at United States Airports', report for the Director General for Transport of the Commission of the European Communities, Putnam, Hayes and Bartlett Ltd., London.

Annex

NERA STUDY on Market Bases Slot Allocation: Executive Summary[1]

All of the market mechanisms we have examined have the potential to increase the proportion of slots at congested airports that are allocated to the airlines that value them most, and to improve the utilization of slots. Passenger numbers at congested airports are likely to increase as a result of four types of impact:

- a shift in the mix of services using congested airports, notably an increase in the proportion of long haul services;
- within each category of service, a general shift to services with higher load factors. Within short haul services, for example, some regional services and services operated by full service carriers other than the hub carrier will be withdrawn, and more services will be operated by low cost carriers. Some of the least profitable long haul services will also be withdrawn;
- where possible, airlines will shift services to off-peak times or to uncongested airports. This is most likely to affect charter services and perhaps some long haul services, and will free up peak capacity for other services. For many services, however, shifting to off-peak times or uncongested airports will not be a realistic option;
- slot utilization will also improve, as the fact that airlines have to pay for slots (or can sell any unwanted slots to other airlines) will reduce the number of slots at congested airports that remain unused.

The illustrative calculations presented in Chapter 6 suggest that, in the medium to long term, such changes could increase the number of passengers at congested airports by approximately 7 per cent, equivalent to an extra 52 million passengers per year[2].

While each of the specific market mechanisms we have examined is likely to deliver many of these benefits, none is likely to be able to function perfectly:

1) *secondary trading*, based on bilateral negotiations between airlines, would be relatively easy to implement. But some potentially compatible buyers and sellers might be unable to identify each other, and some airlines might ignore the potential proceeds from selling their slots and continue instead to run services that fail to make the best use of scarce capacity;

2) *higher posted prices* would, we assume, be introduced only gradually, as it is likely to be quite difficult to establish the market clearing level of prices, and therefore any efficiency improvements may be delayed, perhaps quite significantly. Even in the long run, it is unlikely that higher posted prices will completely clear the market (and therefore administrative primary allocation

---

1   Re-printed with the permission of the EU-Commission and NERA.
2   Based on forecast passenger numbers in 2007.

criteria will still be needed). Some residual inefficiencies are therefore likely to remain;

3) a combination of *higher posted prices and secondary trading* might have the greatest potential of any of our options to achieve the allocation of slots under the ideal market mechanism. The ability to engage in secondary trading may help to address residual inefficiencies that result because higher posted prices do not clear the market. But secondary trading is most effective when there are large differences between the buyer's and the seller's valuation of a slot, and it may therefore only be partially successful in "fine tuning" the allocation of slots among those airlines willing to pay high posted prices;

4) a combination of *auctions of pool slots and secondary trading* also has the potential to achieve a substantial improvement in the allocation of slots. For existing slots, the impact is simply that of allowing secondary trading. But for newly created slots, auctions have the advantage of achieving a more efficient initial allocation of these slots, and also avoid the problem of "giving away" new slots (eg through administrative mechanisms) that can then be sold at a significant profit;

5) *auctions of 10 per cent of slots, combined with secondary trading* could, in theory, achieve the most efficient allocation of slots possible. But in practice, many of the auctions are likely to be so complex, both for auction organizers and for airlines bidding for slots, that it is probably unlikely that an efficient allocation of slots will emerge from this process.

We have carried out some high level calculations to illustrate the potential scale of the impact of each mechanism. The results are summarized in the first few rows of Table 19.1 below. Inevitably, these calculations rely on a large number of assumptions and estimates, as there is very little evidence available about the likely impact of either specific market mechanisms or the use of market mechanisms in general. Nevertheless, we believe these illustrative calculations provide an appropriate (though very approximate) indication of the likely medium to long term effectiveness of each mechanism.

There are also significant differences in the speed at which each mechanism will affect the allocation of slots. Both secondary trading and higher posted prices may have a rather gradual impact. Airlines that might sell their slots will face an opportunity cost rather than a cash cost and therefore may take a while to make the decision to sell. Under posted prices we expect airport co-ordinators and operators to increase prices only gradually, as there is likely to be considerable uncertainty about the impact of price increases.

Perhaps equally importantly, however, is that the impact of some of these mechanisms could be delayed, potentially quite significantly, by difficulties introducing the policy in the first place. Any attempt to introduce higher posted prices or auctions of existing slots, in particular, is likely to be strongly opposed by both EU and non-EU airlines. The resulting disputes and challenges could significantly delay the implementation of either of these mechanisms, and if implemented then they might also provoke retaliatory measures by non-EU states.

Market mechanisms will lead to increased service levels in certain markets, particularly long-haul ones, and service cuts in others, such as regional, short haul and marginal long haul services. Where service levels increase, we expect fares to fall, whereas we expect fares to rise as a result of service reductions. For the industry as a whole, service levels will increase and therefore, on average, we might expect fares to fall. Fares can also be expected to fall as a result of full service short-haul services being replaced by low-cost services, and increased competition on some long haul routes. Apart from these effects, we do not as a general rule expect other impacts on fares. In particular, we do not expect airlines to be able to pass on increases in fixed airport charges at congested airports. Airline price setting at these airports involves matching demand with the available capacity, neither of which will normally change as a result of market mechanisms.

Since market mechanisms will increase the number of flights to and from EU airports and increase aircraft size, they will have negative impacts on the environment (though there may be offsetting factors, such as delaying the need for new airport capacity, and we note that the predicted change in the traffic mix and increase in load factors will lead to lower environmental costs *per passenger km*). In the absence of other measures, market mechanisms may also have a negative impact on the accessibility of regional airports.

All of the options examined are probably likely to lead to increased slot holdings by the major hub carriers. This may be particularly likely under the auction options (especially the auction of 10 per cent of slots), as hub carriers will be better placed than other airlines to cope with the uncertain outcomes of an auction. The fact that these carriers may hold an increased proportion of slots is not necessarily a sign of inefficiency or abuse of a dominant position, as it may simply reflect the benefits of economies of scale and density, and passengers themselves may benefit from the increased journey opportunities offered by improved hub and spoke networks. And increased concentration at airport level does not automatically imply a reduced level of competition, as competition takes place on the basis of routes rather than airports. Although regulatory mechanisms might be used to restrict hub carriers' slot holdings, they appear likely either to be difficult to implement, or to introduce undesirable rigidities in slot use. These considerations perhaps suggest that specific regulations to limit concentration should only be considered if problems occur in practice that cannot be addressed through existing EU competition law.

More generally, however, we believe that market mechanisms will have positive impacts on the degree of competition in the industry. They will remove important entry barriers for low-cost and competing long haul services, which will increase competition on key routes. In part, this is due to the fact that certain other routes will no longer be served from highly congested airports, and may migrate to secondary airports in major city regions. This process will free up slots at the busiest airports that can be used by entrants to compete with incumbents on other routes.

In addition, the specific market mechanisms we have examined are associated with a wide range of implementation costs and other practical issues:

- *secondary trading* is likely to have low implementation costs and is unlikely to interfere with existing slot allocation and scheduling procedures;
- *higher posted prices* (with or without secondary trading as well) will also have relatively low implementation costs, although airports will need to undertake research on a continuing basis in order to set appropriate price levels. Provided prices change only gradually, this option is unlikely to cause widespread disruption to schedules or to the existing slot allocation (at other airports) and scheduling process;
- *auctions of pool slots* may be quite costly to organize and participate in initially, though we would expect large auctions to be relatively rare. Because the primary allocation mechanism (ie the auction) applies only to pool slots, this option will not disrupt existing schedules or slot allocation processes at other airports;
- *auctions of 10 per cent of slots*, in contrast, are likely to be expensive to implement. Costs (including substantial management time) will be incurred by both auction organizers and participants. This option does not fit well alongside existing slot allocation processes, and could also lead to frequent and potentially destabilizing changes in schedules.

Our assessment of the impact of each option does not make specific assumptions in relation to the additional revenues raised from either higher posted prices or auctions. There are a number of possible approaches, including excess revenues being paid to government or ringfenced to fund future airport expansion or improvement projects. We note, however, that if airport operators were to retain these revenues, this could act as a strong disincentive for future airport expansion (as this would reduce the proceeds from higher posted prices or auctions). Even in the case of auctions applied only to pool slots, this will reward airports for providing new capacity, but only if they provide insufficient additional capacity and therefore slot shortages remain.

Table 19.1 below summarizes our main findings in relation to each option.

Table 19.1 Summary of main properties of market mechanisms

	Secondary trading	Higher posted prices	Higher posted prices and secondary trading	Auctions of pool slots and secondary trading	Auctions of 10% of slots and secondary trading
Approximate estimate of impact on passenger numbers					
low case	2.2%	3.8%	4.1%	2.4%	0.4%
central case	4.0%	4.3%	5.0%	4.2%	4.1%
high case	4.8%	5.2%	5.8%	5.0%	4.6%
Implementation costs	very low	low	moderate	moderate	very high
Other factors					
– potential for instability in airline schedules	very low	low	low	low	high
– likelihood of increased concentration of hub airports	moderately high	moderately high	moderately high	high	very high
– consistency with existing scheduling procedures	good	moderately good	moderately good	moderately good	poor
Risk of international disputes, challenges and retaliation	low	high	high	low	very high

Chapter 20

Airport Slots: Perspectives and Policies

Peter Forsyth

Introduction

Excess demand for airport facilities is now a global phenomenon. Increasing the capacity of airports so that they can handle more traffic is difficult, partly because it is expensive, and partly because such expansion often encounters environmental problems. As a consequence, there are many airports which face excess demand, some for part of the day, and several, like London's Heathrow and Frankfurt, for all of the day. Outside the US, capacity at airports is usually rationed by means of a slot system.

As with many administrative allocation systems, the slot system is based on 'grandfather' rights. Airlines which have been using the airport are allocated slots, and they can keep them from year to year. Even though they may not obtain as many slots as they would like to have, the slots that they do manage to obtain can become more and more valuable if demand expands more rapidly than capacity. Over time, slot constraints become effectively tighter, and slots become more valuable, but delays at the airport do not become significantly worse. Unlike in the US, where delays at busy airport can become a serious problem, delays are kept within acceptable bounds. Evidence suggests that the slot system has had an impact as a means of reducing congestion at airports (Janic, Chapter 6 in this volume).

Not surprisingly, the slot system is popular with the established airlines, which possess slots at the preferred airports. Equally unsurprisingly, it is not popular with newer, less established airlines (see Humphreys, 2003). It can be very difficult or impossible for a new airline to develop services from a slot controlled airport. For some airports, in the UK, it may be feasible to purchase a slot from an incumbent with one to spare, but very often, this option is not available, since slots may not be traded. This is especially a concern in continental Europe and parts of the Asia Pacific.

Thus the issue on which most discussion of slots focuses on is that of slot allocation to airlines. As things exist, there are grounds for believing that the allocation of slots to airlines may be less than ideal, and that some low value users may still be able to access busy airports while high value users are excluded. Furthermore, competition from new airlines may be stifled by the difficulties in obtaining slots. Thus there is much discussion of ways of improving slot allocation, whether by introducing slot auctions, by facilitating trading arrangements, or by reallocating a proportion of slots over time. The EC has been reviewing its policy towards slots, and while it has not adopted radical reforms, it has moved in the direction of facilitating trading (Sorensen, Chapter 19 in this volume). There are other efficiency issues to do with slots, but the slot allocation problems are the most discussed, and probably the most difficult to resolve, even though they may not be the most significant.

In this book, we have taken a broad perspective on slots, and explored a wide range of aspects of their working. Some of the key issues which are considered include:

1) how well the slot systems in place are working;
2) how the slot system performs relative to other systems, especially the first come first served approach of the US;
3) whether a greater reliance on posted prices could allocate scarce airport capacity better than slots;
4) whether there are ways of improving allocation of slots, by auctions, trading arrangements or other means;
5) how slot systems interact with price regulation and cost recovery by airports;
6) how slot limits for particular airports are determined;
7) what implications a slot system has for signals for investment to increase airport capacity in the long run; and
8) how the slot system impacts on the performance of the airline industry.

We do come to some clear conclusions in the light of the studies. The slot system does provide the *basis* for an efficient practical solution of the excess demand problem, and it is effective in reducing the costs of congestion. However the attention given to the slot allocation problem is warranted, and evidence suggests that considerable improvement is possible, especially if complementary reforms to airport pricing are made. Some of the gains from limiting congestion costs may well be being squandered. Some improvements are relatively straightforward, though others are likely to be difficult to effect.

This chapter seeks to round off the discussion in the book, and to pick up on some issues not covered in earlier chapters. It begins by identifying a range of efficiency problems which need to be resolved in the context of busy airports, and follow this by making an assessment of how well slots systems as are in place handle these. Then some of the choices between options are reviewed: slots versus prices, a slot system versus the first come first served delay rationing system, and alternative ways of improving slot allocation. The slot system has significant implications for the ways in which airline and airport markets work, and these are sketched out in a section on the political economy of slots. In the final sections some unresolved issues, warranting further study, are identified, and then some key policy conclusions are outlined.

Efficiency and Equity Issues in Airport Capacity Allocation

There are a number of distinct efficiency issues which arise with the allocation of scarce airport capacity to users. Some of these have received a lot of attention, while some have only received scant attention in the literature. In addition, equity, or distributional issues often figure largely in airport capacity management policy.

Key efficiency issues are:

1) *Choice of the slot capacity to be made available to airlines.* Authorities concerned with the airside operations of the airport must determine what level of capacity to make available to airlines. In principle, if the intention of the slot limits is to lessen congestion, the authorities will seek to choose a slot limit which results in an acceptable level of delays (Forsyth and Niemeier, Chapter 5; Czerny, Chapter 7). The value of a slot, if available, will be an indication of the value of additional capacity. In practice, the slot limit will often be set according to arbitrary criteria (e.g. a level of delay). The slot limit may also be set according to environmental criteria – for example, it may be judged that a maximum of 40 movements per hour may be all that can be allowed because of the noise created. In the light of this, one question that must be asked is whether the authorities have set the slot limit at an efficient level?

2) *Allocation of available slots.* Once the number of slots available to airlines is known, they must be allocated to airlines. The slot allocation outcome depends, to a comparable degree, on two factors:

   (a) the slot allocation mechanisms in place, for example grandfathering, auctions or secondary trading (Menaz and Mathews, Chapter 3; Kasper, Chapter 15); and
   (b the prices charged for using the airport (Forsyth and Niemeier, Chapter 8).

   Which users will use the airport, and when they will use it, will depend on these. If the downstream market, the airlines, is competitive, an efficient allocation of slots is one which results in the users with the highest willingness to pay obtaining the slots. When there is market power at the airline stage, this must be qualified.

3) *Minimizing delay costs when capacity is ample. A* slot system does not simply ration capacity when it is scarce. It may have an important role to play in reducing congestion costs at *any* given output. Slot coordination systems play an understated, though very important role in evening up the flow of traffic, and thereby reducing delays or congestion. How slot systems smooth the flow of traffic is discussed in Ulrich, Chapter 2 and Gillen, Chapter 4. Thus, an important efficiency question concerns whether congestion costs have been minimized for the output which is chosen (Forsyth and Niemeier, Chapter 5).

4) *Handling uncertainty.* The demand to use an airport is variable and uncertain. The use of the airport will be determined by what instruments, such as quantity (slot) limits are in place, what prices are in place, and whether congestion is used as a rationing device. Typically, prices and quantity limits must be set in advance, in conditions of imperfect information – thus they can be inappropriate for the demand which eventuates. Some mechanisms are better able to handle uncertainty than others – sometimes prices are better than quantities, and sometimes it is the reverse (Weitzmann, 1974; Czerny, Chapter 7). The problem for airports is to choose options to manage uncertainty at as low a cost as possible.

5) *Signals and incentives for investment.* Prices can act as signals and as incentives for investment, and one of the efficiency objectives for airports is that they invest when the benefits from investment exceed the costs, so that allocative efficiency is achieved in the long run. However slot systems mean that prices are no longer used to ration demand, and the potential market value of the airport capacity is not all received by the owner. In addition, airports are often price regulated. The issue then is whether prices received by the airport provide much guidance for investment, and furthermore, whether the airport has an incentive to expand capacity when warranted.

6) *Competition at the airline level.* Efficiency losses can come about as a result of less competition at the level of airlines. Thus it is desirable that the system of allocating capacity to airlines is such that it does not weaken competition (competition issues are considered by Gillen and Morrison, Chapter 10 and Starkie, Chapter 11). Clearly there are cases where this can happen – for example, if new entrants are unable to obtain slots with which to compete with the incumbent airlines. While airport arrangements may not be able to actively promote airline competition, they should be, if possible, not reducing it.

7) *Rent dissipation.* Rights such as slots which are granted to airlines are valuable, and they enable the recipients to make profits above the norm. Efficiency requires that these profits be used effectively, and not dissipated. Ideally, the profits which accrue to airlines from gaining slots at busy airports should end up as additions to overall airline profits for distribution to the airline's owners. However, these profits can be wasted, for example if the airline cross subsidises unprofitable services with them, or uses them to fund operations which are more costly than necessary.

8) *Achieving additional objectives at minimum efficiency cost.* Often airports are constrained in various ways – one of the most important constraints on busy airports is that they are often required to set prices which do not yield above normal profits. This may be by direct regulation, or publicly owned airports may be expected not to use their market power (Forsyth and Niemeier, Chapter 8). Airports may also have to meet environmental constraints. If these additional requirements are to be imposed on airports, it is desirable that they can be met at minimum cost in terms of efficiency.

9) *Information and transactions costs.* Different mechanisms, such as prices, slot grandfathering and slot auctions, impose different transactions and

information costs on the users, and these costs must be factored in when determining which mechanism is most efficient.

Equity or Distributional Aspects

Distributional aspects can be important for policy choice, alongside efficiency aspects. Different ways of allocating an airport's capacity will have different impacts on the various groups associated with the airport – the airport and its owners, the airlines, their passengers, and the local community surrounding the airport. In practical policy making, who gains and who loses is often critical. Thus in any assessment of the workings of a chosen mechanism, be it slots, prices or congestion, it is necessary to be aware of who gains and loses from it. Reforms which can improve the efficiency of the allocation of airport capacity may not go ahead primarily because some groups lose, and they are able to block the change. It can also be the case that distributional objectives, such as keeping prices low for airlines by using price regulation, or by ring fencing a proportion of slots for use by specific airlines (e.g., regional or international carriers), will often be done only at some cost in terms of efficiency.

Slot Systems – An Efficiency Assessment

Of the various mechanisms of handling excess demand at airports, some perform well in terms of some aspects of efficiency, while others perform well in others. Slot systems perform very well in several aspects, though they have some real limitations. In assessing how slots perform in terms of the nine aspects identified above, it is useful to distinguish three categories. In the first of these, the assessment is straightforward, and slots perform relatively well. In the second category, the assessment is problematic. With the third category, there is reason to believe that slots often do not perform well, though the assessment is complex. These categories are considered in turn.

Straightforward Cases (these cover aspects 1, 3, 4, 8 and 9)

1 Choice of slot limits On the face of it, the choice of slot limits might be a source of considerable inefficiency, since this choice is not made in a particularly scientific manner. Ideally, an assessment of the congestion costs created by additional output should be compared to the benefits of allowing this output (as measured by the value of a slot). In practice, decisions are taken on a more arbitrary basis, and at best a slot limit is declared which will give rise to an 'acceptable' average level of delay. The rules of thumb could be quite arbitrary.

In reality, this ad hoc approach may well be getting things about right (Forsyth and Niemeier, Chapter 5; Janic, Chapter 6). This is because congestion costs are likely to be very sensitive to the utilization rate when this is close to the maximum theoretical capacity of the airport. This means that the optimum utilization rate, or output, is likely to lie quite close to the theoretical capacity. The problem then becomes almost like a fixed capacity allocation problem. It is possible that the authorities may set an inefficient slot limit intentionally. They may be under pressure from airlines to allow more slots than justified, or they may be under pressure from competition authorities to make slots easy for all airlines to obtain. It may be relatively easy for authorities to determine a tolerably efficient slot limit, though this does not mean that they will always implement it. This said, however, the issue has not been explored very rigorously, and more thorough analysis of the slot limit choice should be a priority for future investigation.

3   Minimizing congestion costs when capacity is ample   Slot systems are likely to perform very well in this respect. One of the major sources of delay is the variability of movements *during* the slot period. This aspect has not received much attention in the literature. Slot systems not only limit the number of movements during a period, such as an hour or half-hour, but they also redistribute movements throughout that period. Typically, movements would seek to use the airport at popular times, say at the hour, and would not be distributed evenly throughout the hour – this would lead to many short delays. Slot coordination systems even out the flow of movements, and this results in less delay. Many airports which do not have an excess demand problem still implement a slot coordination process to even out flows and thus reduce delays (Ulrich, Chapter 2; Gillen, Chapter 4; Forsyth and Niemeier, Chapter 5).

4   Handling uncertainty   Slot systems work well, especially in comparison with alternatives such as prices, in handling uncertainty in the airport context. Here the optimum quantity is known, to a good approximation, in advance. Whatever the level of demand at the busy time, the optimum quantity will be a little less than the maximum feasible capacity (possibly with some allowance for no-shows). The price which clears the market at this quantity is not known in advance – it depends on the level of demand which eventuates, and this demand could be quite uncertain and also quite variable from day to day. Slots work by setting a quantity and then allowing the price to adjust. If the market for slots works well, an efficient output, and allocation of capacity, will be the result. The market for slots may not work well, but at least the level of utilization of the airport will be efficient. A slot system creates the potential for a very efficient solution to the uncertainty problem.

8   Price regulation   Many airports, especially privatized ones, are now subject to price regulation (for a discussion of airport regulation, see Forsyth et al., 2004). Slots work well with price regulation or situations where the allowable revenues from the airport are constrained to the level of costs. In situations of excess demand, price regulation precludes the use of prices to ration capacity. This creates a real practical problem, because busy airports are likely to be subjected to revenue or price constraints, as, for example, London Heathrow and Gatwick are (on regulation of these airports, see Hendriks and Andrews, 2004). A slot system is a necessary complement to price regulation. The advantage of slot systems is that they replace prices as the market clearing mechanism with alternative mechanisms. Thus efficient allocation of capacity is made consistent with keeping prices down, the objective of regulation.

Overall, a slot system has the potential to make price regulation consistent with efficient allocation of the airport's capacity. For a busy airport with excess demand, the only feasible means of keeping prices down while allocating capacity efficiently will involve slots.

9   Transactions costs It is generally argued that slots systems impose moderately low transactions and information costs. Slot allocation at a range of complementary airports involves solution of a complex problem, but this is achieved through getting all the parties together to jointly determine a solution. This process entails costs in preparation, evaluating options and in bringing the parties together. The outcome may not be a first best, though it will be feasible. Obtaining a more efficient solution, e.g. through an auction, could be much more complex and costly. However, subject to this caveat, the slot system is not expensive to operate.

Problematic Cases (these include 5, 6, 2 and 7)

5   Investment signals and incentives Slot systems do pose a problem for investment in capacity expansion. Typically, the prices that the airport charges are held below market clearing levels, and slots take over the rationing function. This will mean that the prices received by the airport will be no indication of the value of additional capacity. If the airport is not regulated, it still may not have good information on the value of additional capacity. If the airport is regulated, it has an incentive to invest which depends on how regulated prices relate to the cost of expansion. Very often, regulated prices are such as to just cover costs (possibly with capital costs determined on a historical basis), but additional capacity will often result in higher per unit costs. Unless regulated prices are adjusted, the airport will not have an incentive for investment even though such investment is well worth while. This is more an issue for price cap types of regulation, with pre set maximum prices, than for cost plus types of regulation, where allowable prices are set to recover all costs.

This problem can be regarded as a consequence of price regulation or of setting a requirement that the airport charges prices which are not above costs. If regulation works well it can be overcome. If there is an active slot market, the regulator can observe slot prices, and these will give a signal as to whether investment is warranted. The regulator can then structure the regulation such that the airport has an incentive to invest. In practice, this normally means that the regulator takes on the task of investment evaluation, not the airport. Thus, if regulator is well informed, it can lead to efficient investment under a slot system. However, in actual slot systems, regulation may not be that well informed. For some airports (e.g. London Heathrow) slot prices exist, but it is not clear how reliable or representative they are. For other busy airports (Frankfurt), explicit slot trading for cash does not take place, and there is little or no reliable information on the value of a slot. Even a well intentioned regulator seeking to promote efficient investment decisions will be unable to structure regulation to induce efficient levels of investment because the price signals are not present.

6   Impacts on airline competition Slots systems can have an efficiency cost through reducing competition at the airline level. If there is no slot market, or the market is thin, it may be very difficult for new airlines to enter markets from the airport. Potential entrants simply cannot obtain the slots. This can be a particular problem for international services and low cost carriers.

If there is a slot market, even when it is a fairly open and well functioning one, it may be possible for incumbents with a large share of the slots to game the market (Gillen and Morrison, Chapter 10). Such airlines could hoard slots so as to prevent them from falling into the hands of competitors. If an airline uses its slots from what seem to be low value flights, when it could sell the slots at high prices, it may make sense if it able to keep fares high through discouraging competition. This result might be seen not so much as an inherent weakness of slot systems as a weakness in the presence of dominance of an airport by a small number of carriers. However, there are also positives to slot concentration (Starkie, Chapter 11).

2 and 7   Slot allocation and rent dissipation It makes sense to treat these possible sources of inefficiency together, since it is often difficult or impossible to attribute observed data to one or the other. The objective is to examine available data to determine whether slot systems lead to an efficient allocation of capacity, and whether, if they generate rents at the airline level, these rents are dissipated. Some general stylized facts about slot markets and airlines can be suggested (some of these will be challenged later). These are:

- there are high slot premia at several busy slot constrained airports;
- there are different prices for the same service to different users at these airports;
- most of the airlines using them are private, profit oriented airlines;
- there is little systematic evidence of especially high profitability of airlines with strong positions at slot constrained airports;
- for most slot constrained airports, there is either only limited trading or no officially sanctioned trading in slots; and
- for many slot constrained airports, there are airlines which have a high proportion of the slots.

It is useful to take as a starting point a slot system which is highly efficient in that slots are allocated efficiently and there is no rent dissipation. For popular airports, the slot premia could be high. Abstracting from any attempt to set airport prices to counteract market power issues at the airline level (see Brueckner, 2002; Pels and Verhoef, 2004), to allocate slots efficiently, it would be necessary that airport charges were differentiated only to the extent that they reflected differences in costs due to different passenger or freight loads. All airlines using the airport would be factoring the slot price into their decisions whether to operate a route.

Assuming that they are not making losses on routes other than those using the slot constrained airport, the airline would be making high profits, reflecting the slot premia.

There do not seem to be any slot constrained airports which closely fit this description. In one way or another, airports diverge from this description, suggesting the presence of one or both aspects of inefficiency.

For a start, most airports operate heavily differentiated price structures – one flight may pay ten times the charge that is levied on another, for essentially the same service. If the problem is one of setting prices to make the most efficient use of scarce airport capacity, a uniform, not differentiated one, is needed. Busy slot constrained airports still use a system of pricing which was appropriate for airports with excess capacity, even though these airports have not had excess capacity for decades. The current pricing system in operation at most capacity constrained airports results in high value users being excluded in favour of low value users (Forsyth and Niemeier, Chapter 8). This is a major source of inefficiency even if all other aspects of the slot system are working well.

If a slot system is performing well, the available slots will end up with the users which have the highest willingness to pay for them, which is an efficient solution as long as there is not significant market power at the airline level. If this is the case, high slot prices should be being translated into high profits for the airlines which have strong positions at the slot constrained airports. The airlines which have strong positions at slot constrained airports do not appear to be especially profitable. This suggests that either:

- airlines are not allocating slots to flights very efficiently; or
- they are allocating flights efficiently, but then they are dissipating the rents generated.

Consider first the problem of allocation of slots between airlines, and in particular, the issue of allocation of slots between airlines. Here there are legitimate concerns that inefficiency may be present. For many airports, especially within Europe, there is no market for slots, and trading is difficult if not prohibited. In such a situation, an inefficient allocation of slots is likely – some airlines may value slots much more than other airlines, yet they may not be able to buy them off them. For some airports, slots are traded, but only in very thin markets. The fact that there are very few trades, in an industry which is quite dynamic and changing, suggests that the market is not working very well. There are many airlines which would like to buy slots if they had the opportunity. Thus the between-airline allocation of slots is not efficient. Another source of inefficiency in the allocation of slots stems from the fact that some potential bidders are prevented, by bilateral air services agreements, from bidding.

The allocation of slots *within* airlines may be better (Bauer, Chapter 9; Spitz, Chapter 13). If an airline is seeking to maximize profits, it has a strong incentive to allocate the slots it has efficiently, and to make sure that all flights which use slots are sufficiently profitable to pay the value of the slot. Even if it is not clear what an airline could sell a slot at, the airline would have a minimum internal valuation (or shadow price) of a slot which it would use in its planning, and would allocate its slots so as to maximize their value to the airline. At most busy airports, there are one or two airlines which possess a large share of the slots – several airlines have more than 50 per cent of the slots at highly slot constrained airports. Thus a significant share of the slots should be allocated within airlines efficiently.

If slots are valuable and slot rents are not dissipated, we would expect to see evidence of high profitability for airlines with large slot holdings. The aggregate values of slots held by airlines with large shares of slots at popular airports may approach the values of their fixed assets. To this extent, their slot rents should be obvious in the profit and loss accounts. These airlines should be much more profitable than others. Some legacy airlines, such as British Airways and Lufthansa, appear to be weathering competition from Low Cost Carriers (LCCs) better than others, especially the US airlines which do not have large slot holdings. This is an issue which calls for more study.

This in turn raises the question of whether airlines are indeed making effective use of their slots. It is possible that they are not allocating them very efficiently internally. Since slots mostly have not been paid for, and since they often do not have any clear market value, airlines may be underestimating the value of the slots they hold, and they may not be making good use of them.

Alternatively, airlines may be generating substantial slot rents, but they may then be dissipating them. They may be earning high profits on routes into slot constrained airports, but then using these profits to cross subsidise other routes. This is not the behaviour of an efficient profit maximizing airline, however. Another possibility is that they may be dissipating rents through allowing costs to be higher than the minimum. There is evidence of productive inefficiency in many airlines, especially in Europe. Slot rents may be enabling airlines which are not otherwise cost competitive to survive. In both of these cases, there is an efficiency cost which is measured by the slot rents which are dissipated. A final possibility is that they are sharing the rents with their workforces, by paying them above market wages. This is possible, especially when the workforce unions have some market power over the airline. If this is where slot rents are ending up, at least some of the rents will be shifted, not dissipated. The cost in terms of efficiency will be less.

Slot limits create the potential for slot rents, which one would expect to end up with the airlines which possess the slots. However, they seem to disappear. To an extent, this is because the allocation of slots is imperfect, and thus it does not generate rents to their full potential. In addition, even those rents which are generated are not always kept – they are, to an extent dissipated within the airlines which gain them. Together, poor allocation of slots, and dissipation of slot rents could amount to a considerable inefficiency problem, and represent the least efficient aspect of the slot system. At its worst, a slot system may perform the very useful function of limiting use of the airport to manageable levels, thereby converting wasteful delays into potential rents, but poor allocation of the slots combined with rent dissipation may result in these gains made being wasted.

How large these sources of inefficiency are will depend on how imperfect the slot allocation process is, and how much of the slot rent is dissipated. They will also depend on how high slot values are. If slot values are high, the rents will be large and the costs of misallocation and dissipation will be large. If slot values are moderate or low, the inefficiency costs will be likewise.

The evidence on slot values is very limited (Gillen and Tudor, Chapter 12). Some slot sales suggest high values. Slot pairs for London Heathrow have sold for as high as £10m. If this were annualized at 10 per cent, it would be equivalent to £1m per year, and would translate into about £1,370 per movement, or £10 per passenger (given the convenience and lower access costs of Heathrow over the next best unconstrained alternative, Stansted, this valuation does not seem out of order). All up, this would imply that slots would have an annual value of (i.e. would add to airline profits) £633m or €941 per annum at Heathrow. If slots were similarly valued at Frankfurt, they would be worth £630m or €937m.

As against this, slots may not be valued as highly as these suggest. The slots traded may have been highly preferred slots, and the average price may be much less. In addition, these prices were recorded in very thin markets, and some bidders may have been prepared to pay significantly more than the norm for the slots. The average value could be less if the major holders of slots are withholding slots from sale.

This is illustrated using Figure 20.1. The demand curve for the major holder of slots is given as D_1. Suppose that the major holders of slots are prepared to use slots as long as they add to profits, (essentially costing their slots at zero) though they are not prepared to sell them. Faced with a price of P_a to use the airport, they would use S slots, leaving K-S slots for other users. If the other users have a demand curve of D_2, resetting the zero quantity at S, the market price would be P_1 – which is high. If the major users were to put their slots on the market, the new demand curve would be D_3, the lateral sum of the two demand curves, and the price P_2, much lower than the previous price. Thus if there is significant slot hoarding, the realized market price may be well above the free market slot valuation.

If this were happening, then the efficiency cost of slot misallocation and rent dissipation would be much less than would be implied by the high slot price. Rent dissipation would be smaller, because the rents to be dissipated would be smaller. There are some efficiency costs as a result of hoarding and misallocation, but they are not as large as might be thought. The welfare gains from the major users ceasing to slot hoard would be shown by the triangles ABC and DEF. In addition, slots may not be being allocated efficiently within airlines, but because the slot values are not so high, the efficiency cost of this happening would not be so large.

Slots and Efficiency: Assessment and Options for Improvement

The performance of slot systems in efficiency terms is summed up in Table 20.1. Slot systems handle several dimensions well – they are consistent with the choice of an efficient throughput, and congestion level, of the airport. By coordinating independent users and smoothing out the flow, they can lessen congestion costs.

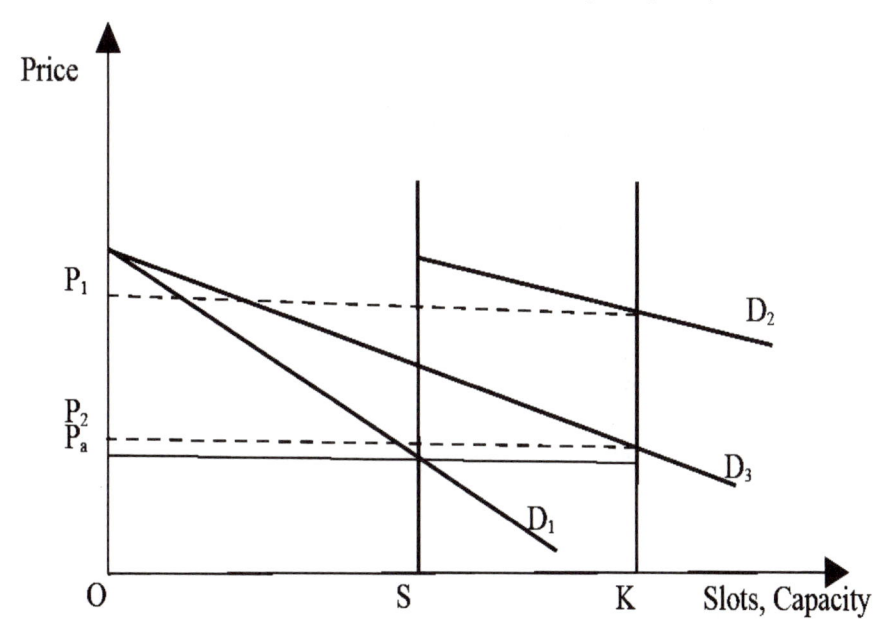

Figure 20.1 Effect of demand on slot values

Since uncertainty at airports is such that approximately efficient quantities are
known in advance but efficient prices are not, they make for an efficient handling
of uncertainty. Slots enable efficient allocation of capacity when price regulation
precludes the use of market clearing prices. Finally, they involve low transactions
costs.

In other respects they are problematic. They break the nexus between the price
received by the capacity provider and the users' valuation of additional capacity.
Along with price regulation, they do not provide good incentives for investment.
As is inevitable with price regulatory systems, the regulator takes on the role of
decision maker on investment – this is evident in the role of the UK CAA in
setting BAA's investment programme (Hendriks and Andrews, 2004). If there is a
well functioning market for slots, the regulator will have sufficient information to
make good decisions. However, if the value of slots is unclear because slot trading
is hindered, well intentioned regulators will struggle to make good decisions.

The performance of slots in terms of impacts on competition is also
problematic. Prohibition of trading and thin markets make it difficult for new
competitors seeking to challenge incumbents with slots. Open trading will help,
as will auctions, which should mean that all competitors are on an equal footing.
Schemes which recycle slots, by taking a proportion of slots off incumbents and
selling them in a broader market, could help.

The weakest aspect of slot systems concerns slot allocation and rent dissipation.
There is evidence that many, probably most, slot systems do not perform well in this
respect. Just how serious an efficiency problem these are is an empirical question,

Table 20.1   Slots: assessment and improvement options

Dimension	Assessment	Options for improvement
Capacity choice	Mostly good	Need for explicit evaluation
Slot allocation	Possibly poor	Efficient airport pricing; Incentives to trade; Auctions
Minimization of congestion cost	Very good	–
Uncertainty	Very good	–
Investment incentives	Problematic	Good regulation
Competition	Problematic	Incentives to trade
Rent dissipation	Possibly poor	Efficient airport pricing; auctions
Additional objectives/ constrain profit	Good	–
Transactions costs	Good	–

but where slots are much in demand, as they are at some airports, there is scope for considerable inefficiency.

There are some improvements which are straightforward to make, though not necessarily easy to make. Where trading in slots is not permitted, opening up the market will be an obvious improvement. Airport pricing structures lead to slot misallocation, and in all airports these can be improved, often significantly. These are reforms which allow markets to work.

In theory, secondary trading, with profit oriented airlines without market power, should be sufficient to ensure an efficient allocation of slots without risking rent dissipation. However, it is recognized that simply allowing markets to work may not be sufficient (see NERA, 2004), especially where single airlines may possess more than half the slots at an airport, and airlines may be under pressure from governments to serve uneconomic routes. Trading may not be as open as desirable. Auctions constitute the most thorough reform, though they would be the hardest to implement (Button, Chapter 16). They would involve considerable shifting of rents, and thus they are strongly opposed. They may encounter problems in terms of information and transactions costs. Auctions would mean that airlines would be required to pay the opportunity cost of the slots they use, and thus they would be faced with the strongest incentive to make good use of them. In addition, since the slot rents would not accrue to the airlines, there would be no problem of rent dissipation by the airlines (though whoever received the rents, be it the airport or government, might dissipate them).

Other mechanisms to encourage trading and reallocation (NERA, 2004), such as slot recycling schemes are likely to be easier to implement than auctions, because they do not have the same distributional consequences. The more they facilitate open trading, the more they will enable airlines without slots to obtain them. With clearly established prices, airlines which possess slots will be more directly faced with the value of the slots they are using, and they will be under more pressure to use them efficiently, and to avoid dissipating the rents.

All in all, slot systems constitute a moderately efficient solution to the airport capacity problem. As they exist at the moment, there are real problems with the allocation of the slots that are made available. However, there are several options, both straightforward and more complex, which can improve slot allocation significantly. If trading in slots can be made open and effective, and airport price structures are reformed, the slot systems would constitute the best of all practical solutions to the airport capacity problem.

Choosing Between Options

Slots versus Prices

When there is a problem of congestion or of rationing excess capacity, economists usually recommend a pricing solution, rather than a mixed quantity/ price solution such as slots. Much of the US literature on airport congestion recognized the externality and recommended price solutions. Price solutions have rarely been adopted in airports – sometimes airports have adopted peak pricing, but they have never used prices as the main rationing device for busy congested airports. Either nothing has been done, and congestion has developed as the primary rationing device, or slot rationing systems have developed over time, as demand exceeded capacity and a rationing problem emerged. However a system of pre determined, or posted prices, should be considered as an alternative to congestion (for some discussion, see NERA, 2004).

The relative advantages and disadvantages of slots and prices are considered in Table 20.2.

As can be seen from Table 20.2, both slots and prices have their advantages and disadvantages. The two systems are equivalent in terms of capacity choice – whatever amount of capacity is made available for slots can be used as the target output when prices are set. Prices are better when it comes to allocation, for reasons discussed above. Slots, however, are better when it comes to minimizing the costs of congestion. Slot coordination can spread the flow of traffic such that clustering and random peaks are avoided – many airports with ample capacity use slot coordination as a means of smoothing traffic. While, in principle, prices could be used to do this, it would require extremely detailed pricing, and doing so would be very information intensive. In practice, there is little alternative to using slot coordination. Slots also can handle uncertainty better, because the approximately efficient output is known in advance, but the prices which will limit demand to this level are not, given that demand is variable and to an extent, unknown.

When it comes to investment incentives, prices perform better. The airport will be faced with prices which reflect users' valuation of capacity, and it can compare these with the cost of expansion. If prices exceed the cost of expansion, the airport will face an incentive to expand. With slots, efficient investment decisions can be made, but these depend on the regulator getting it right.

Table 20.2   Performance in terms of efficiency: slots versus prices

Dimension	Assessment	Options for improvement of weaker system
Capacity choice	Equally good	–
Slot allocation	Prices better	Efficient trading in slots or auctions could make equivalent
Minimization of congestion cost	Slots better	Prices unlikely to be feasible
Uncertainty	Slots better	Little scope to improve on pricing outcomes
Investment incentives	Prices better	Good regulation lessens difference
Competition	Prices better	Incentives to trade could make equivalent
Rent dissipation	Prices can be better	Efficient airport pricing; auctions could make equivalent
Additional objectives/ Constrain profit	Slots better	Little scope to improve pricing outcomes
Transactions costs	Slots slightly better	–

Prices also work better in terms of facilitating competition. Since all users pay the same price, one airline is not given an advantage over another. The performance of slot systems in place is sometimes quite poor in this respect, though if slot trading is effective, the adverse impacts of slots on competition can be minimized.

Prices work well through minimizing the opportunities for rent dissipation by airlines. However this observation must be qualified by noting that prices do not eliminate rents – rather, they shift them to the airport. It is quite possible that an airport may dissipate rents – indeed, granted that many airports are publicly or community owned, or privately owned but subject to cost plus regulation, the likelihood of rent dissipation could be greater at the airport level than the level of airlines, most of which are profit oriented private firms these days. If slots are traded, and there is a clear price for slots, or slots are auctioned, airlines will be under pressure to make good use of their slots and not dissipate the rents from them.

If there is an objective to keep airport revenues no higher than costs, whether this is achieved by government directives to public airports, or regulation of private airports, a slot system can reconcile achieving this objective with efficient allocation of slots. The problem with price solutions is that market clearing prices and prices which just cover costs are inconsistent – for busy airports, prices which ration capacity lead to revenues above costs.

Finally, the transactions costs of implementing a price solution are likely to be somewhat higher than those of a slot system. Airports will need to give a lot of attention to their price levels and structures, and getting matters right will require a large amount of information.

Taking all these aspects into consideration, it does not appear that one system is clearly preferable to the other. A price solution would be preferable if the slot system misallocates slots and allows rent dissipation. However, if these aspects can be addressed, by encouragement to secondary trading or auctions, then slots would become superior. With good allocation of capacity, and limited rent dissipation, slot systems can perform very well or at least tolerably well on all aspects of efficiency. In contrast, while prices perform well in some dimensions, they are inferior to slots in minimizing congestion costs, handling uncertainty and reconciling efficient allocation with price restraint. The deficiencies of slots can be addressed – those of prices, in the main, cannot. For example, it is difficult to see how efficient market clearing prices could be made consistent with just covering costs for airports which face very strong demand. It is difficult to think of ways in which the performance of prices can be improved in these dimensions – the relative advantages of slots here cannot be lessened by reforms to pricing. Thus, overall, slot systems have the potential to perform better than price systems, and if slot trading is sufficient to result in efficient allocation of slots, and discourage rent dissipation, a slot system will be clearly preferable to a pricing solution.

Apart from the advantages in terms of efficiency, slot systems score over prices in terms of practicality. There are very many slot systems in place – there appear to be no price based capacity allocation systems in place with airports, though occasionally prices have a secondary role. Implementing a price solution for an airport which is currently slot controlled would necessitate a shift in rents, away from the airlines and to the airport or slot authority. This would be strongly resisted. It may be feasible to implement a price rationing solution over time, starting before the airport becomes subject to excess demand (and therefore before slot rents come into being). It would be easier to replace a congestion solution by a price solution since, on balance, airlines would not lose when this is done. However, some airlines will gain, and others will lose, and even this move has not been successfully achieved.

Congestion versus Slots

There are two systems of rationing airport capacity which are actually implemented – congestion, and slots. In the US, capacity at most busy airports is allocated on a first come first served basis, with delay or congestion acting as the rationing device. The inefficiencies of this solution are well known (see Carlin and Park, 1970), and the option usually advocated is for the US to move to a pricing solution. Slots are another alternative. The comparative advantages of slots and congestion are summed up in Table 20.3.

One well recognized problem with the congestion solution is that it leads to excessive use of the facility. In this respect, a slot system, in which the use of the facility is optimized, would be clearly preferable. The allocation of capacity under a congestion solution is by willingness to bear delays, rather than willingness to pay. While the allocation of capacity under slot systems is often not ideal, slot systems can be improved through encouragement of trading. The congestion solution does not give any scope for the smoothing out of traffic, with a consequent reduction in delays – in this respect slots are again preferable.

Under a congestion solution, congestion varies as demand varies, and this congestion acts as a variable rationing price. In this respect, congestion is similar to slots. The signals to invest under congestion are the reverse of those under a slot system. As congestion builds up, airlines will pressure the airport to expand capacity. This will be so even where capacity is adequate but poorly allocated. If anything, the airport will be under pressure to build excess capacity, not inadequate capacity. In principle, this could be handled by a well informed regulator which could assess the benefits and costs of expansion.

One respect in which congestion could be superior to slots is in the impact on competition. Under congestion, all users pay the same price in terms of delay. Airlines can enter or exit the market freely – they do not need to obtain slots to serve the airport. This advantage would be minimized if there were a well functioning market for slots. This partly explains the antipathy of US authorities to slot systems.

Table 20.3   Congestion versus slots

Dimension	Assessment	Options for improvement of weaker system
Capacity choice	Slots better	–
Capacity allocation	Slots potentially better	Efficient trading in slots or auctions could make slots better
Minimization of congestion cost	Slots better	–
Uncertainty	Equivalent	–
Investment incentives	Both have problems	Good regulation can minimize these
Competition	Congestion better	Incentives to trade slots could make equivalent
Rent dissipation	Slots usually better	Efficient airport pricing; auctions could make slots better
Additional objectives/ Constrain profit	Equivalent	–
Transactions costs	Congestion better	–

Congestion performs very poorly in terms of rent dissipation. The rents that are possible as a result of scarce airport capacity are all dissipated in delays. With slot systems, full rent dissipation is unlikely even if markets are highly imperfect.

Since congestion rations demand, there is no need for market clearing prices. Thus there is no problem in constraining prices to constrain revenues. Finally, congestion works automatically, and the transactions costs of using congestion are minimal.

In summary, there is very little likelihood that a congestion rationing system would be preferable to a slot system. Even if slots were very poorly allocated and rents were extensively dissipated, the other advantages of slots would prevail. It should not be difficult to design and implement a slot system which performs significantly better than a congestion solution (see Kasper, Chapter 15).

Allocations, Trading and Auctions

As discussed above, one of the main limitations of slot systems is they often allocate slots poorly. If market clearing prices are not to be used, it may still be possible to achieve an efficient allocation of slots by means of slot trading or auctions. Auctions represent the bigger step from current systems. In principle, with profit oriented airlines, slot trading should be sufficient to ensure that slots are efficiently allocated, assuming that the airlines do not possess market power that they can use in slot markets (eg slot hoarding). Profit oriented airlines will not wish to use slots ineffectively, and will not dissipate the rents they earn.

In practice, not all airlines may act as though they are efficient profit maximizers. Several airlines allow their costs to rise beyond feasible levels – these airlines cannot be solely profit oriented. Even privately owned airlines which are not subject to capital market disciplines which are as strong as they might be will allow inefficiencies to develop. Thus they may not use their slots effectively, and may dissipate some of the rents.

If slots are auctioned, airlines are under stronger pressure to make good use of their slots (Button, Chapter 16). To obtain slots to use on a route, they have to buy them. In addition, they do not gain slot rents which can be dissipated. Auctions give less opportunity to hoard slots, or to use their holdings of slots in an anti competitive manner.

There are several disadvantages with auctions. Given the complexities of airline networks, it may be difficult to set up auction mechanisms which work smoothly, at least for some time. Their workings will be demanding of information. Auctions pose a problem for price regulation or keeping revenues close to costs, since they will result in revenues which are possibly well in excess of airport costs. One solution might be to have a government owned auction authority which would conduct the auctions and return net revenues to the government. If this were to work, it could be a very efficient solution, since the slot rents would be ending up as government revenues, which in most countries, have a high shadow price (the marginal cost of raising $1 exceeds $1 because of the need to use distortionary taxation). Slot auctions would be a very efficient way of raising government revenue. On the other hand, it would be necessary to guard against the risk of slot auction revenues accruing to bodies, such as government owned airports with conflicting objectives) with the ability and incentive to dissipate them.

Most of the problems discussed can be addressed. The most difficult practical problem with slot auctions arises as a result of the fact that they shift rents. Airlines would be much worse off as a result of slot auctions, and thus they oppose them strongly. It would be difficult to devise compensation schemes without negating the purpose of moving to an auction.

To this end, the case for using auctions depends on how efficient slots markets are. If slot trading can be freed up, and made to work effectively, it will produce all or nearly all of the benefitsof auctions, but it will not face the implementation problems of auctions. If slot trading remainsthin and ineffective in spite of incentives to trade, moves towards auctions could be worthwhile.

The Political Economy of Slots Systems

Slot systems have distributional impacts, and thus they alter how different interests, such as airlines and airports, gain or lose from policy options or changes in the competitive environment (Button, Chapter 18). Where limited airport capacity is allocated to airlines using slots, slot rents are created, and the airlines gain them. These slot rents can be very valuable, and airlines seek to maintain them. Thus they will strongly oppose options which reduce these rents. As a consequence, some potentially good policy options are very difficult to implement because of airline opposition. Some examples of this follow.

Auctions and Prices

Airlines oppose slot auctions or pricing solutions to capacity problems. This is understandable, since airlines would lose out if a grandfathered slot system were replaced wholly or even partly by a slot auction, or if rationing prices were used instead. Slot auctions or prices have rarely, if ever, replaced grandfathered slots systems.

Expansion of Airport Capacity

Increases in airport capacity devalue slots, making them worthless when capacity is ample. Extra capacity also enables more competition for the incumbent airlines. Thus airlines which have slots at busy slot constrained airports have a strong incentive to oppose capacity expansion. Airlines often state that they would like to see more capacity at such airports, but the vigour with which they pursue this stated objective will be moderated by the realization that they would be destroying their own assets. In reality, there has been little expansion of the very busy slot constrained airports, especially in Europe.

Airport Charges and Regulation

When airports with ample capacity put up their charges, airlines are able to pass these on to their passengers, since air fares are set by competition and costs (the extent to which this happens depends on the degree of competition – in oligopoly or monopoly markets, there will be less than complete pass through). With slot constrained airports, air fares are set differently – they are set so as to ration demand for the use of the airport to the capacity which is available. When an airport increases its charges, the airlines are unable to pass these on – effectively, in this situation, the airlines, not their passengers, pay the increase. Airlines have been particularly critical of airport charges in Europe (Bisignani, 2006). This may be because such charges are higher in Europe than elsewhere, but it also reflects the fact that many airports in Europe are slot constrained, and thus the airlines have to pay the higher charges themselves. It makes good sense for airlines to pay especial attention to the regulation and level airport charges for airports which are slot constrained.

Airport Peak Charges

Airlines strongly oppose airport peak pricing (see IATA, 2000), which appears incongruous given their own extensive use of the tool. The working of the slot system explains this apparent contradiction. Consider an airport which has inadequate capacity at the peak, but adequate capacity in the off peak (i.e., one for which peak pricing might be suggested). Introduction of revenue neutral peak pricing would result in a rise in prices at the peak – airlines would bear these higher prices as they would not be able to pass them on to passengers. While off peak prices would fall, airlines would be forced by competition to pass these on to passengers. Thus airport peak pricing would lower the profits of the airlines, and it makes good sense for them to oppose it.

Slots and LCC Competition

The slot system has helped the legacy airlines or FSCs to withstand competition from LCCs. The legacy airlines have access to the slots at the constrained airports, for which passengers are prepared to pay a substantial premium to use. A new airline, such as a LCC, will need to buy slots at a high price to compete with the legacy airlines (unless it is prepared to wait a very long time to build up slots using the various mechanisms available to it). For many airports it will not be possible to even buy the necessary slots. Mostly, LCCs will use less attractive airports. Thus the legacy carriers will be able to operate at higher costs than the LCCs because they will enjoy higher yields, as a result of operating out of preferred airports (even with higher operational costs at the constrained airports, it is still profitable for them to operate at them). Thus access to slots, especially in Europe, has helped the legacy airlines to adjust to competition from lower cost competitors – US legacy carriers do not have the same advantage, and they are generally less profitable.

Competition on International Routes.

Slot availability can be a major consideration for international air routes. Slots can be used to protect incumbents. While Air Services Agreements may permit more airlines and flights to operate on a route to an airport, slot limitations may prevent them from doing so. If slots were readily traded, this would not be a problem, but for many airports, it is very difficult to obtain slots even if an airline is willing to pay a high price for them (Humphreys, 2003). Thus the slot system as it operates, limits competition on some international routes.

Slots as Currency in International Negotiations

The slot system gives nations a form of currency to be used in international air services negotiations. Slots are valuable, and they can be traded for other rights, such as fifth freedom rights. Access to London Heathrow has been a major issue in the US-EU aviation negotiations.

Slots – Some Unanswered Questions

This book sums up the situation on a range of issues surrounding slots. There remain a number of unanswered questions. It is differing views on these questions which can give rise to different evaluations of how well slot systems are working. Some of the key questions which remain to be answered are:

How Effective is the Slot System in Reducing Delays?

The primary objective of the slot system is to reduce delays, but there is little by way of examination of how effective it is in achieving this. There is some evidence that delays at non slot controlled US airports are higher than slot controlled European airports (Janic, Chapter 6)). There are some mechanisms which lessen airport delays at US airports (See Brueckner, 2002; Mayer and Sinai, 2003) and to an extent, delays are a cost of hub operations. However there is nothing by way of a rigorous study of how delays and delay costs differ between the two systems, and thus how large the economic gains are from implementing a slot system.

How Effective Can Secondary Slot Trading Be?

While there are many criticisms of the results of trading as it is, do these stem from limits imposed by authorities on trading, and would a freeing up of trading remove all problems and lead to an efficient allocation of slots to the users with highest willingness to pay? If this is not the case, there may be other problems present, such as market power at the airline level, and dominance at airports which might lead to trading being constrained, for example, by slot hoarding. If this is so, additional measures, such as recycling of slots, might improve outcomes. Slot trading might need to be backed up by auctions. These questions are essentially practical ones, and it is necessary to see how freer slot trading systems actually work to find the answers.

Are Slots Efficiently Allocated within Airlines, and If So, What Happens to the Slot Rents?

At this stage it is not clear how effectively airlines allocate the slots at their disposal – if they are allocating them efficiently, should not the airlines with strong slot endowments be earning significantly higher profits? This is a question on which more research is feasible – it might be that airlines with strong slot endowments are atypically profitable. Some airlines may gain from more rigorous assessment of how effectively they use their slots, and whether any slot hoarding is taking place.

How Effective Can Slot Auctions Be?

Opinions differ sharply on whether slot auctions would be effective or not. The key question is how complex they would be in the airport case. It is difficult to see how this question could be resolved without practical applications – for example, experiments at airports.

How Well Does the Willingness to Pay for a Slot Reflect the Value to the Community of the Slot?

The demand for airport use and for slots is a derived demand, and when there is market power at the downstream (airline) level, the willingness to pay for inputs will be less than the overall welfare gain from greater use of the inputs. Market power does exist, to an extent, at the airline level, but how serious an issue for slot allocation is it likely to be? This is an issue which more empirically based research would yield results.

How Do Different Slot and Alternative Allocative Mechanisms Impact on Competition at the Airline Levels?

Different mechanisms, such as grandfathering or auctions, might impact on competition at the airline level – this explains some policy makers' preferences for one mechanism over an other. This is an issue on which further empirically based research might pay dividends.

How Efficiently are Slot Limits Set Now?

Slot limits tend to be set according to arbitrary criteria. These may, or may not, result in slot limits at airports which are tolerably efficient. Testing this is a relatively straightforward research issue.

How Well Do Slot Systems Perform in the Investment Context?

Slot systems override market signals for investment, as does price regulation which often accompanies slot allocation. It is possible for well informed regulators to set in train reasonably effective investment incentives. However, whether this happens is another matter. The various parties, airlines, passengers and airports, have diverging interests, and investment outcomes may reflect these rather than what is in the overall public interest. This is essentially a political economy problem, on which empirical research would be interesting.

Reforming the Slot System

In terms of being a means of achieving efficient utilization of scarce airport capacity, the assessment of slot systems is a positive, though not unqualified one. Slot systems do lessen congestion at airports in a way that is consistent with meeting other objectives. In many slot systems, the allocation of slots is poor – this constitutes the main practical problem with slots. It is feasible to improve the working of slot systems considerably by implementing a number of reforms. Several of these reforms are straightforward to implement, though not all are.

Replace First Come First Served Congestion Allocation by a Slot System

If an airport is currently rationing scarce capacity by congestion, an obvious reform is to implement a slot allocation system or to charge rationing prices. The latter have advantages over the former, but also several disadvantages – they are also much more difficult to implement. A well designed slot system has the potential to be an efficient solution.

Clarify Property Rights in Slots

To have effective trade in slots, it is necessary that they be defined clearly and property rights be established (von den Steinen, Chapter 18).

Permit, Facilitate and Perhaps Encourage Slot Trading

A priority is to make slot trading possible. With many slot systems, especially in Europe, slot trading is difficult or prohibited (Killian, Chapter 14; Sorensen, Chapter 19). It is very unlikely that this prohibition serves any purpose. There is no evidence that it lessens problems of airline market power, especially since airline dominance at hubs exists side by side with non traded slots.

When slot markets are not very efficient, there may be a role for interventions to stimulate trading. One suggestion is for slot recycling. This would be at some cost to the airlines and it would be difficult to compensate them. Slot reservation charges would induce airlines to lessen slot hoarding, and make more capacity effectively available at the airport.

Introduce Auctions for Slots if trading Does Not Lead to Efficient Markets

Auctions are potentially the most efficient way of allocating slots, but they are also the most difficult to implement. If it is possible to achieve efficient operation of the slot market, the auctions would not be necessary. Essentially, auctions would only be needed if efficient slot markets do not develop.

While auctions may not bear the main allocation burden for airport capacity, they can be used effectively in more limited ways. Airports periodically have to allocate new or unused slots, and typically they use administrative means to do this. This gives rise to a likelihood of poor allocation, since the administrators have little idea of which airline has the highest willingness to pay. It would be preferable to sell or auction the extra slots. This would also eliminate rent seeking, whereby airlines lobby for free slots. It would also lessen the scope to use slots as an international currency in air services negotiations, a situation in which they are likely to be misallocated.

Reform Airport Pricing

If airport capacity is to be allocated efficiently, the structure of airport pricing will be of comparable importance to that of slot allocation. Thus reform of airport pricing structures is also a priority. At virtually all busy slot constrained airports the price structure is one designed for the days of ample capacity several decades ago. Weight or passenger based charges discriminate against the aircraft which can make effective use of the scarce capacity at airports. Some busy airports have made tentative moves towards price structures which ration their capacity effectively; however these moves have been rare, and not always permanent. Some inertia is to be expected granted that some airlines will gain from price reform while others will lose, though on balance airlines will be unaffected as long as the level of prices remains the same. Achieving efficient allocation of slots through trading or other means is solving only part of the problem – prices have to induce airlines to use the airports' capacity more efficiently.

Set Slot Limits on a Systematic Basis

The amount of capacity which is made available to airlines to use under slot systems is mostly determined on an ad hoc, quasi engineering basis. Nevertheless, given the nature of the congestion function, it is possible that the slot constraint has been set at about the right level at many airports. Slot constraints are subject to lobbying, and authorities may bend to pressure and set slot constraints which are too slack to achieve their full potential in reducing congestion. Thus there is a need for rigorous economic evaluation of slot constraints to determine whether they are being set efficiently.

Review Ring Fencing of Slots for Preferred Users

Many airports allocate slots on a preferential basis to specific users, such as regional airlines. If slot trading is to emerge, the status of these slots must be determined. If tradability between slot categories is to be permitted, an efficient allocation of slots (according to willingness to pay) will come about, but the regional airlines will probably cash out their slots and reduce services. Regional airlines will gain, but regional communities will suffer. Alternatively, regional airlines might be prohibited from selling their slots to non regional airlines. This would preserve the slots for regional services, but at the cost of lower value services using the airport. In addition, such ring fencing of regional slots would weaken the incentive for operators of small aircraft to develop and use reliever airports, which would be more cost effective than using the constrained busy airport. Once a preferred category of user has been granted slots it is then difficult to reform the system, because of resistance from the beneficiaries. However with segregated markets for different users of slots, the opportunity costs of ring fencing would become apparent.

Unbundle Slot Allocation from International Negotiations

The solution to the airport capacity problem is achieved most efficiently when slots are not tied up with other considerations. Ideally slots should be unbundled from international aviation negotiations, and traded separately. This would remove the inefficiency inherent in preventing an airline from using an airport because it does not have slots. It would also lessen the inefficient allocation of slots by air services negotiators who are granted slots free of charge, and who are responding to external pressures when they dispose of the slots.

These policy reforms do not directly address the problems of slot allocation systems affecting airline competition and rent dissipation. It is not clear how different systems of slot allocation impact on airline competition, nor whether any effects are significant or systematic. If anything, it might be thought that ready availability of slots at going market prices would be helpful to competition – such a situation would come about if the slot market develops effectively.

As recognized, airlines may not use their slots well in generating rents, or, if they do, they may not use the rents efficiently (e.g. cross-subsidising loss making services). This problem, if it is a significant one, is less easy to address through the design of the slot system since it is a problem which develops within the airline. Again, having well functioning slot markets producing going market prices for slots is likely to help, since this will make the opportunity cost of using slots very clear to the airline. If evidence indicates that it is a serious efficiency issue, a greater reliance on prices or on auctions can lessen the scope for rent dissipation. They will do this by shifting rents to the airport or slot administrator/ auctioneer – achieving such a rent shift will be difficult. In addition, there needs to be some mechanism in place to ensure that the new recipient of the rents does not dissipate them instead.

Conclusions – Achieving Reform of Slot Systems

In many respects, slots are like the taxi licences of the air. They are valuable because they restrict supply of a service and enable their owners to make profits. They differ from taxi licences in that slot limits do serve a valuable purpose in lessening congestion at airports. However, they are similar in that once created, they are very difficult to change. In Europe, moving from the existing system to one in which the development of slot markets is possible has been a slow and uncertain process. The slot system has achieved its objectives at the expense of creating a group of stakeholders who are resistant to any reforms which might make the system perform better. However, unlike the rents from taxi licensing, there are serious questions concerning how effectively the rents are used.

There are several reforms which can result in the airport capacity problem being solved better. Some of these are relatively straightforward, such as allowing markets to develop, reforming price structures at airports, and evaluating slot limits more scientifically. At another stage, steps might be taken to encourage trading, perhaps through recycling of a small proportion of slots, and by auctions of slots which become available. If these work well, there is no need to go further. If these are not sufficient, auctions for slots might be implemented, though this would be difficult to achieve and perhaps costly to operate. An essential step is the reform of airport pricing – no matter how effective the slot allocation process is, slots will not be allocated efficiently if airport price structures are not reformed.

The significance of the slot system goes well beyond the obvious task of rationing demand to capacity. Slots can be important in reducing delays at airports with ample capacity, through evening out traffic flows – hence many such airports have slot coordination even though they have no problem of reducing demand to capacity. Slots enable regulators to keep airport charges down while still making efficient allocation of the scarce capacity feasible. However, this means that airport prices are not an accurate measure of the value of additional capacity. Thus regulators need to give special attention to the investment incentive problem. Slots can impact on airline competition. Finally, slots have major implications for who gains the rents from scarce airport capacity, and this affects in the interests of the different parties, airlines and airports, in policy options. A slot system means that airlines have a stronger than normal interest in keeping airport charges low, that airlines will oppose peak pricing by airports, and that airlines will not press for airport capacity expansion even when it is manifestly inadequate. Slots systems have major implications for the way the air transport market operates.

References

Bisignani, G. (2006), Remarks at Brussels Press Lunch, 13 February, IATA website.
Brueckner, J. (2002), 'Airport Congestion when Carriers have Market Power', *American Economic Review*, 92 (5), pp. 1357–75.

Carlin, A. and Park, R.E. (1970), 'Marginal Cost Pricing of Airport Runway Capacity', *American Economic Review*, 60, pp. 310–19.

Forsyth, P., Gillen, D., Knorr, A., Mayer, O., Niemeier, H.-M. and Starkie, D. (eds) (2004), *The Economic Regulation of Airports: Recent Developments in Australasia, North America and Europe*, Aldershot: Ashgate.

Hendriks, N. and Andrew, D. (2004) 'Airport Regulation in the UK', in Forsyth, P., Gillen, D., Knorr, A., Mayer, O., Niemeier, H.-M. and Starkie, D. (eds), *The Economic Regulation of Airports: Recent Developments in Australasia, North America and Europe*, Aldershot: Ashgate.

Humphreys, B. (2003), 'Slot Allocation: A Radical Solution', in K. Boyfield, *A Role Market Airport Slots*, London: IEA, Readings 56, pp. 51–79.

International Air Transport Association (IATA) (2000), 'Peak/Off-peak Charges', Ansconf Working Paper No. 82, ICAO, Montreal.

Mayer, C. and Sinai T. (2003), 'Network Effects, Congestion Externalities, and Air Traffic Delays: Why Not All Delays Are Evil', *American Economic Review*, 93 (4), pp. 1194-1215

National Economic Research Associates (NERA) (2004), 'Study to Assess the Effects of Different Slot Allocation Schemes', Final Report for the European Commission, DG Tren, London January.

Pels, E. and Verhoef, E. (2004), 'The Economics of Airport Congestion Pricing', *Journal of Urban Economics*, 55, pp. 257–77.

Weitzman, M. (1974), 'Prices vs Quantities', *Review of Economic Studies*, 41 (4), pp. 477–91.

Slots, Pricing and Airport Congestion: Some Key References

Introduction and Overview

Pels, E. and Verhoef, E. (2004), 'The Economics of Airport Congestion Pricing', *Journal of Urban Economics*, 55, pp. 257–77.

Starkie, D. (2005), 'Making Airport Regulation Less Imperfect', *Journal of Air Transport Management*, 11, pp. 3–8.

Toms, M. (1994), 'Charging for Airports', *Journal of Air Transport Management*, 1, pp. 77–82.

Zhang, A. and Zhang, Y. (2006), 'Airport Capacity and Congestion when Carriers have Market Power', *Journal of Urban Economics*, 60, pp. 229–47.

Airport Congestion and Pricing

Brueckner, J.K. (2002), 'Airport Congestion When Carriers Have Market Power', *American Economic Review*, 92, pp. 1357–75.

Basso, L. and Zhang, A. (2007), 'A Survey of Analytical Models of Airport Pricing', in Lee, D. (ed.), *Advances in Airline Economics, Volume 2: The Economics of Airline Institutions, Operations and Marketing*, Amsterdam: Elsevier.

Carlin, A. and Park, R.E. (1970), 'Marginal Cost Pricing of Airport Runway Capacity', *American Economic Review*, 60, pp. 310–19.

Daniel, J.L. (1995), 'Congestion Pricing and Capacity of Large Hub Airports: A Bottleneck Model with Stochastic Queues', *Econometrica*, 63, pp. 327–70.

Levine, M.E. (1969), 'Landing Fees and the Airport Congestion Problem', *Journal of Law and Economics*, 12, pp. 79–109.

Mayer, C. and Sinai, T. (2003), 'Network Effects, Congestion Externalities, and Air Traffic Delays: Or Why Not All Delays Are Evil', *American Economic Review*, 93, pp. 1194–215.

Morrison, S.A. and Winston, C. (1989), 'Enhancing the Performance of the Deregulated Air Transportation System', in Bailey, M.N. and Winston, C. (eds), *Brookings Papers on Economic Activity*, Washington, DC: Brooking Institution.

Slot Systems

Abeyratne, R.I.R. (2000), 'Management of Airport Congestion through Slot Allocation', *Journal of Air Transport Management*, 6, pp. 29–41.
Council of the European Communities (1993), Council Regulation (EEC) No. 95/93 on Common Rules for the Allocation of Slots at Community Airports, 18 January: http://europa.eu.int/eur-lex/lex/LexUriServ/LexUriServ.do?uri= CELEX:31993R0095:EN:HTML.
International Air Transport Association (IATA) (2005), *Worldwide Scheduling Guidelines*, 11th edn, Montreal: IATA.
Airport Council International, Air Transport Action Group and International Air Transport Association (2003b) *Airport Capacity/Demand Profiles 2003*, Geneva: ACI, ATAG and IATA.
National Economic Research Associates (NERA) (2004), *Study to Assess the Effects of Different Slot Allocation Schemes A Final Report for the European Commission, D G Tren*, London, January: http://europa.eu.int/comm/ transport/air/rules/doc/2004_01_24_nera_slot_study.pdf.

Reforming the Slot System

Boyfield, K. (ed.) (2003), *A Market in Airport Slots*, London: IEA, Readings 56, (includes papers by K. Boyfield, T. Bass, B. Humphreys and D. Starkie).
Civil Aviation Authority (2001), *The Implementation of Secondary Slot Trading*, London, November: http://www.caa.co.uk/docs/5/ergdocs/slotsnov01.pdf.
Civil Aviation Authority and Office of Fair Trading (2005), *Competition Issues Associated with the Trading of Airport Slots*, paper produced for DG TREN by the CAA and OFT, CAA, June, London.
Commission of the European Communities (2004), Commission Staff Working Document: *Commercial Slot Allocation Mechanisms in the Context of a Further Revision of Council Regulation (EEC) 95/93 on Common Rules for the Allocation of Slots at Community Airports*, 17 September.

Airport Slots: International Experiences and Options for Reform

Jones, I., Viehoff, I. and Marks, P. (1993), 'The Economics of Airport Slots', *Fiscal Studies*, 14, pp. 37–57.
Council of the European Communities (2004), *Regulation (EC) No. 793/2004 of the European Parliament and of the Council*, Official Journal of the European Union, 21 April: http://europa.eu.int/eur-lex/pri/en/oj/dat/2004/l_138/ l_13820040430en00500060.pdf.
International Air Transport Association (IATA) (2004), *Comments on the European Commissions's Staff Working Document regarding Commercial Slot Allocation Mechanisms*, November: http://www.iata.org/NR/ContentConnector/CS2000/

Siteinterface/sites/whatwedo/file/IATA_response_to_EC_Consultation_Paper_on_Slots.pdf.
Langner S.J. (1995), 'Contractual Aspects of Transacting in Slots in the United States', *Journal of Air Transport Management*, 2, pp. 151–61.

Auctions and Trading

Grether, D.M., Isaac, R.M. and Plott, C.R. (1981), 'The Allocation of Landing Rights by Unanimity Among Competitors', *American Economic Review, Papers and Proceedings,* 71, pp. 166–71.
Grether, D., Isaac, M. and Plott, C. (1989), *The Allocation of Scarce Resources*, Boulder, CO: Westview Press.
Klemperer, P. (1999), 'Auction Theory: A Guide to the Literature', *Journal of Economic Surveys*, 13, 227-286
Milgrom, P.R. (2004), *Putting Auction Theory to Work*, Cambridge: Cambridge University Press.
Rassenti, S.J., Smith, V.L. and Bulfin, R.L. (1982), 'A Combinational Auction Mechanism for Airport Time Slot Allocation', *Bell Journal of Economics*, 13, pp. 369–84.
Sentance, A. (2003), 'Airport Slot Auctions: Desirable or Feasible?', *Utilities Policy*, 11, pp. 53–7.
Starkie, D. (1998), 'Allocating Airport Slots: A Role for the Market?', *Journal of Air Transport Management*, 4, pp. 111–16.

Appendix

Characteristics of Slot Coordinated Airports, 2003–2004[1]

Nathalie McCaughey; Peter Forsyth and David Starkie

In the following two tables, we present data for a range of airports, most of which are subject to slot coordination. The objectives are to provide a statistical background to the chapters of the book, and to be of use as a more general reference. Data were collected from a range of sources, though the main source is IATA, ACI and ATAG Airport Capacity/Demand Profiles Report (2003). Airport websites and Annual Reports were also consulted. Since data have had to be collected from different sources, caution must be exercised in using them as problems of non comparability can exist.

Table A1 summarises information on the tightness of, and rationale for, slot coordination. One important indicator concerns the level of slot coordination. The level of coordination is the result of an administrative decision. While very busy, constrained airports are almost certainly fully coordinated (if slots are used), the fact that an airport is fully coordinated need not imply that it is very busy. Information is provided on the stated reason for coordination- noise limitations, runway congestion etc. With several airports, there are multiple reasons. Noise curfews also impact on an airports ability to handle demand, and airports' subject to noise curfews are identified.

Where available, information is provided on the airports' capacities. The number of runways is given as an indication, along with the declared capacity for movements- for arrivals, departures and for total movements. Terminal capacity, in passengers, is also listed for most airports.

Table A2 summarises information on traffic at airports and the utilisation of capacity. The total aircraft movements are shown. Airlines which are using airports subject to tight capacity constraints might be expected to schedule larger aircraft so as to enable them to serve more passengers. Therefore average aircraft size, in terms of passengers per movement, is listed. In addition, the percentage of flights with 1 to 49 passengers is given – as it might be expected that busy airports would have a small percentage of small aircrafts if their capacity is being rationed effectively. Finally an indicator of the concentration in slot use is given by the percentage of movements accounted for by the main three carriers.

---

1   Funding from the Australian Research Council for preparation of this appendix is gratefully acknowledged.

In the final column, an indicator of utilisation is presented. An estimate of annual capacity, derived by assuming that the airport is operated for 18 hours per day for airports subject to a curfew, and 24 hours per day for airports not subject to a curfew, is made. Total actual movements is then related to this, and expressed as a percentage. Given the arbitrariness of the assumptions underlying it, this indicator should not be interpreted as a precise measure. However it provides a broad indication of which airports are subject to heavy demand pressure and which are not.

The intention of this appendix is to provide an overview of the characteristics of slot coordinated airports. The situation at individual airports is changing, as demand builds up, and in some cases, as more capacity is provided. Thus the data on individual airports should be used with care and updated where necessary. For example, in the case of Madrid Barajas Airport, a substantial addition to runway and terminal capacity has been made since the data for this appendix were collected.

Reference

International Air Transport Association (IATA), Airports Council International (ACI) and Air Transport Action Group (ATAG) (2003) *Airport Capacity/Demand Profiles, 2003 Edition*, Geneva-Montreal: IATA.

Table A1   Slots and airport capacity

Country	City	Airport	Grading of tightness of slot constraints	Rationale for slot constraints	Night curfew	Terminal capacity (million passengers per year)	Number of runways	Declared runway capacity (available slots per 60 min.)
Australia	Brisbane	Brisbane International Airport	SCR Level 3	N/A	None	N/A	2	N/A
Australia	Melbourne	Tullamarine Airport	SCR Level 3	1	None	N/A	2	N/A
Australia	Sydney	Kingsford Smith Airport	SCR Level 3	1 and NSAP 2, 4	√	25	3	80
Austria	Vienna	Vienna International	SCR Level 3	1 and NSAP for 2, 3, 4	None	24 (by 2008)	2	66
Belgium	Brussels	Brussels International Airport	SCR Level 3	1, 5 and NSAP for 3, 4	None	N/A	3	74
Canada	Montreal	Montreal Dorval Airport	SMA Level 2	1	√	16	3	40
Canada	Vancouver	Vancouver International Airport	SCR Level 3	1	√	21	3	75
China	Beijing	Beijing Capital International Airport	SCR Level 3	1 and NSAP for 2, 3, 4	None	27	2	60
China	Hong Kong	Chep Lap Kok International	SCR Level 3	N/A	None	45	2	49
China	Shanghai	Hong Qiao International Airport	SCR Level 3	5	None	N/A	1	30
Czech Republic	Prague	Ruzyne Airport	SCR Level 3	N/A	None	10	3	45

Country	City	Airport	Grading of tightness of slot constraints	Rationale for slot constraints	Night curfew	Terminal capacity (million passengers per year)	Number of runways	Declared runway capacity (available slots per 60 min.)
Denmark	Copenhagen	Kastrup International Airport	SCR Level 3	1 and NSAP for 3, 4	None	25	3	83
Finland	Helsinki	Vantaa Airport	SCR Level 3	N/A	None	15	3	50
France	Lyon	Saint Exupery Airport	SCR Level 3	1 and NSAP 2, 3, 4	None	8	2	51
France	Marseille	Provence Airport	Not slot constraint	NSAP for 2, 4	None	11.5	2	35
France	Nice	Cote d'Azur Airport	SMA Level 2	1 and NSAP 2, 3, 4	√	13	2	44
France	Paris	Ch. de Gaulle	SCR Level 3	1, 5	√	47	4	101
France	Paris	Orly	SCR Level 3	1	None	30	3	76
Germany	Berlin	Tegel	SCR Level 3	N/A	√	10	4	35
Germany	Cologne - Bonn	Konrad Adenauer Airport	SMA Level 2	1, 5	None	12	3	52
Germany	Düsseldorf	Düsseldorf International	SCR Level 3	1 and NSAP for 3, 4	√	22	3	38
Germany	Frankfurt	Frankfurt Main International	SCR Level 3	1 and NSAP for 2, 3, 4	√	56	3	78
Germany	Hamburg	Hamburg Fuhlsbüttel Airport	SMA Level 2	NSAP for 2, 3	√	15	2	52
Germany	Munich	Franz Josef Strauß Airport	SCR Level 3	1 and NSAP for 2, 4	√	45	2	86
Germany	Stuttgart	Stuttgart Airport	SCR Level 3	1 and NSAP for 2, 3, 4	None	13	1	36
Greece	Athens	Eleftherios Venizelos Airport	SMA Level 2	N/A	None	16	2	52

Country	City	Airport	Grading of tightness of slot constraints	Rationale for slot constraints	Night curfew	Terminal capacity (million passengers per year)	Number of runways	Declared runway capacity (available slots per 60 min.)
Hungary	Budapest	Ferihegy International	SMA Level 2	1 and NSAP for 2, 4	√	2.5	2	40
India	Mumbai	Chhatarpati Shivaji International Airport	SCR Level 3	1 and NSAP for 2, 4	None	18.5	2	N/A
Indonesia	Denpasar - Bali	Ngurah Rai Airport	SCR Level 3	1, 5 and NSAP for 2, 3, 4	None	10	1	N/A
Indonesia	Jakarta	Soekarno-Hatta International	SCR Level 3	NSAP for 2	None	N/A	2	74
Ireland	Dublin	Dublin Airport	SCR Level 3	N/A	None	18	3	44
Italy	Milan	Linate Airport	SCR Level 3	N/A	None	N/A	1	18
Italy	Milan	Malpensa Airport	SCR Level 3	N/A	None	24	2	70
Italy	Naples	Capodichino Airport	Not slot constraint	N/A	N/A	N/A	1	N/A
Italy	Rome	Ciampo	SCR Level 3	N/A	√	4		22
Italy	Rome	Fiumicino - Leonardo da Vinci	SCR Level 3	5 and NSAP for 3, 4	None	37	4	90
Japan	Osaka	Kansai International Airport	SMA Level 2	5 and NSAP for 2, 3, 4	None	N/A	1	30
Japan	Tokyo	Haneda International Airport	SCR Level 3	NSAP for 2	None	N/A	3	N/A
Japan	Tokyo-Narita	New Tokyo International Airport	SCR Level 3	1, 5 and NSAP for 2, 3, 4	√	N/A	2	44

Country	City	Airport	Grading of tightness of slot constraints	Rationale for slot constraints	Night curfew	Terminal capacity (million passengers per year)	Number of runways	Declared runway capacity (available slots per 60 min.)
Kingdom of Bahrain	Bahrain	Bahrain International Airport	SMA Level 2	N/A	None	10	2	N/A
Malaysia	Kuala Lumpur	Kuala Lumpur International	SCR Level 3	NSAP for 3	None	25	2	50
Netherlands	Amsterdam	Amsterdam Airport Schiphol	SCR Level 3	1, 5 and NSAP for 3, 4	√	45	6	106
New Zealand	Auckland	Auckland International Airport	SCR Level 3	N/A	None	8	1	N/A
New Zealand	Wellington	Wellington International Airport	SCR Level 3	1	√	5	1	N/A
Norway	Oslo	Gardermoen Airport	SCR Level 3	1	None	17	2	80
Poland	Warsaw	Frederic Chopin Airport	SMA Level 2	1, 5 and NSAP for 2, 3, 4	None	6.5	2	32
Portugal	Lisbon	Lisbon International	SCR Level 3	NSAP for 3, 4	√	12.3	2	30
Russia	Moscow	Domededovo Airport	SCR Level 3	5 and NSAP for 2, 3, 4	None	14	2	25
Russia	Moscow	Sheremetyevo Airport	SCR Level 3			11	2	N/A
Russia	St.Petersburg	Pulkovo Airport	SMA Level 2	1, 5 and NSAP for 2	None	N/A	2	28
Singapore	Singapore	Changi International	SCR Level 3	5 and NSAP for 4	None	44	2	66
Slovakia	Bratislava	M R Stefanik Airport	SMA Level 2	N/A	None	N/A	2	N/A

Country	City	Airport	Grading of tightness of slot constraints	Rationale for slot constraints	Night curfew	Terminal capacity (million passengers per year)	Number of runways	Declared runway capacity (available slots per 60 min.)
Slovenia	Ljubljana	Ljubljana Airport	SMA Level 2	1, 5 and NSAP2, 3	None	N/A	1	15
South Korea	Seoul - Gimpo	Gimpo International	SCR Level 3	1, 5 and NSAP for 3	√	34.7	2	32
South Korea	Seoul - Incheon	Incheon International Airport	Not slot constraint	N/A	None	30	2	N/A
Spain	Alicante	Alicante Airport	SCR Level 3	N/A	None	15	1	N/A
Spain	Barcelona	Barcelona Airport	SCR Level 3	N/A	None	20	2	N/A
Spain	Ibiza	Ibiza Airport	SCR Level 3	N/A	√	N/A	1	N/A
Spain	Madrid	Aeropuerto de Madrid/Barajas	SCR Level 3	1, 5 and NSAP for 2, 3, 4	√	70	3	78
Spain	Palma de Mallorca	Palma de Mallorca Airport	SCR Level 3	1 and NSAP for 2	√	>20	3	60
Sweden	Stockholm	Arlanda Airport	SCR Level 3	1 and NSMD 2,3 and NSAP 4	√	~25	3	76
Switzerland	Geneva	Geneva International Airport	Not slot constraint	1 and NSAP 2, 3, 4	√	10	2	36
Switzerland	Zürich	Zurich Ariport	SCR Level 3	1, 5 and NSAP 2,3,4	√	N/A	3	66
Thailand	Bangkok	Don Muang International Airport	SCR Level 3	5	None	35	2	60
Turkey	Istanbul	Ataturk Airport	SCR Level 3	1 and 5	√	~21	3	36
Ukraine	Kyiv	Boryspil State International Airport	SCR Level 3	N/A	None	N/A	2	N/A

Country	City	Airport	Grading of tightness of slot constraints	Rationale for slot constraints	Night curfew	Terminal capacity (million passengers per year)	Number of runways	Declared runway capacity (available slots per 60 min.)
United Arab Emirates	Dubai	Dubai International Airport	Not Slot constraint	N/A	None	N/A	1	N/A
United Kingdom	Glasgow	Glasgow Airport	SMA Level 2	N/A	None	N/A	2	N/A
United Kingdom	London	Gatwick Airport	SCR Level 3	1 and NSMD 2, 4	None	40	1	50
United Kingdom	London	Heathrow Airport	SCR Level 3	1, 5 and NSAP 2, 4 and NSMD 3	√	90	3	86
United Kingdom	London	Stansted Airport	Not Slot constraint	N/A	√	25	1	42
United Kingdom	Manchester	Manchester Airport	SCR Level 3	1, 5 and NSAP2, 3, 4	√	30	2	59
United States	Atlanta	Hartsfield-Jackson International Airport	SCR Level 3	NSMD for 3, 4	None	~85	4	184
United States	Chicago	O'Hare International Airport	SCR Level 3	NSMD for 2, 3, 4	None	N/A	7	N/A
United States	Denver	Denver International	SCR Level 3	NSAP for 3	None	N/A	5	200
United States	Los Angeles	Los Angeles International	SMA Level 2	NSAP for 2, 3, 4	None	79	4	153
United States	New York	J F Kennedy	SCR Level 3	1 and NSAP for 2, 3, 4	None	N/A	4	98
United States	New York	Newark International	SMA Level 2	1, 5 and NSAP for 2, 3, 4	None	N/A	3	108
United States	New York	La Guardia Airport	SCR Level 3	1, 5 and NSAP for 2, 3, 4	None	N/A	2	81

Country	City	Airport	Grading of tightness of slot constraints	Rationale for slot constraints	Night curfew	Terminal capacity (million passengers per year)	Number of runways	Declared runway capacity (available slots per 60 min.)
United States	San Francisco	San Francisco International Airport	SMA Level 2	1 and NSAP for 3	√	N/A	4	120
United States	Washington	Dulles International Airport	Not slot constraint	1	None	55 (after extension)	3	N/A
United States	Washington	Ronald Reagan National Airport	Not slot constraint	1	None	N/A	3	62
United States	Washington	Baltimore-Washington Int'l Airport	Not slot constraint	1	None	N/A	4	N/A

SCR Level 3:   Fully Slot-coordinated Airport
SMA Level 2:   Slot Facilitated Airport

Rationale for slot constraint
- Noise Restriction = 1
- Terminal Capacity = 2
- Runway Capacity = 3
- Apron Capacity = 4
- ATC Considerations = 5

NSMD :   Near saturated most of the day
NSAP: Near saturated at peak times

Table A2   Airport utilization

Country	City	Airport	Total movements (2004)	Proportion of movements accounted for by main carriers (2004)	Average aircraft size (passengers per movement) (2004)	% of movements accounted by aircrafts of 1–49 seats (2004)	Declared runway capacity (available slots per 60 min.)	Night curfew (*)	Annual capacity	Capacity utilization indicator (total movements/ total capacity)
Australia	Brisbane	Brisbane International Airport	72,377	45%	99	N/A	N/A	None	N/A	N/A
Australia	Melbourne	Tullamarine Airport	165,300	42%	116	N/a	N/A	None	N/A	N/A
Australia	Sydney	Kingsford Smith Airport	266,746	45%	95	28%	80	√	525,600	51%
Austria	Vienna	Vienna International Airport	230,900	59%	65	31%	66	None	578,160	40%
Belgium	Brussels	Brussels International Airport	253,255	30%	74	23%	74	None	648,240	39%
Canada	Montreal	Montreal Dorval Airport	208,329	N/A	N/A	55%	40	√	350,400	79%
Canada	Vancouver	Vancouver International Airport	322,986	44%	51	35%	75	√	657,000	66%
China	Beijing	Beijing Capital International	472,000	36%	114	1%	60	None	525,600	90%

Country	City	Airport	Total movements (2004)	Proportion of movements accounted for by main carriers (2004)	Average aircraft size (passengers per movement) (2004)	% of movements accounted by aircrafts of 1–49 seats (2004)	Declared runway capacity (available slots per 60 min.)	Night curfew (*)	Annual capacity	Capacity utilization indicator (total movements/ total capacity)
China	Shanghai	Hong Qiao International Airport	149,477	39%	100	7%	30	None	262,800	57%
China	Hong Kong	Chep Lap Kok International	210,112	26%	186	N/A	49	None	429,240	49%
Czech Republic	Prague	Ruzyne Airport	160,213	N/A	67	N/A	45	None	394,200	41%
Denmark	Copenhagen	Kastrup International Airport	272,518	47%	67	31%	83	None	727,080	37%
Finland	Helsinki	Vantaa Airport	166,286	61%	65	N/A	50	None	438,000	38%
France	Lyon	Saint Exupery Airport	122,000	N/A	50	45%	51	None	446,760	27%
France	Marseille	Provence Airport	115,110	N/A	81	26%	35	None	306,600	38%
France	Nice	Cote d'Azur Airport	119,854	N/A	47	58%	44	√	289,080	41%
France	Paris	Ch. de Gaulle	516,425	58%	101	10%	101	√	663,570	78%
France	Paris	Orly	218,798	61%	107	12%	76	√	665,760	44%
Germany	Berlin	Tegel	143,000	35%	79	N/A	35	√	229,950	62%

Country	City	Airport	Total movements (2004)	Proportion of movements accounted for by main carriers (2004)	Average aircraft size (passengers per movement) (2004)	% of movements accounted by aircrafts of 1–49 seats (2004)	Declared runway capacity (available slots per 60 min.)	Night curfew (*)	Annual capacity	Capacity utilization indicator (total movements/total capacity)
Germany	Cologne - Bonn	Konrad Adenauer Airport	152,700	N/A	56	30%	52	√	455,520	45%
Germany	Düsseldorf	Düsseldorf International	200,584	39%	87	12%	38	√	249,660	80%
Germany	Frankfurt	Frankfurt Main International	477,475	60%	114	9%	78	√	512,460	93%
Germany	Hamburg	Hamburg Fuhlsbüttel Airport	151,434	N/A	72	24%	52	√	341,640	44%
Germany	Munich	Franz Josef Strauß Airport	383,110	64%	74	24%	86	√	565,020	68%
Germany	Stuttgart	Stuttgart Airport	156,000	N/A	70	28%	36	√	315,360	66%
Greece	Athens	Eleftherios Venizelos Airport	191,000	46%	74	10%	52	None	455,520	42%
Hungary	Budapest	Ferihegy International Airport	111,723	N/A	67	15%	40	√	350,400	43%
India	Mumbai	Chhatarpati Shivaji International Airport	217,800	32%	104	N/A	N/A	None	N/A	N/A

Country	City	Airport	Total movements (2004)	Proportion of movements accounted for by main carriers (2004)	Average aircraft size (passengers per movement) (2004)	% of movements accounted by aircrafts of 1–49 seats (2004)	Declared runway capacity (available slots per 60 min.)	Night curfew (*)	Annual capacity	Capacity utilization indicator (total movements/total capacity)
Indonesia	Denpasar - Bali	Ngurah Rai Airport	N/A	N/A	124	13%	N/A	None	N/A	N/A
Indonesia	Jakarta	Soekarno-Hatta International	241,846	31%	96	N/A	74	None	648,240	37%
Ireland	Dublin	Dublin Airport	182,175	31%	101	N/A	44	None	385,440	47%
Italy	Milan	Linate Airport	122,221	N/A	N/A	29%	18	None	157,680	78%
Italy	Milan	Malpensa Airport	214,365	59%	87	N/A	70	None	613,200	35%
Italy	Naples	Capodichino Airport	N/A	N/A	N/A	N/A	N/A	N/A	N/A	N/A
Italy	Rome	Ciampo	44,269	N/A	N/A	N/A	22	√	192,720	31%
Italy	Rome	Fiumicino - Leonardo da Vinci	295,963	45%	94	9%	90	None	788,400	38%
Japan	Osaka	Kansai International Airport	102,571	26%	177	2%	30	None	262,800	39%
Japan	Tokyo	Haneda International Airport	N/A	45%	217	0.40%	N/A	None	N/A	N/A

Country	City	Airport	Total movements (2004)	Proportion of movements accounted for by main carriers (2004)	Average aircraft size (passengers per movement) (2004)	% of movements accounted by aircrafts of 1–49 seats (2004)	Declared runway capacity (available slots per 60 min.)	Night curfew (*)	Annual capacity	Capacity utilization indicator (total movements/ total capacity)
Japan	Tokyo-Narita	New Tokyo International Airport	185,243	22%	233	3%	44	√	289,080	64%
Kingdom of Bahrain	Bahrain	Bahrain International Airport	73,891	55%	109	N/A	None	None	N/A	N/A
Malaysia	Kuala Lumpur	Kuala Lumpur International	164,483	53%	139	N/A	50	None	438,000	38%
Netherlands	Amsterdam	Amsterdam Airport Schiphol	431,456	51%	98	16%	106	√	928,560	62%
New Zealand	Auckland	Auckland International Airport	154,812	N/A	72	N/A	N/A	None	N/A	N/A
New Zealand	Wellington	Wellington International Airport	100,480	N/A	46	N/A	N/A	√	N/A	N/A
Norway	Oslo	Gardermoen Airport	196,355	54%	78	14%	80	None	700,800	28%
Poland	Warsaw	Frederic Chopin Airport	126,870	N/A	56	39%	32	None	280,320	45%
Portugal	Lisbon	Lisbon International	112,200	N/A	88	15%	30	√	262,800	57%

Country	City	Airport	Total movements (2004)	Proportion of movements accounted for by main carriers (2004)	Average aircraft size (passengers per movement) (2004)	% of movements accounted by aircrafts of 1–49 seats (2004)	Declared runway capacity (available slots per 60 min.)	Night curfew (*)	Annual capacity	Capacity utilization indicator (total movements/total capacity)
Russia	Moscow	Domededovo Airport	N/A	25%	72	2%	25	None	219,000	N/A
Russia	Moscow	Sheremetyevo Airport	N/A	57%	N/A	N/A	N/A	N/A	N/A	N/A
Russia	St. Petersburg	Pulkovo Airport	N/A	N/A	69	6%	28	None	245,280	N/A
Singapore	Singapore	Changi International	184,932	36%	161	N/A	66	None	578,160	32%
Slovakia	Bratislava	M R Štefánik Airport	27,133	N/A	33	N/A	N/A	None	N/A	N/A
Slovenia	Ljubljana	Ljubljana Airport	35,502	N/A	30	60%	15	None	131,400	27%
South Korea	Seoul - Gimpo	Gimpo International	N/A	53%	147	N/A	32	√	210,240	N/A
South Korea	Seoul - Incheon	Incheon International Airport	105,923	36%	161	N/A	N/A	None	N/A	
Spain	Alicante	Alicante Airport	76,109	N/A	N/A	N/A	N/A	None	N/A	N/A
Spain	Barcelona	Barcelona Airport	291,369	46%	85	N/A	N/A	None	N/A	N/A

Country	City	Airport	Total movements (2004)	Proportion of movements accounted for by main carriers (2004)	Average aircraft size (passengers per movement) (2004)	% of movements accounted by aircrafts of 1-49 seats (2004)	Declared runway capacity (available slots per 60 min.)	Night curfew (*)	Annual capacity	Capacity utilization indicator (total movements/ total capacity)
Spain	Ibiza	Ibiza Airport	49,603				N/A	√	N/A	N/A
Spain	Madrid	Aeropuerto de Madrid/ Barajas	401,503	57%	96	11%	78	√	683,280	78%
Spain	Palma de Mallorca	Palma de Mallorca Airport	182,028	24%	122	11%	60	√	394,200	46%
Sweden	Stockholm	Arlanda Airport	245,360	43%	70	31%	76	√	665,760	49%
Switzerland	Geneva	Geneva International Airport	166,631	46%	68	15%	36	√	236,520	70%
Switzerland	Zürich	Zurich Ariport	266,660	51%	77	15%	66	√	433,620	61%
Thailand	Bangkok	Don Muang International Airport	208,556	29%	246	1%	60	None	525,600	40%
Turkey	Istanbul	Ataturk Airport	N/A	64%	83	1%	36	√	236,520	N/A
Ukraine	Kyiv	Boryspil State International Airport	N/A	N/A	N/A	N/A	N/A	None	N/A	N/A

Country	City	Airport	Total movements (2004)	Proportion of movements accounted for by main carriers (2004)	Average aircraft size (passengers per movement) (2004)	% of movements accounted by aircrafts of 1–49 seats (2004)	Declared runway capacity (available slots per 60 min.)	Night curfew (*)	Annual capacity	Capacity utilization indicator (total movements/ total capacity)
United Arab Emirates	Dubai	Dubai International Airport	217,165	36%	111	N/A	N/A	None	N/A	N/A
United Kingdom	Glasgow	Glasgow Airport	93,000	N/A	N/A	N/A	N/A	None	N/A	N/A
United Kingdom	London	Gatwick Airport	245,000	44%	129	8%	50	None	438,000	56%
United Kingdom	London	Heathrow Airport	470,000	42%	134	1%	86	None	753,360	83%
United Kingdom	London	Stansted Airport	177,000	57%	N/A	N/A	42	√	275,940	64%
United Kingdom	Manchester	Manchester Airport	212,000	31%	103	37%	59	√	387,630	55%
United States	Atlanta	Hartsfield-Jackson International Airport	964,858	75%	87	20%	184	None	1,611,840	60%

Country	City	Airport	Total movements (2004)	Proportion of movements accounted for by main carriers (2004)	Average aircraft size (passengers per movement) (2004)	% of movements accounted by aircrafts of 1–49 seats (2004)	Declared runway capacity (available slots per 60 min.)	Night curfew (*)	Annual capacity	Capacity utilization indicator (total movements/ total capacity)
United States	Chicago	O'Hare	992,427	48%	80	29%	N/A	None	N/A	N/A
United States	Denver	Denver International	560,198	55%	81	29%	200	None	1,752,000	32%
United States	Los Angeles	Los Angeles International	655,097	29%	90	25%	153	None	1,340,280	49%
United States	New York	J F Kennedy	349,518	24%	112	20%	98	None	858,480	41%
United States	New York	New Ark International	437,828	66%	78	30%	108	None	946,080	46%
United States	New York	La Guardia International	398,957	33%	62	42%	81	None	709,560	56%
United States	San Francisco	San Francisco International Airport	353,231	56%	97	19%	120	√	1,051,200	45%
United States	Washington	Dulles International Airport	469,634	47%	48	N/A	N/A	None	N/A	N/A
United States	Washington	Ronald Reagan National Airport	22,891	45%	59	N/A	62	None	543,120	4%
United States	Washington	Baltimore-Washington Int'l Airport	311,454	43%	74	N/A	N/A	None	N/A	N/A

Index